ADVANCES IN CARBENE CHEMISTRY

Volume 1 • 1994

ADVANCES IN CARBENE CHEMISTRY

Editor: UDO H. BRINKER
Department of Chemistry
State University of New York at Binghamton
Binghamton, New York

VOLUME 1 • 1994

JAI PRESS INC.

Greenwich, Connecticut *London, England*

55 Old Post Road, No. 2
Greenwich, Connecticut 06836

JAI PRESS LTD.
The Courtyard
28 High Street
Hampton Hill, Middlesex TW12 1PD
England

ISBN: 1-55938-545-6

Manufactured in the United States of America

CONTENTS

LIST OF CONTRIBUTORS

James E. Jackson	Department of Chemistry Michigan State University East Lansing, Michigan
Maitland Jones, Jr.	Department of Chemistry Princeton University Princeton, New Jersey
Wolfgang Kirmse	Fakultät für Chemie Ruhr-Universität Bochum, Germany
Holger Memmesheimer	Fachbereich Chemie Universität Kaiserslautern Kaiserslautern, Germany
Robert A. Moss	Department of Chemistry Rutgers, The State University of New Jersey New Brunswick, New Jersey
Matthew S. Platz	Department of Chemistry The Ohio State University Columbus, Ohio
Manfred Regitz	Fachbereich Chemie Universität Kaiserslautern Kaiserslautern, Germany

INTRODUCTION TO THE SERIES: AN EDITOR'S FOREWORD

The JAI series in chemistry has come of age over the past several years. Each of the volumes already published contains timely chapters by leading exponents in the field who have placed their own contributions in a perspective that provides insight to their long-term research goals. Each contribution focuses on the individual author's own work as well as the studies of others that address related problems. The series is intended to provide the reader with in-depth accounts of important principles as well as insight into the nuances and subtleties of a given area of chemistry. The wide coverage of material should be of interest to graduate students, postdoctoral fellows, industrial chemists, and those teaching specialized topics to graduate students. We hope that we will continue to provide you with a sense of stimulation and enjoyment of the various sub-disciplines of chemistry.

Department of Chemistry
Emory University
Atlanta, Georgia

Albert Padwa
Consulting Editor

PREFACE

Beginning as chemical curiosities, carbenes are now solidly established as reactive intermediates with fascinating and productive research areas of their own. Five decades of divalent carbon chemistry have provided us with a vast repertoire of new, unusual, and surprising reactions. Some of those reactions, once classified as exotic, have become standard methods in organic synthesis. These highly reactive carbene species have been harnessed and put to work to achieve difficult synthetic tasks other reactive intermediates cannot easily perform.

The fruitful relationship between experiment and theory has pushed carbene chemistry further toward the direction of reaction control; that is, regio- and stereoselectivity in intra- and intermolecular addition and insertion reactions. The interplay between experiment and modern spectroscopy has led to the characterization of many carbenes that are crucial to both an understanding and a further development of this field.

Monographs devoted entirely to carbenes appeared in the late sixties, such as the classic works of Hine and of Kirmse. Second-generation surveys published in the seventies consisted of the series of three volumes edited by Jones and Moss. At the end of the eighties, the status of such knowledge was collected into the two imposing volumes of Houben-Weyl (*Methoden der Organischen Chemie*; Regitz, editor).

Our understanding of carbene chemistry has advanced dramatically, especially in the last decade, and new developments continue to emerge. Some of the recent exciting findings are collected in this first volume of *Advances in Carbene*

Chemistry. The purpose of this volume is to launch a series for the periodic coverage of carbene chemistry in its broadest sense—leading into the next century.

Udo H. Brinker
Series Editor

CARBENES AND THE O–H BOND

Wolfgang Kirmse

Advances in Carbene Chemistry
Volume 1, pages 1–57.

ISBN: 1-55938-545-6

I. INTRODUCTION

The insertion of carbenes into X–H bonds is arguably the most characteristic reaction of divalent carbon species. Although X may be varied widely, the mechanism of insertion responds sensibly to the polarity of the X–H bond. Singlet carbenes insert into the nonpolar C–H bond in a concerted, single-step process, with retention of configuration.[1] In contrast, "insertion" into the polar O–H bond has been discussed primarily in terms of stepwise mechanisms: (a) electrophilic attack of the carbene on the oxygen atom, followed by proton transfer (ylide mechanism), and (b) protonation of the carbene to give a carbocation (ion pair), followed by nucleophilic substitution (carbocation mechanism) (Scheme 1).[2,3] A concerted process (c) cannot be excluded *a priori* but has not received definitive experimental support. Discrimination between reaction paths (a)–(c) will be a major theme of the present account. After having reviewed some basic premises, tools for probing the mechanism of O–H insertion reactions will be discussed. The various classes of carbenes will then be surveyed for their behavior toward O–H bonds.

II. SINGLET VERSUS TRIPLET REACTIVITY

Two electronic states of methylene, $^1A_1(\sigma^2)$ and $^3B_2(\sigma^1p^1)$,[4] can each be detected spectroscopically,[5] and their chemistry is distinct.[1] Methylene is a ground state triplet; experimental and theoretical estimates of the singlet–triplet energy gap now converge at $\Delta H_{ST} = 9 \pm 1$ kcal/mol.[6,7] The effects of substituents on the magnitude of ΔH_{ST} may be dissected into steric, primarily bond angle, and electronic components. Both contraction of the carbene–carbon bond angle and electron donation into the vacant *p* orbital favor the singlet state. For example, many carbenes with heteroatoms attached directly to the central carbon have singlet ground states, the excited triplet states being experimentally inaccessible.

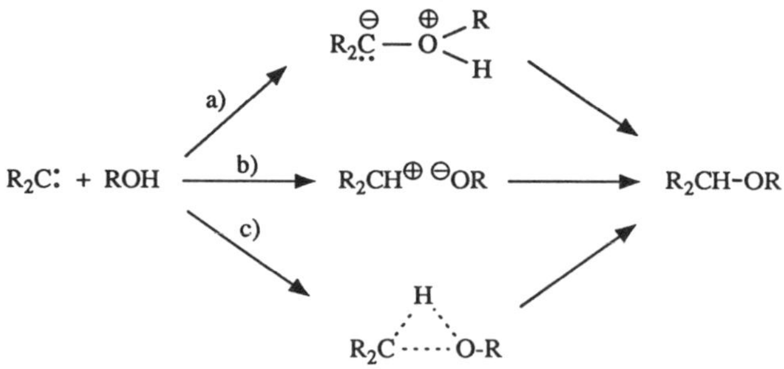

Scheme 1.

It is widely believed that spin-state-specific mechanisms can be assigned to the reactions of carbenes. Singlet carbenes undergo concerted, stereospecific cycloadditions with alkenes and insert readily into O–H bonds.[1] Stepwise, nonstereospecific addition to alkenes and hydrogen abstraction from alcohols is characteristic of carbenes that react from their triplet ground state; for example, anthronylidene.[8,9]

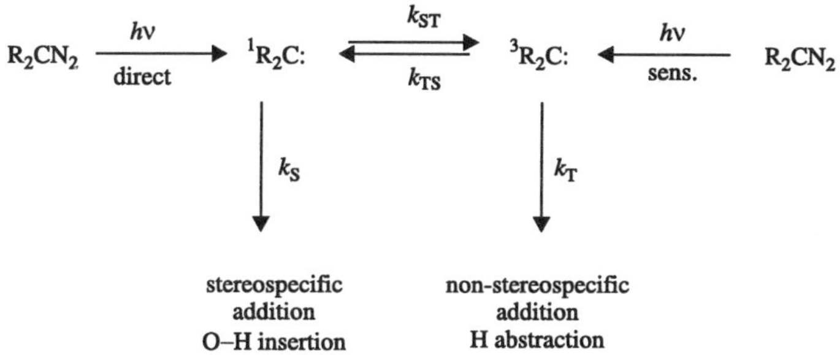

(1)

The behavior of aryl- and diarylcarbenes [(di)arylcarbenes] seemingly violates these simple rules. Although the triplet has been shown to be the ground state of phenylcarbene by low-temperature electron paramagnetic resonance (EPR),[10] the reaction of phenylcarbene with alkenes gives cyclopropanes with greater than 95% stereospecificity.[11,12] Diphenylcarbene has been exhaustively studied by spectroscopic techniques. Electron paramagnetic resonance spectroscopy clearly established the ground state of this carbene as the triplet.[10] The optical spectrum of the triplet was recorded first in a 1,1-diphenylethylene host[13] and later in rigid matrices.[14] Flash photolysis of diphenyldiazomethane in a variety of solvents produces triplet diphenylcarbene, whose transient absorption is readily monitored ($\lambda_{max} = 314$ nm).[15] Although singlet diphenylcarbene remains invisible to this day, the rate of intersystem crossing (k_{ST}) could be measured by picosecond time-resolved laser-induced fluorescence.[16] Upon addition of alcohols in the laser flash photolysis (LFP) experiments, quenching of both singlet diphenylcarbene ($k_s \approx 10^{10}\ M^{-1}\ s^{-1}$)[16,17] and triplet diphenylcarbene ($k_T \approx 10^7\ M^{-1}\ s^{-1}$)[15,17,18] is observed. The former effect leads to a decrease in yield, the latter to a decrease in lifetime of the triplet carbene (whose absorption is monitored).

While k_S of diphenylcarbene is estimated by a Stern–Volmer-type analysis, direct measurements are available for xanthylidene (**1**)[19] and 3,6-dimethoxyfluorenylidene (**2**).[20] In contrast to diphenylcarbene, both **1** and **2** are ground state singlets which react with alcohols at a diffusion-controlled rate.

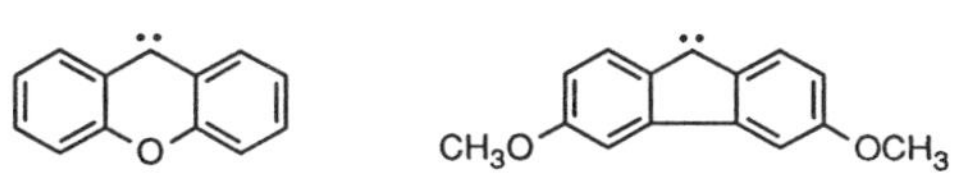

1 **2**

Pre-equilibrium and *surface crossing* mechanisms have been proposed for the relatively slow reaction of triplet (di)arylcarbenes with O–H bonds. Certainly any process leading from a triplet reactant to a singlet product must involve surface crossing. It is the timing of spin inversion and of chemical events that distinguishes the mechanistic alternatives. According to the pre-equilibrium mechanism, triplet to singlet interconversion (k_{TS}) occurs *prior to* O–H insertion of the singlet carbene (k_S), Eq (1). On the other hand, Griller et al.[18] suggested that triplet (di)arylcarbenes can react directly with alcohols, with surface crossing occurring *after* the carbene has begun to interact with the O–H bond.

The mechanistic alternatives are not readily distinguished by experiment. According to Eq. (1), the observed rate constant for the reaction of a triplet carbene with alcohol is

$$k_{obsd} = \frac{k_{TS}k_S}{k_{ST} + k_S[\mathrm{ROH}]} \tag{2}$$

At relatively low concentrations of alcohol, with $k_{ST} > k_S[\mathrm{ROH}]$, k_{obsd} reduces to

$$k_{obsd} = k_S K \qquad \text{where} \qquad K = \frac{k_{TS}}{k_{ST}} = \frac{[\mathrm{S}]}{[\mathrm{T}]} \tag{3}$$

As a consequence of Eq. (3), a "limiting" value of k_{obsd} is expected for all substrates that react with the singlet carbene at a diffusion controlled rate. In fact, triplet diphenylcarbene in acetonitrile was found to react with methanol ($k_{obsd} = 2.4 \times 10^7\ M^{-1}\ s^{-1}$)[17,18a] and with diphenyldiazomethane ($k_{obsd} = 3.4 \times 10^7\ M^{-1}\ s^{-1}$)[21] at virtually the same rate. Different rates would be anticipated for surface crossing with participation of the reactant. More experiments of this type are clearly desirable to exclude a fortuitous coincidence.

Equation (3) also correlates k_{obsd} with the magnitude of the singlet–triplet energy gap, $\Delta G_{ST} + RT\ln K$. Electron-donating substituents in the *para* position(s) of diphenylcarbene should reduce the magnitude of ΔG_{ST} while electron-accepting substituents should enhance it; k_{obsd} varies accordingly[22] (Table 1). Widening of the carbene–carbon bond angle also enhances ΔG_{ST}. This may be achieved by sterically crowding the *ortho* positions of diphenylcarbene.[23] A glance at Table 1 reveals that k_{obsd} decreases in the predicted way (**3a** > **4** > **5** > **6**). Conversely, the bond angle may be contracted by shortening the "tether" of cyclophanylcarbenes (**7a–d**), with a concomitant increase of k_{obsd} (Table 1).[24] Although these data are nicely compatible with the pre-equilibrium mechanism, it may be argued that any decrease in ΔG_{ST} will also lower the activation barrier of nonequilibrium surface crossings, and *vice versa*.

The pre-equilibrium mechanism requires that the activation energy of the reaction of triplet diphenylcarbene with methanol be equal to the singlet–triplet enthalpy gap (ΔH_{ST}) plus the apparent activation energy of the diffusion-controlled reaction of singlet diphenylcarbene with methanol (≈ 2 kcal/mol).

$$E_a(\text{triplet}) = \Delta H_{ST} + E_a(\text{singlet}) \tag{4}$$

ΔH_{ST} was calculated from $\Delta G_{ST} = -RT\ln K$ (Eq. 3) on the assumption that the entropy difference between the spin states of diphenylcarbene is solely due to the

Table 1. Bimolecular Rate Constants for the Reaction of Triplet Diarylcarbenes with Methanol at 298–300 K

Carbene	*Solvent*	$k_Q(M^{-1}s^{-1})$	*Ref.*
3a	cyclohexane	1.2×10^7	22
3b	cyclohexane	1.0×10^8	22
3c	isooctane	3.6×10^7	22
3d	isooctane	6.5×10^6	22
3e	cyclohexane	2.9×10^6	22
3f	cyclohexane	1.3×10^6	22
3g	cyclohexane	8.7×10^5	22
4	isooctane	8.2×10^4	23
5	isooctane	2.1×10^4	23
6	cyclopentane	$< 2.0 \times 10^3$	24
3a	tetrahydrofuran	6.3×10^6	24
7a	tetrahydrofuran	2.0×10^8	24
7b	tetrahydrofuran	2.4×10^8	24
7c	tetrahydrofuran	6.8×10^8	24
7d	tetrahydrofuran	2.0×10^9	24

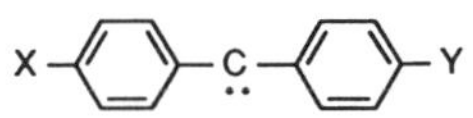

3

	3		**7**
	X	Y	*n*
a	H	H	12
b	CH_3	CH_3	11
c	H	CH_3	10
d	Br	Br	9
e	Cl	Cl	
f	H	CO_2CH_3	
g	H	CN	

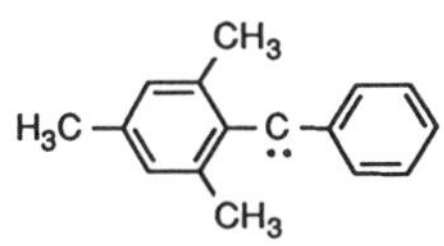

4

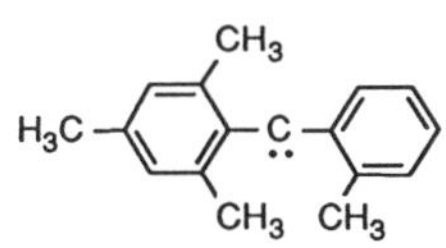

5

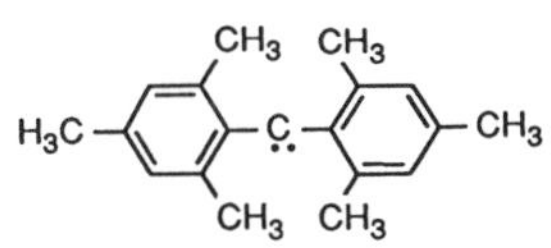

6

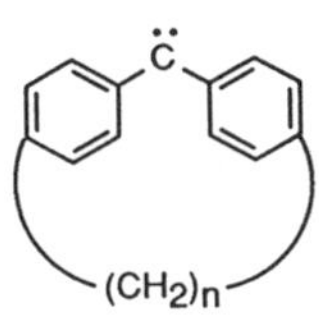

7

electronic degeneracy of the triplet; that is, $\Delta S_{ST} = R\ln 3$, or 2.2 eu.[17,18] The experimental values of E_a (triplet), 1.2–3.6 kcal/mol, were found to be smaller than those predicted by Eq. (4); 5–7 kcal/mol.[18] Similar inconsistencies were observed in the reaction of triplet diphenylcarbene with pyridine.[25] The fact that E_a (triplet) is less than the calculated ΔH_{ST} argues in favor of the surface crossing mechanism. The discrepancy would be resolved, however, if ΔS_{ST} is much larger than $R\ln 3$. Restricted internal rotation[26] and enhanced solvation of the polar singlet state may contribute to ΔS_{ST}. These and other complications have been extensively discussed elsewhere.[27]

A strong argument in favor of surface crossing comes from the observation that triplet diphenylcarbene in frozen methanol matrices reacts at 77 K to give the ether.[28] With $\Delta H_{ST} \geq 2$ kcal/mol, the singlet population would be too small to account for quenching of the triplet carbene within minutes. Whether or not the matrix results have any bearing on reactions in solution at ambient temperature remains questionable.

In conclusion, it can be honestly said that the reaction of triplet (di)arylcarbenes with O–H bonds is not yet fully understood. The effort expended on this problem is due, at least in part, to the advent of LFP techniques by which triplet (di)arylcarbenes are readily monitored.[27] The wealth of kinetic data for triplets should not obscure the fact that singlet carbenes are much more reactive toward O–H bonds than triplets. Conditions must be chosen very carefully in order for the triplet state to contribute significantly to ether formation. Any direct photolysis of diazo compounds in neat alcohols will proceed predominantly, if not entirely, by quenching of the initially formed singlet carbenes. The mechanism of this process will be the central theme of the following sections.

III. ELECTROPHILIC VERSUS NUCLEOPHILIC REACTIVITY

Singlet carbenes have been experimentally classified as electrophiles, nucleophiles, or ambiphiles depending on their selectivity toward alkenes.[29] Electrophilic carbenes are described as being more reactive toward electron-rich alkenes such as tetramethylethylene, whereas nucleophilic carbenes react faster with electron-poor alkenes such as ethyl acrylate. Ambiphilic carbenes react more rapidly with olefins substituted with *either* electron-donating or electron-withdrawing groups than with 1-pentene or 1-hexene. Moss has developed a method of quantifying the olefinic selectivity of carbenes with a free energy relationship that yields a selectivity index m_{CXY}. The relative rates of addition of the carbene under investigation (CXY) with a standard set of alkenes are plotted against the same measurements for CCl_2 ($m_{CCl_2} \equiv 1.00$) to obtain m_{CXY} as the slope of a log–log correlation. Multiple linear regression analysis of the dependence of m_{CXY} (obsd) on σ_R^+ and σ_I afforded the dual parameter equation, Eq. (5), in which Σ_{XY} represents the sum of the appropriate σ constants for the substituents of CXY.[30]

Table 2. Selectivity Indices and Differential Orbital Energies for Carbene–Alkene Additions

Carbene	m_{CXY} *(obsd)*	$m_{CXY}{}^{a}$ *(calcd)*	*p-π[b] (eV)*	*π*-σ[b] (eV)*	
CH_3CCl	0.50	0.58			
PhCCl	0.83	0.71			electrophilic
CCl_2	1.00	0.97	10.82	13.22	electrophilic
CF_2	1.48	1.47	12.40	15.16	electrophilic
CH_3OCCl		1.59	12.97	12.60	ambiphilic
CH_3OCF		1.85			ambiphilic
$C(OCH_3)_2$		2.22	14.60	12.59	nucleophilic

Notes: [a]Eq. (4).
[b]Calculated for ethylene.

$$m_{CXY} = -1.10\ \Sigma_{X,Y}\ \sigma_R^+ + 0.53\ \Sigma_{X,Y}\ \sigma_I - 0.31 \qquad (5)$$

Frontier molecular orbital theory has been successful in accounting for the selectivity relationships exemplified by the m_{CXY} values in Table 2.[31] An electrophilic carbene has a relatively low-lying lowest unoccupied molecular orbital (LUMO) (p) and the dominant interaction is carbene LUMO (p) with an alkene highest occupied molecular orbital (HOMO) (π). If electron donation from the substituents X and Y is strong enough, the dominant interaction will become carbene HOMO (σ) with alkene LUMO (π^*); the carbene is now nucleophilic (Table 2). The "border" of electrophilicity and ambiphilicity appears to be experimentally located at $m_{CXY} \cong 1.5$. Recently, absolute rate constants for quenching of arylhalocarbenes by alkenes have been obtained using LFP techniques.[32] These studies revealed "latent ambiphilicity" of PhCX; that is, the electron-withdrawing cyano or methoxycarbonyl substituents make $CH_2{=}CHR$ more reactive toward PhCF or PhCCl than does the butyl substituent of 1-hexene. These data, supplemented by absolute rate constants for addition to substituted styrenes, caution against the previous characterization of arylhalocarbenes as electrophilic.

The mechanism by which carbenes insert into O–H bonds will also be influenced by their "philicity". Ylide formation is initiated by electrophilic attack at the nonbonding electrons of oxygen. On the other hand, nucleophilic reactivity prevails as the carbene accepts the proton of alcohols to give a carbocation or ion pair (Scheme 1). We should not expect, however, that the philicity scale defined by the addition of carbenes to alkenes applies to O–H insertion reactions as well. Electrophilic and nucleophilic cyclopropanation reactions differ in the early stages of reactant approach but lead ultimately to the same type of product. Since the transition state is reactant-like and the reaction is strongly exothermic, the effect of substituents on product stability may be neglected. In contrast, the stepwise insertion of carbenes into O–H bonds produces highly reactive intermediates (ylides or carbocations) whose relative stability may be pivotal to the choice of mechanism. Therefore, it will not be appropriate to focus exclusively on electronic effects in the carbene, as described by m_{CXY} values. The simplistic view that "the majority of carbenes are electrophiles" has inhibited the discussion of O–H insertion mechanisms for many years.

IV. MECHANISTIC PROBES FOR O–H INSERTION REACTIONS

A theme that runs through many areas of mechanistic research is the distinction of concerted and stepwise reactions. Theory solves the problem, at least in principle, by computation of the potential energy surface on which saddle-points correspond to transition states and local minima represent intermediates. Unfortunately, conclusive results have not been obtained even for the simplest case of O–H insertion—the reaction of methylene with water to give methanol. Although the methyleneoxonium ylide CH_2OH_2 corresponds to a minimum on the HF/6-31 G* surface, the barrier for rearrangement disappears when single-point calculations that include electron correlation are carried out.[33] At their highest level of calculation (MP4/6-31 G**), Radom et al. again found a barrier for rearrangement, but it is only 1 kcal/mol.[34]

Valuable experimental criteria of concertedness include stereochemistry and energetics. A concerted reaction is bound to proceed stereospecifically, by way of a transition state that is lower in energy than any conceivable intermediate. The stereochemical criterion is clearly inapplicable to O–H insertion reactions, owing to the achiral nature of divalent oxygen. The thermochemistry of carbene reactions is still in its infancy. Time-resolved photoacoustic calorimetry can be utilized to estimate energies and heats of reaction of short-lived transients.[35] Some data for arylhalocarbenes have been reported,[36] including O–H insertion reactions with methanol ($\Delta H_r \approx 50$ kcal/mol). Activation energies for the reaction of diphenylcarbene with methanol have been measured[18] but cannot be interpreted unequivocally, as discussed in Section II. Even with reliable data for uncomplicated carbene reactions at hand, the problem of estimating the energies of *solvated* ylides or carbocations would remain. As a consequence, the mechanistic evaluation of O–H insertion reactions to date rests mostly on the identification of intermediates. In reviewing these efforts we should keep in mind that the failure to detect intermediates may well be due to inadequate methods (scavengers), rather than to concerted mechanisms.

A. Delocalization of Charge

The various pathways envisioned for O–H insertion reactions (Scheme 1) of carbenes are most readily distinguished if the divalent carbon is conjugated to a π system. Stepwise insertion mechanisms will then create a delocalized charge, thus giving rise to characteristic distributions of products and/or labels. Cylopentadienylidene (**9**) and cycloheptatrienylidene (**17**) are illustrative examples.[37] Photolysis of diazocyclopentadiene (**8**) in MeOD afforded a mixture (4:1) of [5-D]-1-methoxycyclopentadiene (**12**) and [5-D]-2-methoxycyclopentadiene (**13**) (Scheme 2). In order to avoid isomerization by way of 1,5-H shifts, both the photolysis of **8** and the cycloaddition of **12** and **13** to hexafluoro-2-butyne were performed at –70 °C. The deuterium was located at C-7 of both adducts, **14** and **15**, thus excluding interconversion of **12** and **13**. No adduct derived from 5-methoxycyclopentadiene (**11**) was observed. Therefore, a concerted insertion of **9** into O–H bonds is unlikely. The data are consistent with electrophilic attack of **9** on the oxygen of methanol, followed by deuteration of the cyclopentadienide part of the ylide **10**.

In contrast to **9**, cycloheptatrienylidene (**17**) should be immediately protonated by alcohols to give the stable tropylium ion (**18**). Photolysis of tropone tosylhydrazone sodium salt (**16**)[38] and base-induced elimination of hydrogen chloride from chlorocycloheptatrienes (**19**)[39] are established routes to **17** (Scheme 3). In the presence of ROD, both methods afforded 7-alkoxycycloheptatrienes (**21**) with random distribution of deuterium among all positions of the seven-membered ring.[37] These results support the intervention of tropylium ions but do not clearly define their origin. According to MNDO computation,[40] infrared (IR) spectroscopy,[41] and trapping experiments,[42] cycloheptatetraene (**20**) appears to be the global

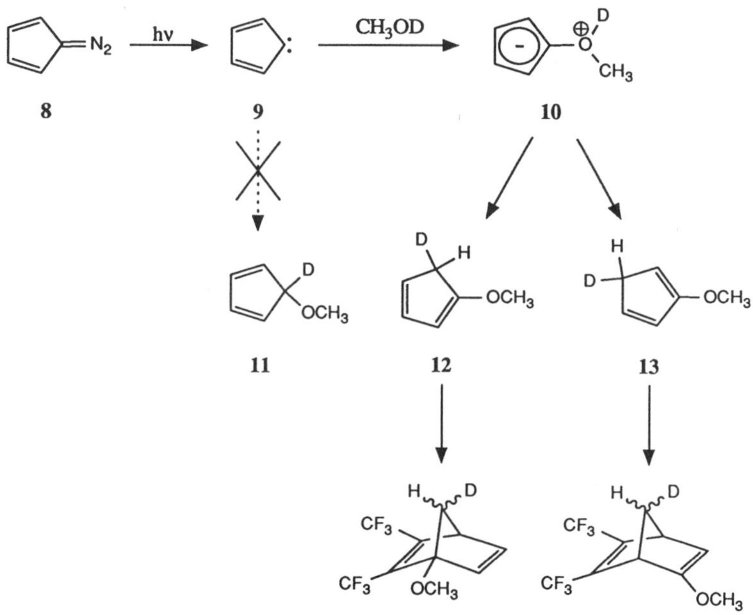

Scheme 2.

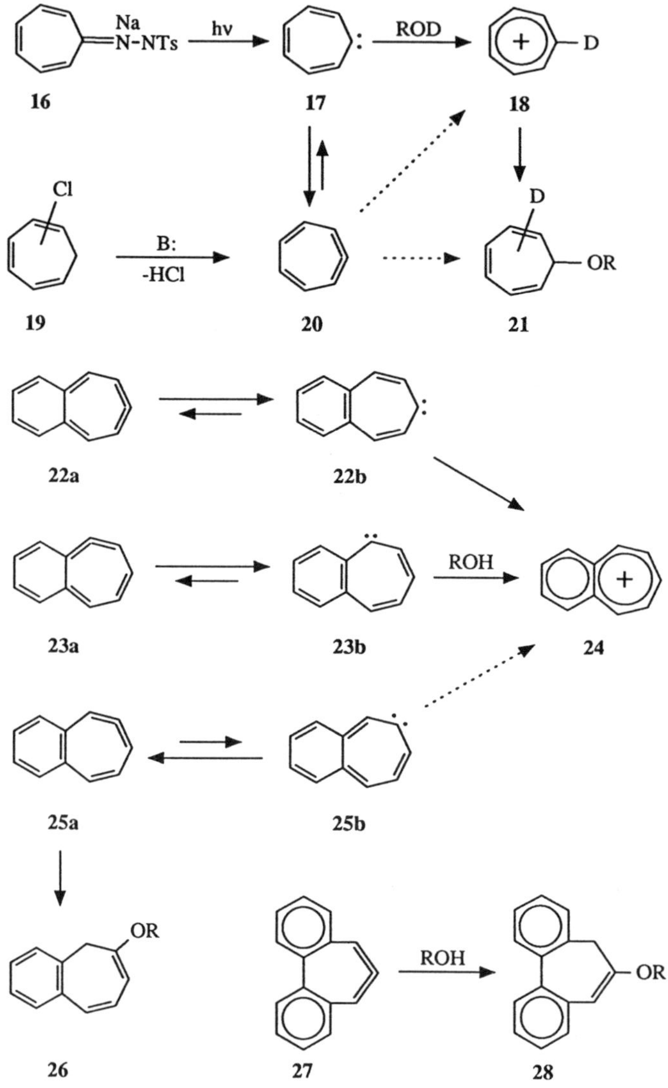

Scheme 3.

minimum on the potential energy surface of C_7H_6. In order to exclude protonation of **20**[40f] as the source of **18**, the effect of benzoannelation on the relative stability of **17** and **20** has been exploited.[43] Inspection of resonance structures as well as INDO calculations[44] indicate that 1,2- and 3,4-benzoannelation stabilize the carbene (**22b, 23b**) but destabilize the tetraene (**22a, 23a**). Both species **22** and **23** were

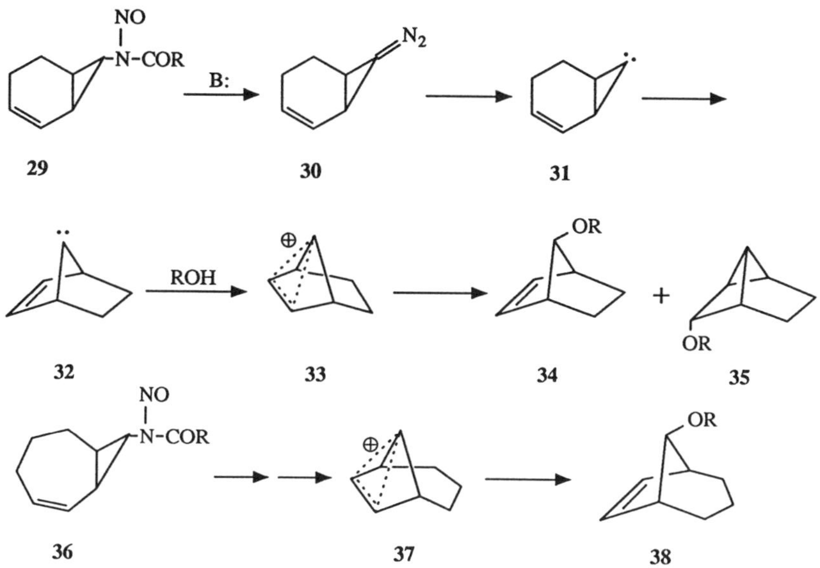

Scheme 4.

found to react exclusively by way of the benzotropylium ion (**24**) to give identical mixtures of benzocycloheptatrienyl ethers. 2,3-Benzoannelation, on the other hand, stabilizes the tetraene (**25a**) and destabilizes the carbene (**25b**). Only a minor fraction of **25** reacted via **24** while the major product was 1,2-benzo-6-alkoxycycloheptatriene (**26**). Annelation of a second benzene ring widens the energy gap between carbene and tetraene even further, the consequence being exclusive formation of **28** from **27**.[43] The observations demonstrate that cycloheptatetraenes accept the alkoxy group, rather than the proton, of alcohols at the *sp*-hybridized carbon. Tropylium ions arise by protonation of cycloheptatrienylidenes unless the carbene–tetraene equilibrium is extremely unfavorable.

Conjugation of the electron-deficient center to a π system may be replaced by homoconjugation. Deacylation of 7-norcarenyl nitrosoamides (**29**)[45,46] in the presence of alcohols afforded *anti*-7-norbornenyl (**34**) and *endo*-6-tricyclo[3.2.0.0]heptyl ethers[2,7] (**35**) in a base-dependent ratio, pointing to the intervention of 7-norbornenyl cations (**33**).[47] Skattebøl rearrangement[48] of 7-norcarenylidene (**31**) is thought to generate 7-norbornenylidene (**32**) which reacts with alcohols by way of **33**. The homologous bicyclo[3.2.1]oct-6-en-8-yl cation (**37**) does not give tricyclic products, but stereospecific formation of the ether **38** in the deacylation of **36**[49] suggests a reaction sequence analogous to that proposed for **29** (Scheme 4).

The effect of charge delocalization in allylic systems may be partially offset by ion pairing. Proton transfer from alcohols to carbenes will give rise to carbocation–

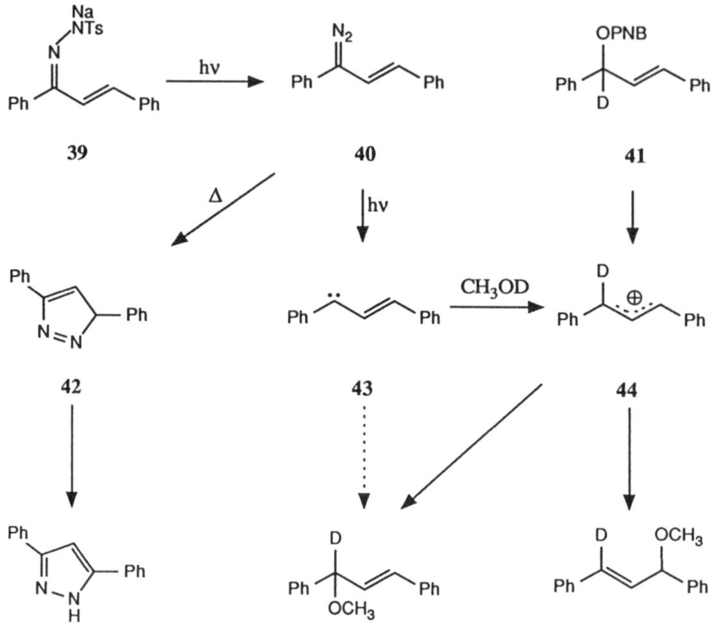

	46a	46b
39:	66	34
41:	52	48

Scheme 5.

alkoxide ion pairs; that is, the counterion will initially be closer to the carbene-derived carbon than to any other site. Unless the ions are rapidly separated by solvent molecules, collapse of the ion pair will mimic a concerted O–H insertion reaction.

Technical problems are also encountered in the investigation of vinylcarbenes. The use of diazoalkenes as precursors is often precluded by rapid cyclization, with the formation of pyrazoles. However, photolysis of 1,3-diphenylpropenone tosylhydrazone sodium salt (**39**) in methanol afforded the pyrazole **45** and 1,3-diphenyl-3-methoxypropene (**46**) in comparable amounts. Obviously, photolysis of the diazoalkene **40**, generated *in situ* from **39**, is competitive with ring closure to give **42** (Scheme 5). On photolysis of **39** in MeOD, deuterium was distributed between C-1 and C-3 of **46** in a 1:2 ratio.[50] Disregarding a small secondary isotope effect, the labeled 1,3-diphenylallyl cation (**44**) should give rise to an even distribution of deuterium, as was confirmed by methanolysis of the *p*-nitrobenzoate **41**. It is not clear whether the uneven distribution of deuterium observed with **39** is due to ion pairing or to competing reactions of the carbene **43**.

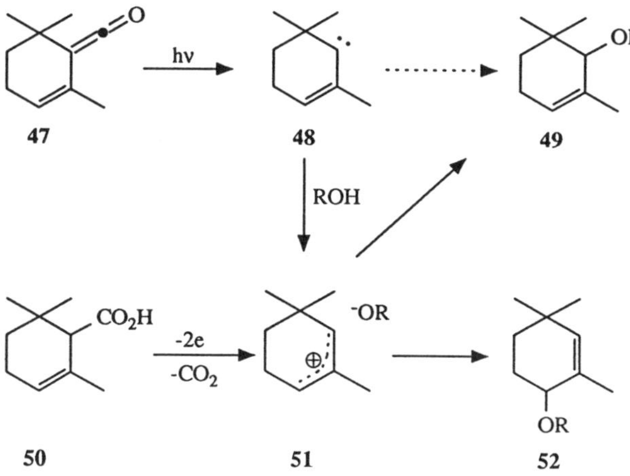

Scheme 6.

Variation of the medium might provide an answer, but the choice of solvents is very limited in the case of **39**. The sterically encumbered ketene **47**[51] is a less sensitive precursor that does not react with OH groups at ambient temperature. Photolysis of **47** gave mixtures of allylic ethers or esters (**49** and **52**) which are indicative of a delocalized intermediate.[52] Oxidative decarboxylation of the carboxylic acid **50** served as an alternative route to the 2,4,4-trimethylcylcohexenyl cation (**51**) (Scheme 6). Although the solvent effect on both reactions was qualitatively similar, the fraction of rearranged product **52** from the ketene **47** was always inferior to that from the carboxylic acid **50**. The increase in **52** with increasing polarity of the solvent supports the intervention of ion pairs. If competing reactions of the carbene **48** were involved, increasing acidity of the solvent should lead to enhanced rearrangement. The opposite effect was observed (cf. MeOH vs. AcOH) (Table 3).

While ion pairing is a complicating factor in the analysis of O–H insertion reactions of carbenes, new opportunities in carbocation chemistry may be envi-

Table 3. Product Ratios **49**:**52** Obtained from Photolyses of **47** and Oxidative Decarboxylation of **50**

	Precursor	
Solvent	**47**	**50**
Acetic acid	82:18	33:67
Methanol	58:42	28:72
Trifluoroethanol	43:57	13:87

sioned. Proton transfer to carbenes should produce carbocation–nucleophile ion pairs of unsurpassed "tightness" which are not accessible by heterolysis of covalent substrates. For example, carbocation–alkoxide pairs cannot be generated from ethers; the C–O bond will not break unless the oxygen is protonated or complexed to Lewis acids.

B. Trapping of Intermediates

Alkenes should be useful scavengers that differentiate between carbenes (cycloaddition) and carbocations (electrophilic addition); a successful application of this technique follows. Treatment of *N*-nitroso-*N*-(2-vinylcyclopropyl)urea (**53**) with sodium methoxide generates 3-cyclopentenylidene (**56**) via Skattebøl rearrangement[48] of 2-vinylcyclopropylidene (**55**) (Scheme 7). In methanol, the products derived from **56** were 4-methoxycyclopentene (**58**) and cyclopentadiene (**59**) (≈ 2:1). When methyl vinyl ether was added as a scavenger, cycloaddition of the carbene **56** to give **60** and electrophilic addition of 3-cyclopentenyl cation (**57**), leading to the acetal **62** by way of the carboxonium ion **61**, were observed in a 1:14 ratio.[53] Since **57** is the major species being trapped by the vinyl ether, it was inferred that **57** should also be the major precursor of **58**. Labeled substrates revealed that the reaction sequence was stereospecific (**53a** → **58a**; **53b** → **58b**). Therefore, nonplanar, homoconjugated structures were assigned to **56** and **57**.

The limitations of carbocation trapping by alkenes become obvious when more selective species are involved. The reaction of phenylcarbene (**63**) with equimolar mixtures of methanol and alkenes afforded predominantly phenylcyclopropanes (**66**) and benzyl methyl ether (**67**) (Scheme 8).[54] Electrophilic addition of benzyl cations (**64**), affording **68** and **69** by way of **65**, was only a minor route (≈ 6%). Since benzyl cation (**64**) from other sources also discriminated strongly in favor of methanol, the intervention of **64** cannot be assessed quantitatively.

Selectivity should not be a problem with scavengers that are closely related to alcohols, such as ethers. Oxetane was chosen because ring strain facilitates the Stevens rearrangement of oxetane-derived ylides (**71** → **72**). The generation of carbenes in neat oxetane led to C–H insertion (predominantly, if not exclusively, in the α position), ring expansion, and ring cleavage (Scheme 9).[55] With oxetane–methanol mixtures, the methyl ethers **74** and 1-alkoxy-3-methoxypropanes (**76**) were obtained as additional products. The immediate precursor of **76** is the oxetanium ion **75** which may arise by two routes: protonation of the ylide **71**, or protonation of the carbene, followed by electrophilic attack of the carbocation on oxetane. If protonation of the ylide **71** is the only route to **75**, the product ratio (**72** + **73** + **76**)/(**70**) will not depend on the concentration of MeOH. Ethoxycarbonylcarbene and bis(methoxycarbonyl)carbene were found to behave accordingly.[56] Protonation of the carbene, on the other hand, opens an additional route to **75**, thus enhancing the product ratio. A step increase of (**72** + **73** + **76**)/(**70**) with increasing MeOH concentration was observed for diphenylcarbene.

A simplified view focuses on the fraction of carbene that is scavenged by methanol in equimolar mixtures of oxetane and methanol (Table 4). The table includes some carbenes to which the previous analysis is inapplicable because they do not insert into C–H bonds of oxetane. Two categories are clearly distinguished. Some carbenes are scavenged by methanol in the order of 40–60%; therefore, they react with methanol and with oxetane at nearly the same rate. Others prefer methanol to the virtual exclusion of oxetane, and give > 90% of methanol-derived

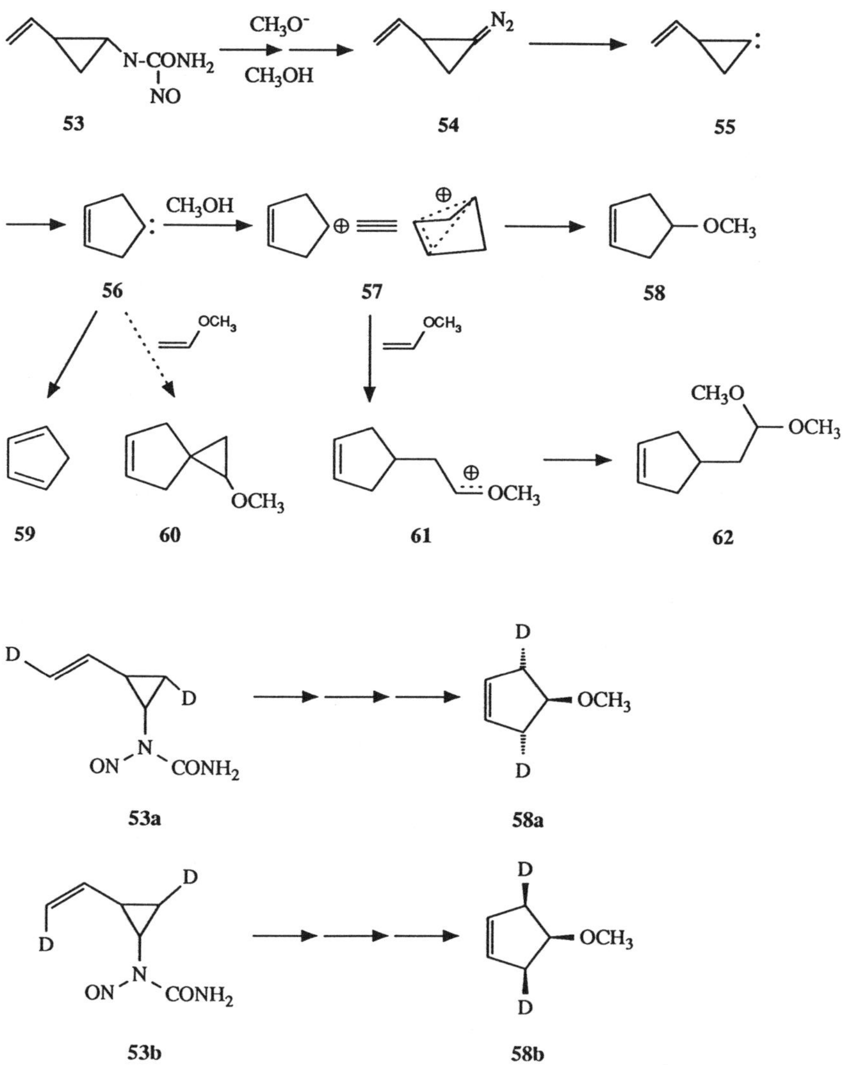

Scheme 7.

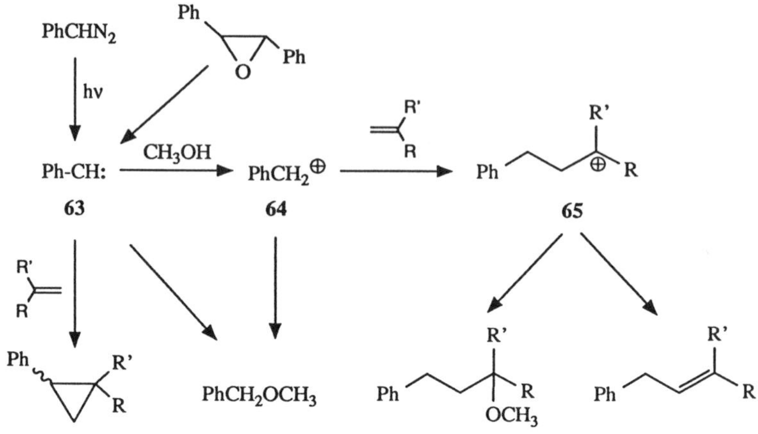

66		67	68	69
49		45	1	5%
60	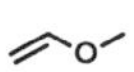	34	5	1
64	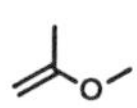	30	4	2

Scheme 8.

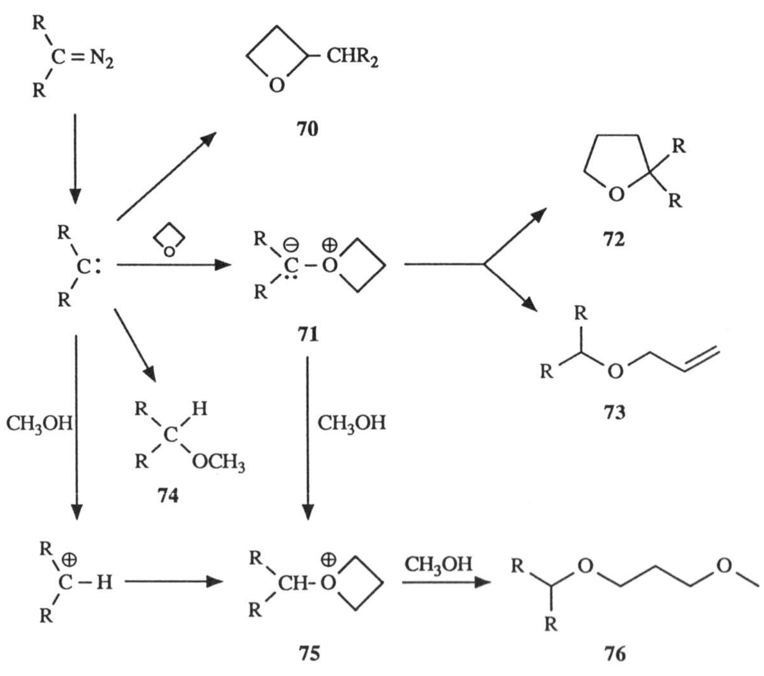

Scheme 9.

Table 4. Fraction of Carbene Scavenged by Methanol at Oxetane–Methanol Molar Ratios of 1:1 (6:1)

Carbene	*Fraction scavenged (%)*
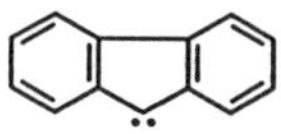	59
Ph-CH:	40
:CH–CO_2Et	59
:C(CO_2Me)$_2$	52
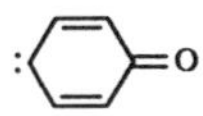	58
:CHPh$_2$	>98 (97)
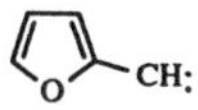	> 98 (95)
	98 (93)

products even at an oxetane:methanol ratio of 6:1.[56] It is obvious that different mechanisms are involved. Approximately equal reactivities toward oxetane and methanol point to electrophilic attack of the carbene at the oxygen of both reactants. On the other hand, protonation is the most likely route for carbenes that discriminate strongly in favor of methanol.

As a rule, carbenes bearing hydrogen vicinal to the divalent carbon undergo 1,2-H shifts with formation of alkenes. In the presence of alcohols, protonation of the carbene with subsequent elimination of a proton from the β position must be considered as an alternative route to alkenes. The carbenic and cationic pathways are readily distinguished in deuterated solvents (Scheme 10). When benzylphenylcarbenes or benzylchlorocarbenes (**77**) were generated in MeOD, the formation of alkenes **78** was found to proceed with incorporation of deuterium (0.3–15%, depending on R and X).[57] Partial, although not exclusive, intervention of the carbocation **80** accounts for these observations.

Occasionally, competitive intramolecular and intermolecular O–H insertion reactions of carbenes can provide mechanistic insight. When 2-(hydroxymethyl)phenylcarbene (**82**) was generated in protic media, the ratio of dihydroisobenzofuran (**83**) to 2-(alkoxymethyl)benzyl alcohols (**87**) was found to

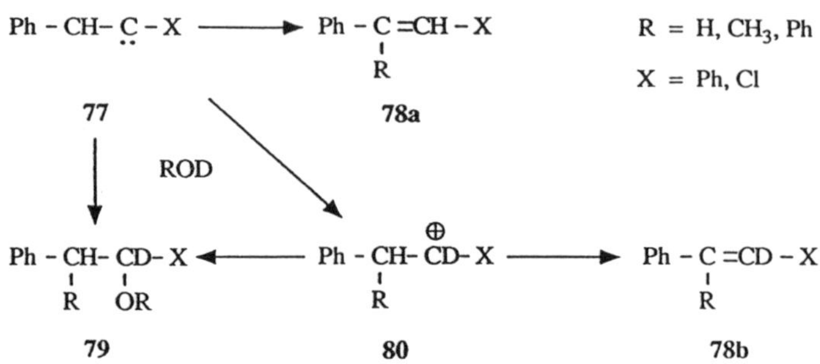

Scheme 10.

depend strongly on the nature of the solvent (Scheme 11).[58] In terms of an electrophilic attack by **82** on both internal and external hydroxyl groups, it might be expected that decreasing solvent nucleophilicity should lead to more of the intramolecular product **83**; however, the opposite was found experimentally. This parallels the increase in solvent acidity, which should increase the rate of protonation of **82**. The observed trend in product ratios suggests that the 2-(hydroxymethyl)benzyl cation (**86**) does not undergo cyclization to give **83**, whereas the carbene **82** does. The divergent intramolecular reactivities of **82** and **86** are explicable in terms of the much higher rotational barrier of benzyl cation relative to singlet phenylcarbene.[59] These views were substantiated experimentally by generating **86** from the diazonium ion **84** or from the bromide **85**. Intermolecular

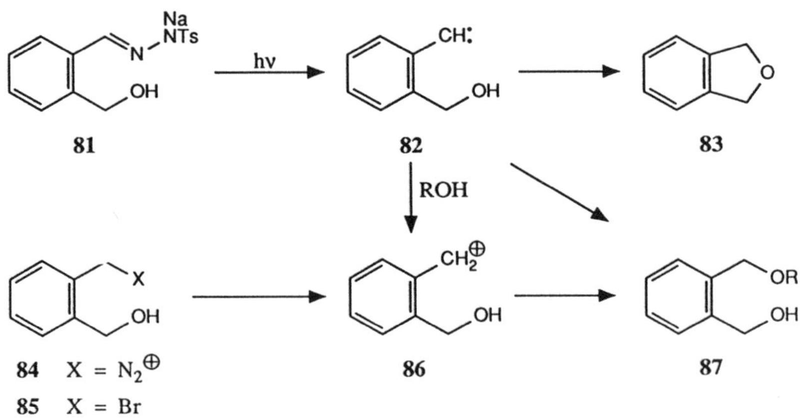

ROH	*t*-BuOH	EtOH	MeOH	H_2O	TFE
83 : **87**	71 : 29	51 : 49	38 : 62	10 : 90	2 : 98

Scheme 11.

products were formed exclusively; the cyclic ether **83** was not obtained. The results with **84** and **85** exclude the cation **86** as a precursor to **83**. Therefore, the formation of **83** from **82** must proceed by way of the ylide mechanism. On the other hand, protonation of **82** by external alcohols is indicated by the solvent-dependent product ratios. It appears that **82** behaves as an ambiphile in its reactions with OH groups.

C. Spectroscopic Detection of Intermediates

The advent of the pulsed laser has strongly affected mechanistic organic chemistry. The short-lived intermediates that are now accessible for time-resolved measurements include diarylcarbenium ions.[60] Laser flash photolysis (LFP) of diarylmethyl halides, acetates, or ethers was found to generate diarylcarbenium ions as well as diarylmethyl radicals, the ratio depending on substrate and solvent (Scheme 12). The diarylcarbenium ions were identified by their UV/VIS spectra, which had previously been measured in strong acids ("stable ion conditions"). The kinetic data gathered from LFP experiments indicate that the lifetimes of many diarylcarbenium ions are in the μs regime, even in the presence of water or alcohols. Since the reaction rates of diarylcarbenes with OH groups are close to diffusion-controlled,[16,17] it should be possible to observe the spectra of the intervening diarylcarbenium ions.

In fact, LFP of 4,4′-dimethoxydiphenyldiazomethane in acetonitrile–water (1:2) gave rise to the same transient-absorption as that obtained by photoheterolysis of Ar_2CHX (Figure 1).[61] The rate constants of pseudo first-order decay were also in

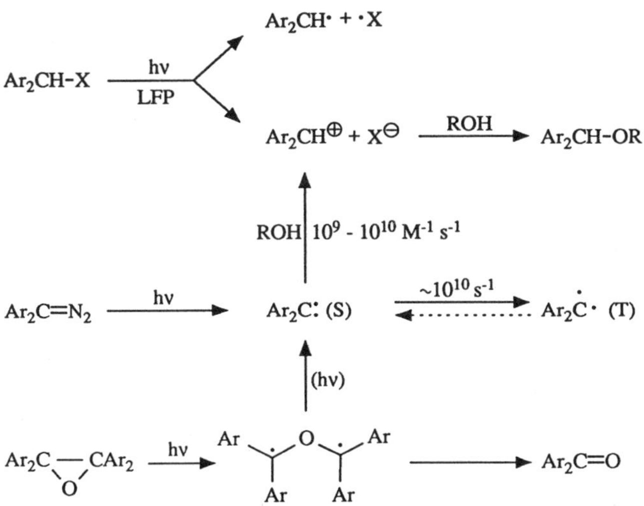

Scheme 12.

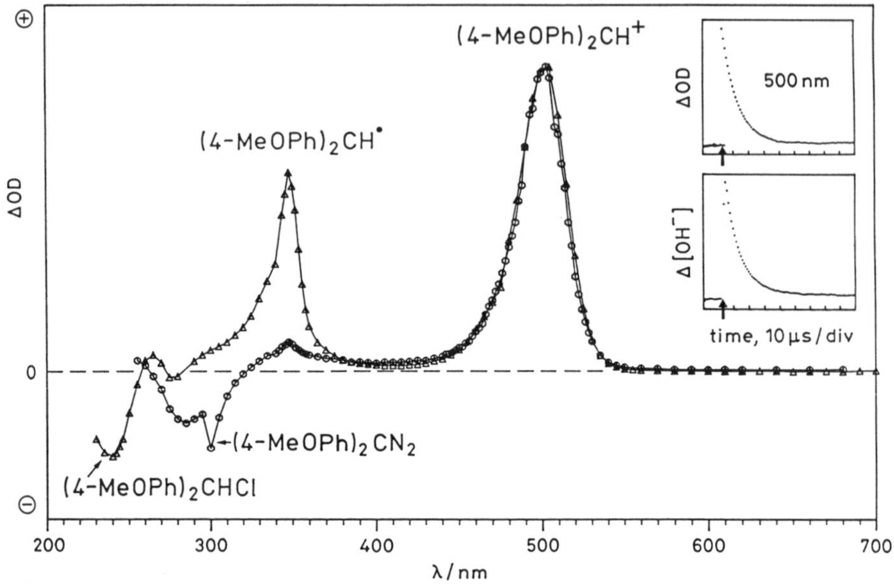

Figure 1. Changes in optical density (OD) observed on 248 nm photolysis of (4-$MeOC_6H_4)_2CN_2$ (○) and (4-$MeOC_6H_4)_2CHCl$ (△) in acetonitrile–water. Insets: Comparison of the decay of Ar_2CH^+, monitored at 500 nm (upper inset), with the decrease of conductance (lower inset).

excellent agreement (Table 5). Additional support was provided by the use of time-resolved conductivity detection in weakly basic solution (Figure 1). The observed steep increase in conductance after the pulse is due to protonation of the carbene by H_2O to give the carbenium ion and OH^-. The subsequent decrease in conductance is caused by reaction of the cation with solvent, neutralizing OH^-.

Table 5. Diarylcarbenium Ions Generated by Protonation of Diarylcarbenes (by Photoheterolysis of Ar_2CHX)

Substituents				
4-	*4′-*	*Solvent*	λ_{max} *(nm)*	*k* (s^{-1})
MeO	MeO	MeCN–H_2O (1:2)	500 (500)	1.2×10^5 (1.3×10^5)
MeO	H	MeCN–H_2O (1:4)	455 (455)	2.3×10^6 (2.1×10^6)
Me	Me	MeCN–H_2O (1:2)	462 (460)	4.2×10^7 (3.2×10^7)
H	H	MeCN–TFE (1:1.6)	430 (435)	6.6×10^6 (6.8×10^6)
		TFE	435 (440)	3.2×10^6 (3.2×10^6)
Cl	Cl	MeCN–TFE (1:1.6)	470 (472)	5.9×10^6 (7.0×10^6)
CF_3	H	TFE	425 (425)	3.7×10^7 (3.8×10^7)

Laser flash photolysis of 4-methoxy- and 4,4′-dimethyldiphenyldiazomethane in acetonitrile–water also generated diarylcarbenium ions whose absorption and reactivity were in good agreement with previous measurements (Table 5). On nanosecond LFP of diphenyldiazomethane in acetonitrile–water, the absorption of the parent diphenylcarbenium ion was detected but decayed too rapidly for kinetic investigation. With enhanced time resolution, using picosecond LFP of diphenyldiazomethane, Ph_2CH^+ was observed in acetonitrile–water (1:1), 2-propanol, ethanol, and methanol with lifetimes of 750, 85, 70, and 40 ps, respectively.[62] Much smaller rates of decay were measured in weakly nucleophilic solvents, such as 2,2,2-trifluoroethanol (TFE) (Table 5).[61] Protonation by TFE was also found for some diarylcarbenes bearing electron-withdrawing groups (e.g., 4,4′-Cl and 4-CF_3), but failed with 4,4′-CF_3 substitution. In each case of diarylcarbenium ion intervention, the absorptions were formed "instantaneously," within the excitation pulse. Obviously, singlet diarylcarbenes are protonated by ROH in strong competition with intersystem crossing to ground state triplets.

Recent reports indicate that O–H insertion reactions of arylmethoxycarbonylcarbenes (**89**) to give the hydroxy esters **92** are strongly influenced by the nature of the aryl group.[63,64] Laser flash photolysis of the diazo precursors **88** generated the carbenes **89** as short-lived transients that were rapidly quenched by water. The

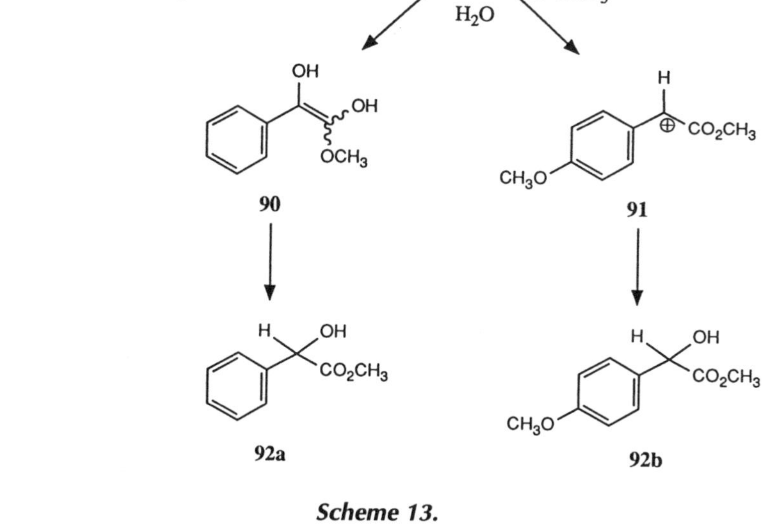

Scheme 13.

second transient obtained from **89a** was assigned as the enol of methyl mandelate (**90**) (Scheme 13). The decay of **90** was subject to general acid and base catalysis, the pH profile being characteristic for ketonization of an enol.[63] The formation of **90** suggests that **89a** reacts with water by way of the ylide mechanism. In contrast, the second transient arising from **89b** was identified as the cation **91** since the rate of decay was strongly affected by nucleophiles ($k_{TFE} = 1.6 \times 10^5\ s^{-1}$; $k_{H_2O} = 2.0 \times 10^7\ s^{-1}$; $k_{Br^-} = 6.0 \times 10^9\ M^{-1}\ s^{-1}$).[64] The change in mechanism is obviously due to the cation-stabilizing effect of the *p*-methoxy group.

Photolysis of 3*H*-pyrazoles (**93**) is known to generate vinylcarbenes (**94**), presumably by way of diazoalkenes.[65] Photolysis of **93a–e** in protic solvents revealed that protonation of **94** is competitive with 1,3-cyclization of **94a** and **94b** to give cyclopropenes **95**, but not with 1,5-cyclization of **94c** and **94d**, yielding indenes **96** (Scheme 14).[66] Thus **93b** (= **93c**) generated a mixture of **94b** and **94c** of which

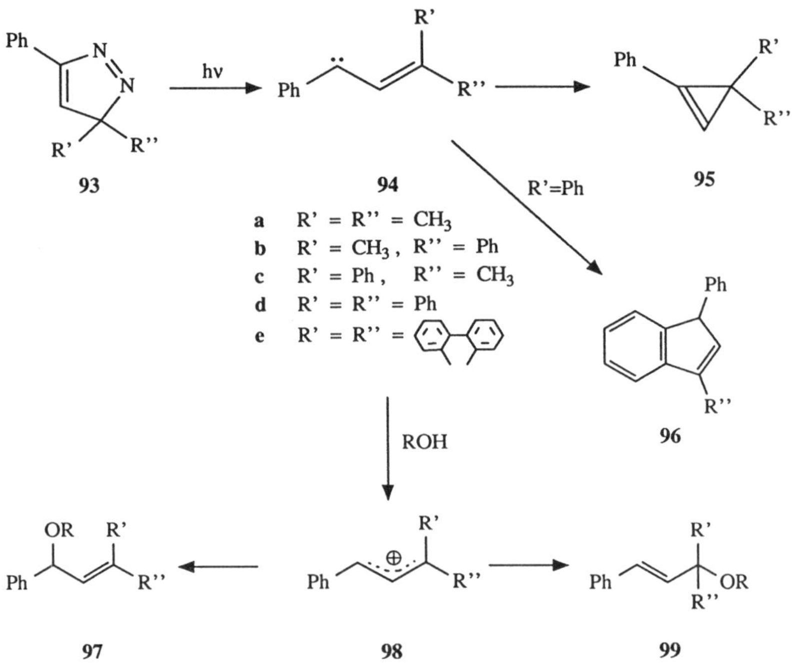

	λ max (nm)	k (s^{-1}) TFE	k (s^{-1}) CH_3CN - H_2O
a	405	$1.2 \cdot 10^5$	$1.1 \cdot 10^8$
b	490	$1.7 \cdot 10^4$	$2.1 \cdot 10^6$
e	550	$3.9 \cdot 10^3$	$1.9 \cdot 10^6$

Scheme 14.

only **94b** was trapped intermolecularly with formation of **97b** and **99b**. The *Z* isomer **97c** was not detected, due to rapid conversion of **94c** into **96c**. Laser flash photolysis of **93a** and **93b** gave rise to transient spectra of the cations **98a** and **98b**, with appropriate absorption maxima and decay kinetics. In contrast, neither ethers nor cationic transients were obtained from **93d**. Somewhat unexpectedly, LFP of **93e** generated the well-behaved cation **98** whose reaction with nucleophiles afforded **97e**. Relatively slow 1,5-cyclization of **94e**, due to ring strain of the resulting cyclopentindene skeleton, is thought to account for the divergent reactivities of **94d** and **94e**.

The protonation of oxacarbenes to give carboxonium ions has also been confirmed by time-resolved spectroscopy. Two processes were found to compete in the LFP of benzocyclobutenedione (**100**).[63] One leads to the formation of the long-lived bisketene **101** (λ_{max} = 380 nm) (Scheme 15). The other process leads to the oxacarbene **102** which could be monitored indirectly, using pyridine and adamantanethione as probes. The much faster reaction of **102** with TFE ($k = 1.2 \times 10^8\ M^{-1}\ s^{-1}$) than with methanol ($k = 3.4 \times 10^5\ M^{-1}\ s^{-1}$) is indicative of protonation, as is a transient absorption ($\lambda_{max} \approx 370$ nm) attributed to the cation **103**. Equilibration of **102** with **103**, and overlapping absorptions of **101** and **103**, complicate interpretation of the data but do not explain the failure to stabilize **103** with TFE.

A different approach to oxacarbenes starts from acylsilanes (**105**). The excited triplet state of the acylsilanes was shown to isomerize to siloxycarbenes (**106**),[68] which can be captured by alcohols to form acetals (**109**)[69] or by pyridine to give transient ylides (**107**) (Scheme 16).[70] The relative reactivity of alcohols toward

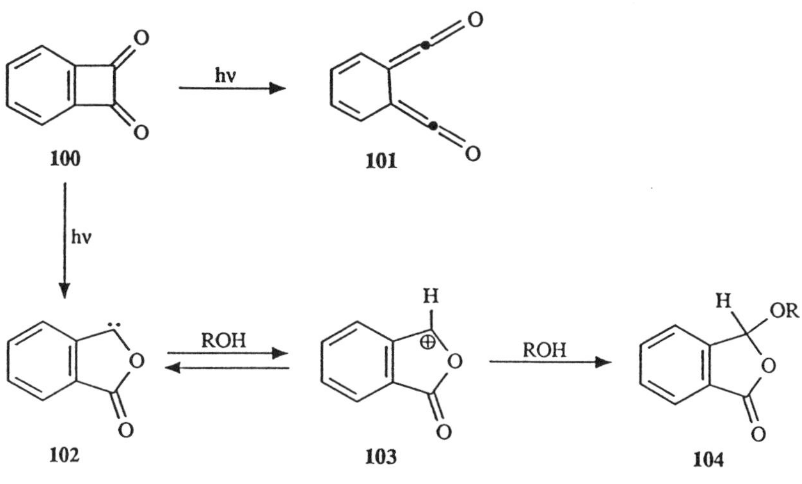

Scheme 15.

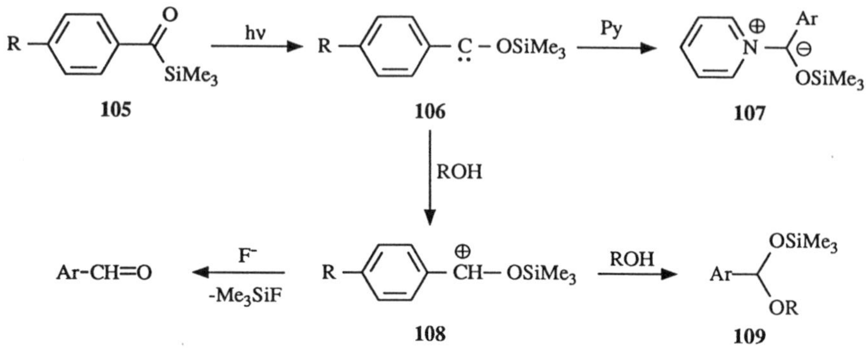

R	λ max (nm)	k (s^{-1}) TFE	HFIP
H	305	$1 \cdot 10^7$	$1 \cdot 10^4$
CH_3	315	$2 \cdot 10^6$	$7 \cdot 10^3$
OCH_3	355	$2 \cdot 10^5$	$2 \cdot 10^2$

Scheme 16.

106 was found to parallel the acidity of the alcohols, suggesting a proton transfer mechanism.[70] Laser flash photolysis of **105** in TFE or HFIP produced transient absorptions of the cations **108**, which were characterized by their reactions with nucleophiles.[71] The relative reactivity of halide ions toward **108**, $F^- >> Cl^- > Br^-$, points to an attack of the fluoride ion at silicon. The formation of aldehydes rather than acetals in the presence of fluoride supports the desilylation route of **108**.

V. SCOPE AND LIMITATIONS OF O–H INSERTION REACTIONS

A. Methylene

Reports on the insertion reaction of methylene with O–H bonds mark the renaissance of carbene chemistry five decades ago. In their pioneering work on the photolysis of diazomethane in the presence of 2-propanol, Meerwein and his students obtained 2-methoxypropane, 2-butanol, and 2-methyl-2-propanol by competitive insertion into O–H and C–H bonds.[72] Palazzo observed the formation of methanol on dropwise addition of water to a solution of diazomethane in diethyl ether under the irradiation of sunlight.[73] In a study of the photolysis of ketene in the presence of methanol in the gas phase, C–H insertion was reported to be about

2.5 times faster than O–H insertion[74]; however, this work suffered from an extensive dark reaction of ketene with methanol. Kerr et al. found by gas chromatographic analysis that the relative rates for the insertion of $^1CH_2$ into an O–H bond as compared with insertion into a primary C–H bond are as follows: methanol, 21.8; ethanol, 21.2; 2-propanol, 14.9; and 2-methyl-2-propanol, 10.9.[75] Similar data (methanol, 33; ethanol, 21) were reported in an independent study.[76]

In the photolysis of diazomethane in 1 atm of dry N_2 and 0–4 torr of water, methanol and ethylene were formed competitively, with the rate constant ratio $k(CH_2N_2)/k(CH_3OH) = 10 \pm 2$.[77] On the basis of these data, rate constants for the insertion of $^1CH_2$ into the O–H bonds of methanol were estimated ($k = 1.8 \times 10^9$–$2.5 \times 10^{10}\ M^{-1}\ s^{-1}$). Absolute values for the removal rates of $^1CH_2$ by water[78] and methanol[79] have been determined in the gas phase at low pressures. Methylene was generated by pulsed laser photolysis of ketene. The measurements were carried out by observing the laser-induced fluorescence of $^1CH_2$ after excitation by a pulsed laser as a function of the time delay between the photolysis and analysis pulses. The overall rate constants were found to be $1.3 \times 10^{11}\ M^{-1}\ s^{-1}$ and $1.6 \times 10^{11}\ M^{-1}\ s^{-1}$ for water and methanol, respectively. These rate constants include physical deactivation of $^1CH_2$, leading to $^3CH_2$. The intersystem crossing channel was investigated by monitoring the formation of $^3CH_2$ with laser magnetic resonance. The branching ratios of physical quenching versus total removal of $^1CH_2$ turned out to be 0.30 and 0.13 for water and methanol, respectively.

Not surprisingly, the O–H insertion reaction of methylene has also been scrutinized by *ab initio* calculation. As was briefly mentioned in Section IV, the depth of the potential well corresponding to the methyleneoxonium ylide H_2COH_2 depends strongly on the basis set used. The barrier for the rearrangement $H_2COH_2 \rightarrow CH_3OH$ virtually disappears when electron correlation is included.[33,34] Analogous calculations for the reaction of methylene with methanol (6-31 G**, no electron correlation)[80] produced a reaction profile very similar to that obtained for water at the same level of approximation. The authors suggested that the intervention of a "stable" intermediate in O–H insertion reactions, as opposed to C–H insertion, should reverse the ratio of O–H to C–H insertion at low temperatures.

B. Dialkylcarbenes and Cycloalkylidenes

Attempts at observing the insertion of (di)alkylcarbenes into O–H bonds (**116** → **117**) are often frustrated by undesired reactions of the carbene precursors. The majority of diazoalkanes (**113**) are protonated even by weakly acidic alcohols to give alkanediazonium ions (**114**) and products derived therefrom (Scheme 17).[81] As a rule, (di)alkylcarbenes (**116**) are not accessible from **113** in protic solvents. Generation of **113** *in situ* does not solve the problem. Photolyses of tosylhydrazones (**110**) in dilute aqueous base and nitrous acid deamination reactions of the analogous amines (**111**) in dilute aqueous acid give virtually identical product distributions.[82] Obviously, both reactions proceed by way of diazonium ions (**114**) as the common

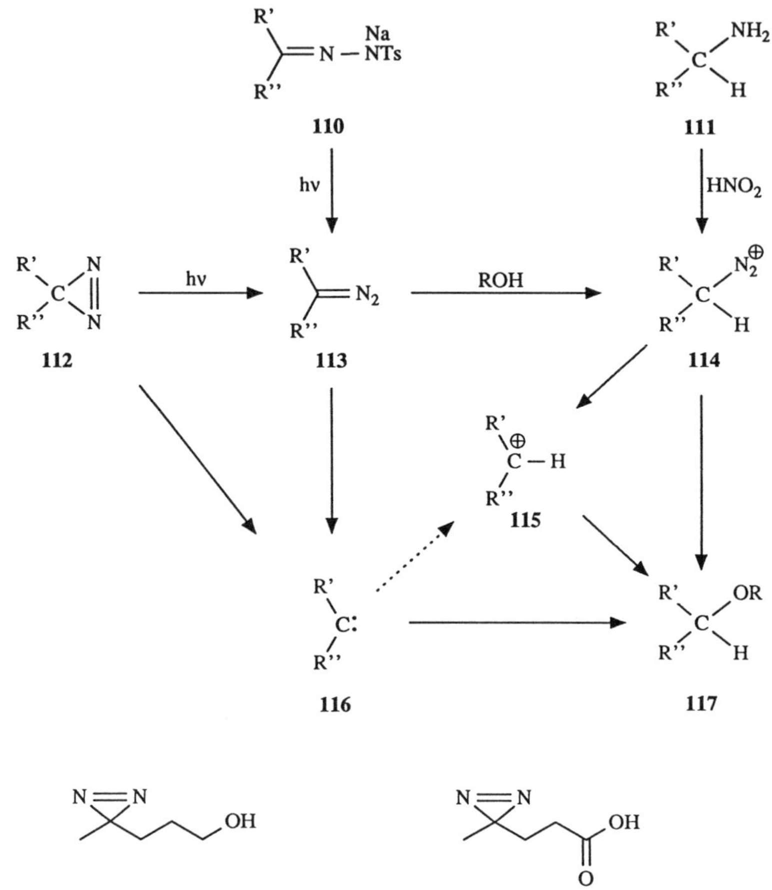

Scheme 17.

intermediates, that is, protonation of **113** is faster than photolysis to give **116**. The protonation of **113** is slowed down, of course, at low concentrations of ROH. Photolysis of diadamantyldiazomethane (**113**; $R^1 = R^2 =$ 1-adamantyl) in *cis*-2-butene with 0.25% of methanol added led to 77% of Ad_2CHOMe.[83] The ether was thought to arise, under these conditions, by O–H insertion of singlet diadamantylcarbene. Laser flash photolysis of Ad_2CN_2 in aerated cyclohexane produced the transient spectrum of a carbonyl ylide which served to monitor the reactions of the invisible carbene with methanol ($k = 2 \times 10^7$–$3 \times 10^7\ M^{-1}\ s^{-1}$) and TFE ($k = 1.5 \times 10^8$–$1.7 \times 10^8\ M^{-1}\ s^{-1}$).[84]

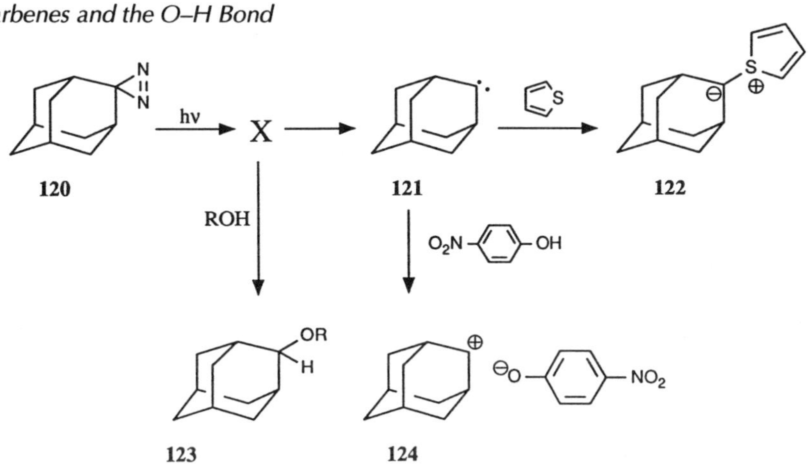

Scheme 18.

Diazirines (**112**) are stable in protic solvents. Photolyses of **112**, however, proceed in part by way of the isomeric diazoalkanes (**113**).[85] With a series of cycloalkanespirodiazirines (**112**; R^1, R^2 = $[CH_2]_n$), the intervention of diazoalkanes was detected by spectroscopic methods as well as by capture with acetic acid.[86] When AcOD was used, ~60% of the products were acetates and deuterated alkenes. During photolyses of the neat diazirines, 30–55% of the analogous diazo compounds were detected by IR spectroscopy. Therefore, the acetates and deuterated alkenes are thought to arise largely from diazoalkanes (**113**) by way of diazonium ions (**114**) and carbocations (**115**) although protonation of the carbenes (**116**) cannot be excluded as a minor reaction path. Surprisingly, thermolyses and photolyses of the diazirines **118** and **119** did not afford any cyclic ethers and lactones, respectively.[87] 1,2-Shifts of hydrogen to give alkenes must either proceed faster than the intramolecular O–H insertion of the carbenes, or in concert with the extrusion of nitrogen.

In an LFP study of the diazirine **120**, 2-adamantylidene (**121**) was trapped by thiophene to produce the intense transient absorption (390 nm) of the ylide **122** (Scheme 18).[88] Alcohols effectively quenched the yield of **122** but had no effect on the rate of formation of **122**. These observations suggest that a precursor X to **121** is intercepted by alcohols. X may be an excited state of either diazirine **120**, the corresponding diazo compound, or carbene **121**. In contrast, *p*-nitrophenol was found to react with **121** at a diffusion-controlled rate, with formation of the ion pair **124**.

In view of the complications encountered with diazo compounds and diazirines in protic media, some effort has been directed toward other carbene precursors. Photolysis of the sterically congested ketene **125** in protic solvents resulted in exclusive formation of the alkene **127** (Scheme 19).[52] The cyclohexylidene **126** is not captured by alcohols, in contrast to the unsaturated analog **48** (Scheme 6). Direct irradiation of alkenes in nonnucleophilic media leads to skeletal rearrangements that appear to involve carbene intermediates.[89] The carbenes are thought to arise from the π,R (3s) Rydberg excited states of the alkenes by retro-1,2 shifts. As a rule, the Rydberg excited states are trapped nucleophilically by alcohols, thus precluding O–H insertion reactions of the carbenes. However, sterically congested alkenes, such as **128**, are exceptional.[90] On irradiation of **128** in pentane, the products **130** and **133** were formed, most probably via C–H insertion of the carbene **129**. In methanol **128** additionally afforded the ether **131** and the rearranged alkenes **134**

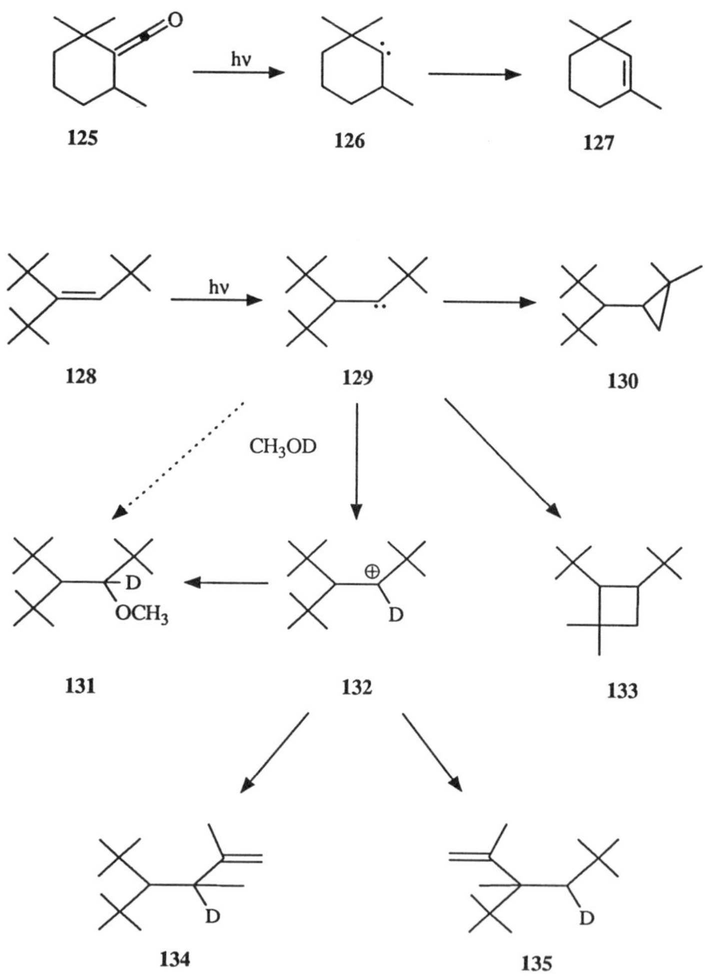

Scheme 19.

and **135**, presumably derived from protonation of the carbene **129**. The proposed mechanism is supported by incorporation of deuterium from MeOD (nucleophlic trapping of Rydberg excited states does not produce deuterated products). At first sight, the reactivity of **126** and **129** toward methanol appears to differ substantially. We note, however, that regeneration of **128** by 1,2-H shift may well be a major, albeit undetected reaction of **129**.

A variant of the di-π-methane rearrangement, also applicable to triptycenes,[91] was used to generate the cyclopropylcarbene **137** from **136** (Scheme 20).[92] Formation of the cyclobutene **138** in aprotic media is in accordance with the established behavior of cyclopropylcarbenes. In the presence of methanol, the methyl ethers **139** and **140** (≈ 1:1) were obtained.[93] The product distribution suggests that **129** is protonated to give a cyclopropylcarbinyl cation.

Carbene–carbene rearrangements may also help to avoid the interference of leaving groups. The protonation of 3-cyclopentenylidenes, generated by the Skattebøl rearrangement, has been discussed in Section IV (cf. Schemes 4 and 7). These species, however, cannot be regarded as "true" cycloalkylidenes, owing to interactions between the divalent carbon and the π bond. Recent studies have shown that homocubylidenes (**146**), generated directly or by rearrangement of cubylcarbenes (**144**), react readily with alcohols (Scheme 21).[94,95] The ratio of iso(topo)meric products, **147:148**, was found to depend on the carbene precursors and on the concentration of R′OD. The skeletal reorganization leading to **147** cannot be taken as evidence for the intervention of carbocations. It was argued that the degenerate rearrangement of the homocubyl cation (R = H) should lead to further distribution

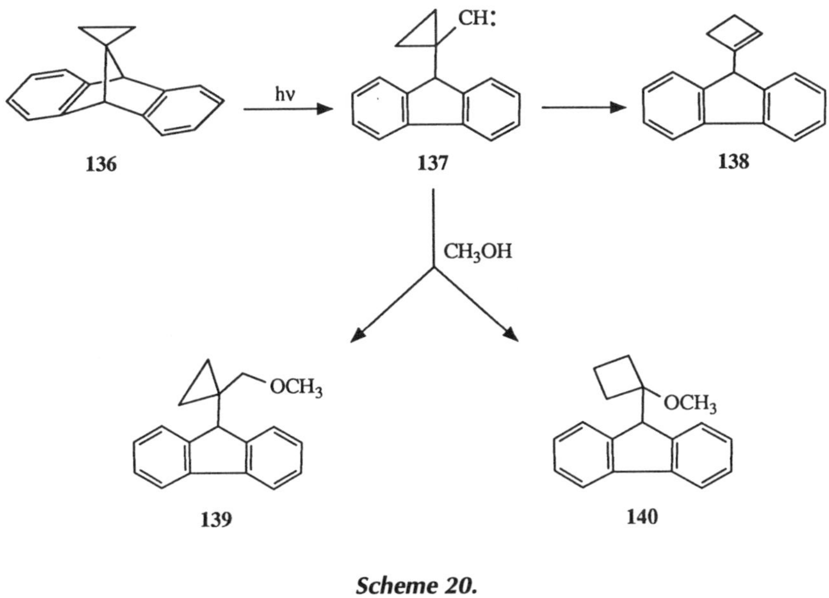

Scheme 20.

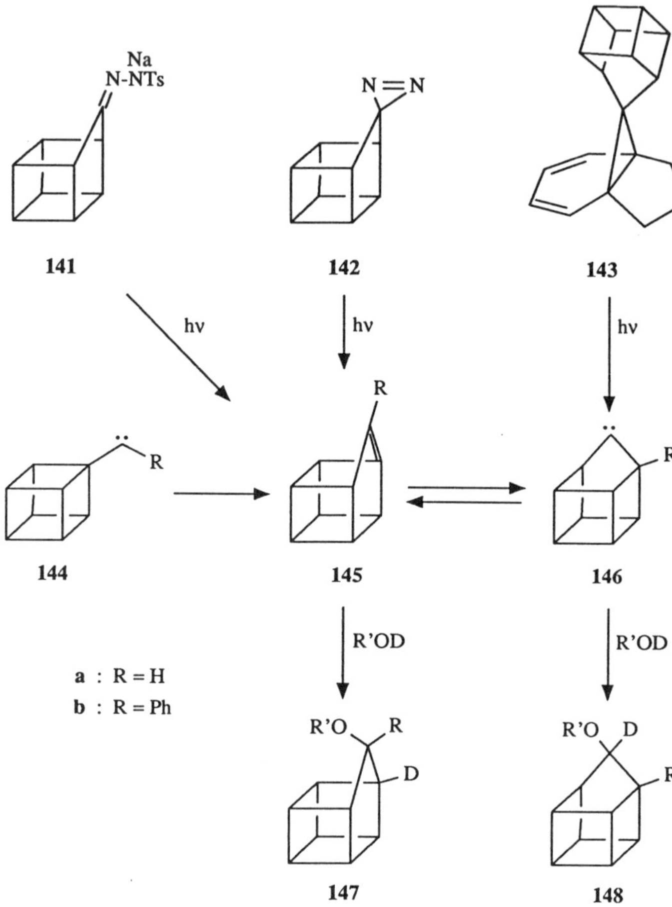

Scheme 21.

of the deuterium. Addition of R′OD to the highly strained alkenes **145** is a plausible route to **147**. Homocub-1(9)-ene (**145a**) and homocub-9-ylidene (**146a**) appear to be in equilibrium, with *K* close to unity.[95] Trapping by R′OD of both **145a** and **146a** is competitive with equilibration. Depending on the precursor, the equilibrium can be approached from either side. Photofragmentation of **143** leads directly to **146a** whereas nitrogen extrusion from **141** and **142** proceeds in concert with alkyl migration to give **145a**.[95]

Intermolecular reactions of cyclopropylidenes, including O–H insertion, are often competitive with ring opening to allenes.[96] In contrast to the majority of diazoalkanes, diazocyclopropanes thermolyze rapidly at ambient temperature. The diazo–diazonium dichotomy discussed above (Scheme 17) applies here, too, but

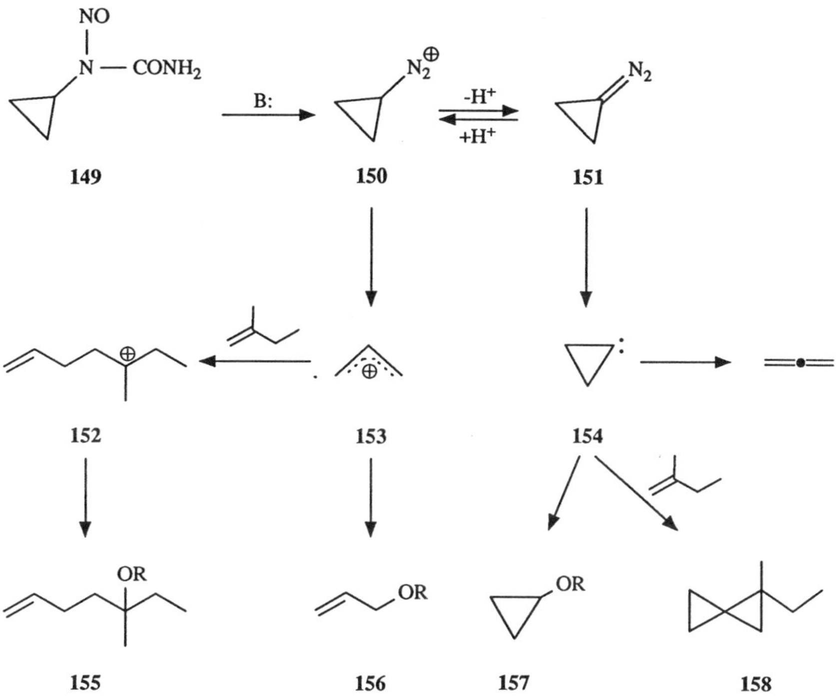

Scheme 22.

the reaction paths are more readily dissected. In the absence of steric constraints, cyclopropanediazonium ions give rise to allylic products whereas cyclopropylidenes produce cyclopropyl derivatives and allenes. Deacylation of *N*-nitroso-*N*-cyclopropylurea (**149**) generates the parent cyclopropanediazonium ion (**150**) in equilibrium with diazocyclopropane (**151**) (Scheme 22). The ratio of cyclopropyl ethers (**157**) to allyl ethers (**156**) was shown to rise steeply with increasing concentration of base, due to displacement of the equilibrium **150** $\rightleftharpoons$ **151** in favor of **151**.[97] Added 2-methyl-1-butene was found to trap the allyl cation (**153** → **152** → **155**) as well as cyclopropylidene (**154** → **158**) but gave no indication of the hypothetical cyclopropyl cation.[98] In neutral methanol, the analogous 9-bicyclo[6.1.0]nonyl system, **159** $\rightleftharpoons$ **160**, afforded 3-methoxy-*E*-cyclononene (**162**) exclusively (Scheme 23). With increasing concentration of methoxide, *exo*-9-methoxybicyclo[6.1.0]nonane (**163**) and 1,2-cyclononadiene (**164**) were formed in increasing amounts, but in a constant ratio (1:15).[99] These data support 9-bicyclo[6.1.0]nonylidene (**161**) as the common precursor to **163** and **164**.

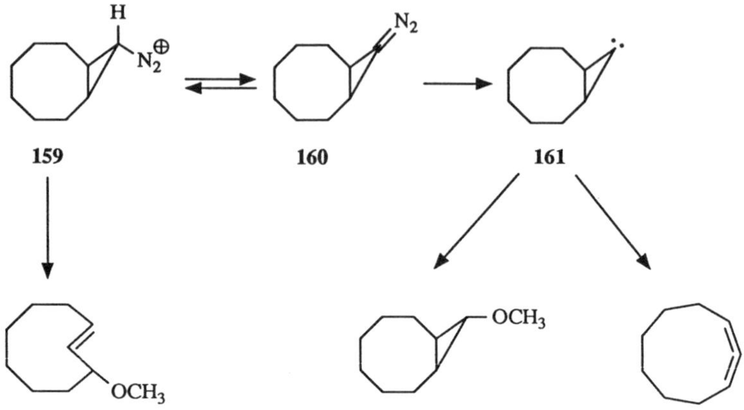

Scheme 23.

A detailed mechanistic study has been performed on the O–H insertion reaction of bicyclo[5.1.0]oct-2-en-8-ylidene (**165**).[100] Trapping of **165** by alcohols is competitive with the Skattebøl rearrangement leading to bicyclo[3.2.1]oct-6-en-8-ylidene (**168**) (Scheme 24). Protonation of this homoconjugated carbene (**168** → **37** → **38**) is sufficiently established (cf. Section IV. A.)[49] to provide a standard for comparison with **165**. Separate lines of evidence, involving temperature-dependent

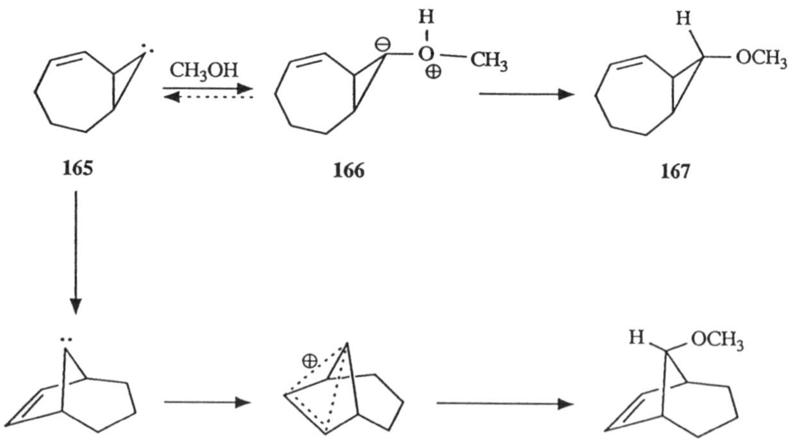

Scheme 24.

isotope effects and product ratios, suggest that **165** reacts with methanol via (reversible) formation of the ylide **166**.

Despite the effort expended on dialkylcarbenes, a consistent mechanistic picture has not yet emerged. It appears that dialkylcarbenes undergo intramolecular 1,2-H shifts in preference to intermolecular O–H insertion. Significant barriers to rearrangement facilitate the reaction with alcohols (cf. **121** and **146**). Suggestive evidence for and against protonation has been adduced (**121**, **129** and **137** vs. **146**, **154** and **165**). The cause of divergent mechanisms is not always clear. It seems plausible that cyclopropylcarbenes should be protonated while cyclopropylidenes should not, but what distinguishes 2-adamantylidene (**121**) from homocubylidene (**146**)? More work is clearly required to correlate the structure and reactivity of dialkylcarbenes.

C. Arylcarbenes

Although arylcarbenes are correctly defined as Ar–CH:, the term will be used here in a broader sense to include diarylcarbenes, arylalkylcarbenes, and arylvinylcarbenes as well. Arylcarbenes bearing acyl groups or hetero atoms as the second substituent will be covered in Sections V. D.–V. F.

Aryldiazomethanes are more stable and less basic than their aliphatic analogs. Therefore, many arylcarbenes can be generated in protic media by photolysis of diazo or diazirine precursors. Among other criteria, the O–H insertion reaction has been taken as evidence for the intervention of arylcarbenes in the photolysis of arylcyclopropanes (**169**),[101] aryloxiranes (**170**),[101] tri- and tetraarylmethanes (**171**),[102] arylallenes (**172**),[103] and triptycenes (**173**)[91,104] (Scheme 25). Laser flash photolysis of **170** (Ar = R = Ph),[61] **172** (Ar = Ph),[66] and **173** (R = Ph)[66] in TFE produced the transient absorption spectra of arylcarbenium ions.

The reaction rates with alcohols for a variety of arylcarbenes have been measured by time-resolved spectroscopy (Table 6; see also Section II, Table 1). The data in Table 6 refer to the triplet ground states of the arylcarbenes; the magnitude of the singlet–triplet gap is thought to correlate inversely with the apparent rate of O–H insertion. Diffusion-controlled rates of singlet arylcarbenes have been derived from a Stern–Volmer-type analysis (Section II). Despite variations in mechanism (Section IV. C), good to excellent yields of ethers were obtained from all arylcarbenes listed in Table 6. In contrast, anthronylidene (**183**),[9] 9-mesityl-9-boraanthrylidene (**184**),[114] and 3,5-diazafluorenylidene (**185**)[115] were found to react with alcohols predominantly by hydrogen abstraction to give radicals and products derived therefrom. Comparison with structurally related carbenes in Table 6 suggests that the deviant reactivity of **183–185** is due to electron-withdrawing resonance effects that destabilize the singlet state. Dianthrylcarbene (**186**) cannot be trapped by methanol, even if generated by direct photolysis of the diazo precursor.[116] The exceptional properties of triplet **186**—self-reaction is some four orders of magni-

Table 6. Reaction Rates of Arylcarbenes with Alcohols (at 25 °C)

Carbene		*Alcohol*	*Solvent*	$k(M^{-1}s^{-1})$	*Ref.*
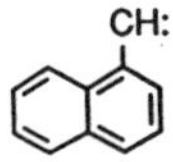	**175**	MeOH	Freon 113	1.6×10^7	105
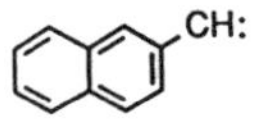	**176**	MeOH	Isooctane	7.3×10^6	106
Ph_2C:	**3a**	MeOH	Benzene	2.0×10^7	18a
		MeOH	Isooctane	6.4×10^6	18a
		MeOH	Acetonitrile	2.4×10^7	17
		i-PrOH	Acetonitrile	1.5×10^7	17
		t-BuOH	Acetonitrile	6.1×10^6	17
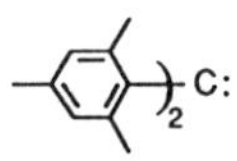	**6**	MeOH	Cyclopentane	$< 2 \times 10^3$	23a
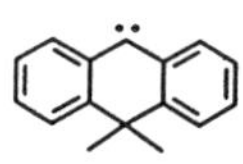	**177**	MeOH		4.6×10^9	27, 107
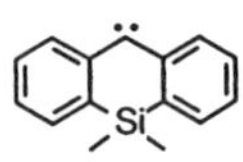	**178**	MeOH	Cyclohexane	6.9×10^6	108
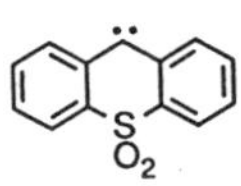	**179**	MeOH	Benzene	9.3×10^6	109
		t-BuOH	Benzene	7.0×10^6	
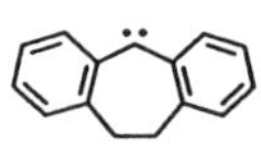	**180**	MeOH	Cyclohexane	1.9×10^6	110
		MeOH	Toluene	2.5×10^6	

(continued)

Table 6. (*continued*)

Carbene		*Alcohol*	*Solvent*	$k(M^{-1}s^{-1})$	*Ref.*
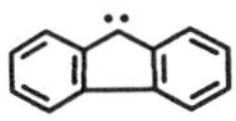	**181**	MeOH	C_6F_6	1.2×10^8	111
		MeOH	Acetonitrile	8.6×10^8	
		EtOH	C_6F_6	2.4×10^8	
		EtOH	Acetonitrile	7.3×10^8	
		i-PrOH	C_6F_6	9.8×10^7	
		i-PrOH	Acetonitrile	5.2×10^8	
		t-BuOH	Acetonitrile	1.6×10^8	112
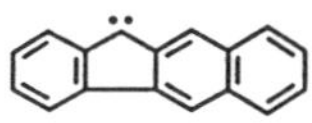	**182**	MeOH	Acetonitrile	3.6×10^9	113
		MeOH	Benzene	6.3×10^9	

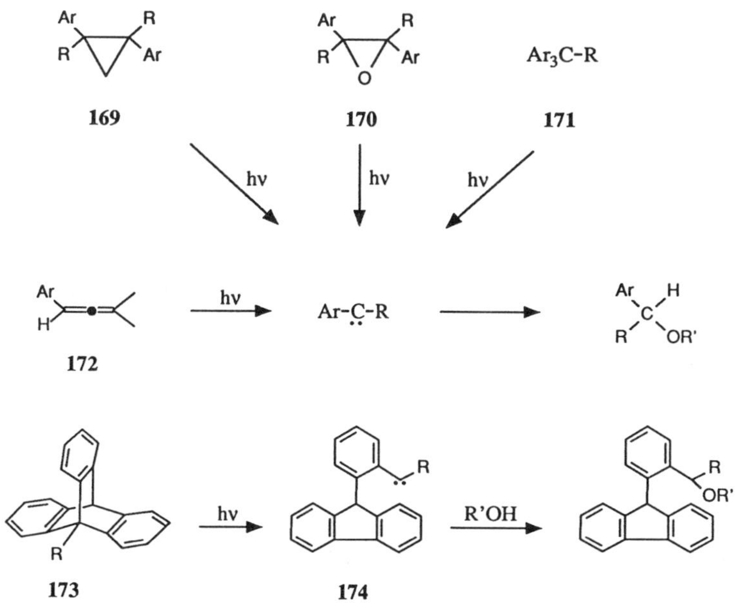

Scheme 25.

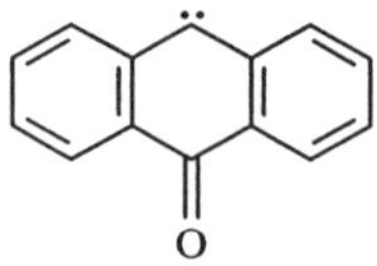

183

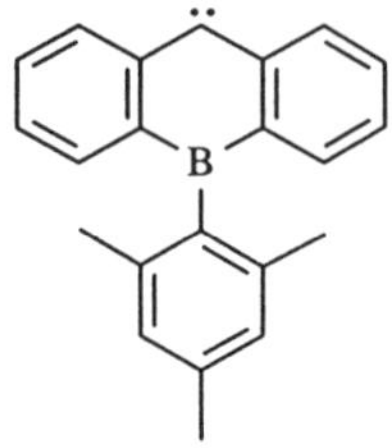

184

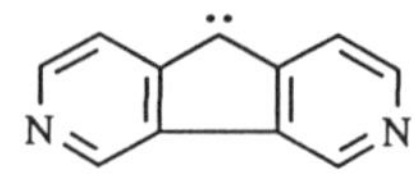

185

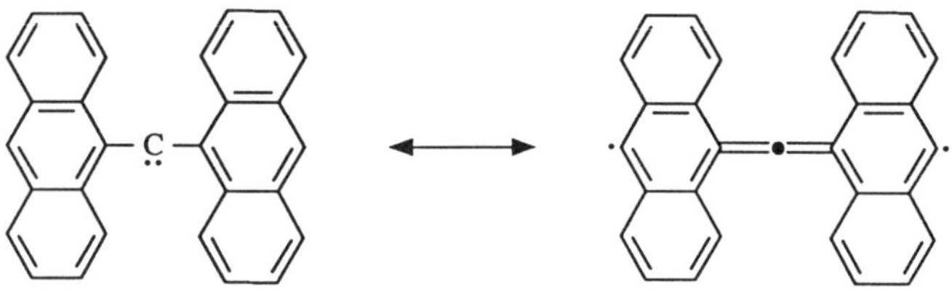

186 **186a**

tude faster than the reaction with oxygen—are consistent with a linear, allenic structure **186a**.

Intermolecular competition revealed that phenylcarbene prefers addition to 3-butene over insertion into the O–H bond of methanol by a factor of 2.1–2.2.[117] For substituted phenylcarbenes, the relative rates of O–H insertion and alkene addition have been correlated with σ^+ to give $\rho = -0.96$ (2-propanol/ethyl vinyl ether)[118] and $\rho = -0.54$ (methanol/*cis*-4-methyl-2-pentene),[119] respectively. The negative ρ values point to enhanced electron demand (= nucleophilicity) in the reaction with alcohols, relative to alkenes. "Spin-equilibrated" diarylcarbenes add very slowly to simple alkenes. Even the more efficient scavengers butadiene, isoprene, and styrene are inferior to methanol; the rate constants differ by at least one order of magnitude.[27] On the other hand, triplet diphenylcarbene was found to react with nitroxyl radicals [2,2,6,6-tetramethylpiperidin-1-oxyl (TEMPO)] ~10 times faster than with O–H bonds.[120] Except for special cases, such as **183–185**, the insertion of arylcarbenes into C–H bonds compares unfavorably with O–H insertion; however, the relative reactivity toward C–H and O–H bonds changes dramatically when arylcarbenes are generated in rigid matrices at low temperatures. The reader is referred to a thoughtful review of the matrix effect.[121]

Intramolecular reaction can divert functionalized arylcarbenes from O–H insertion. Electrocyclic ring closure of (2-vinylphenyl)carbene (**188a**)[122] and (2-phenylphenyl)carbene (**188b**)[123] proceeds by way of **189** to give indene and fluorene, respectively (Scheme 26). When these carbenes were generated in methanol, the product ratio **187:190** was close to unity.[59a] The more remote double bond of

(2-allylphenyl)carbene and (2-vinyloxyphenyl)carbene (**192**) was found to be an efficient scavenger for the triplet but not for the singlet state.[117] Consequently, the product ratio **191:193** varied from ~10 in direct photolyses of the diazo precursors ~0.1 in sensitized photolyses. The reactivity of 1-phenyl-2,*n*-dienylidenes (**195, 198**), arising from photolytic cleavage of appropriate cyclopropenes (**194, 197**), is similarly affected by the position of the second double bond. While **195** was not scavenged by methanol,[124] **198** afforded a mixture of allylic ethers (**199** and **200**) which indicates the intervention of a delocalized carbocation[125] (see Section IV. C and Scheme 14 for the related 1,3-diphenylpropenylidenes).

Hetero atoms can replace π bonds as intramolecular carbene traps. The formation of the ylide **202** from (2-methoxymethylphenyl)carbene (**201**) is quenched in part by alcohols, as testified by an excess of **204** over **206** (82:16).[126] In contrast, the

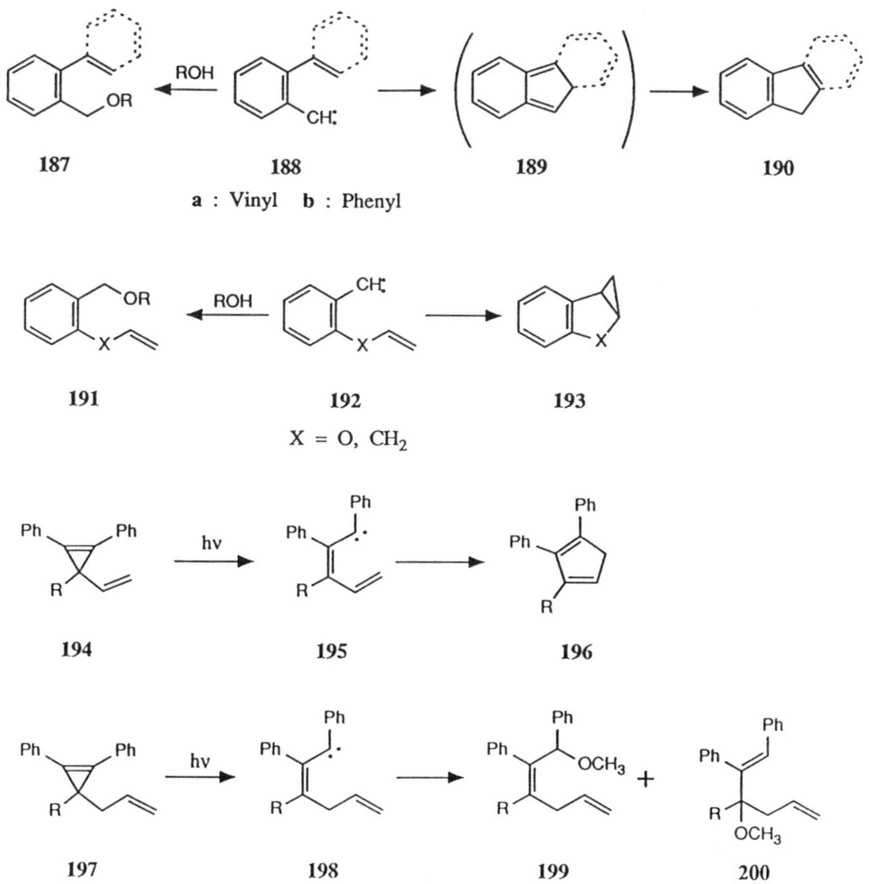

Scheme 26.

carbene **207** appears not to be scavenged en route to the sulfur ylide **208**.[127] The Stevens rearrangement of **208**, leading to **209**, was not affected by the presence of methanol (Scheme 27).

The competition of intramolecular and intermolecular O–H insertion reactions is of particular interest. The case of (2-hydroxymethylphenyl)carbene (**82**) has been reviewed in Section IV. B (Scheme 11). The carbenes **210** feature hydroxyl and alkyl groups α to the divalent carbon. Products arising by intramolecular O–H insertion (**211**) and alkyl migration (**212**) predominate in aprotic solvents (Scheme 28). In the presence of methanol, intermolecular O–H insertion produces **213** and **214** at the expense of **211** and **212**.[128] As a rule, arylalkylcarbenes prefer the reaction with protic solvents to intramolecular rearrangement. Exceptions are found among highly strained systems. Thus cubylphenylcarbene (**144b**; Scheme 21) cannot be intercepted by methanol prior to ring expansion.[94] Thermolysis of the tosylhydrazone salt **215** in refluxing ethanol afforded the products **218** and **219** in a 9:5 ratio.[129] It appears that 1-bicyclo[1.1.1]pentylphenylcarbene (**216**) undergoes O–H insertion competitively with alkyl migration (**216** → **217**).

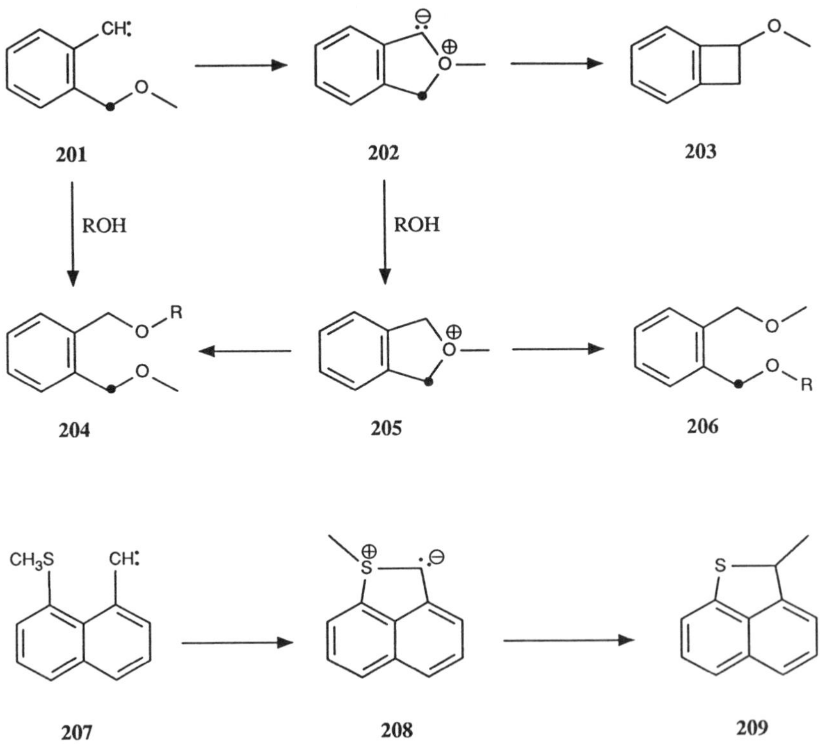

Scheme 27.

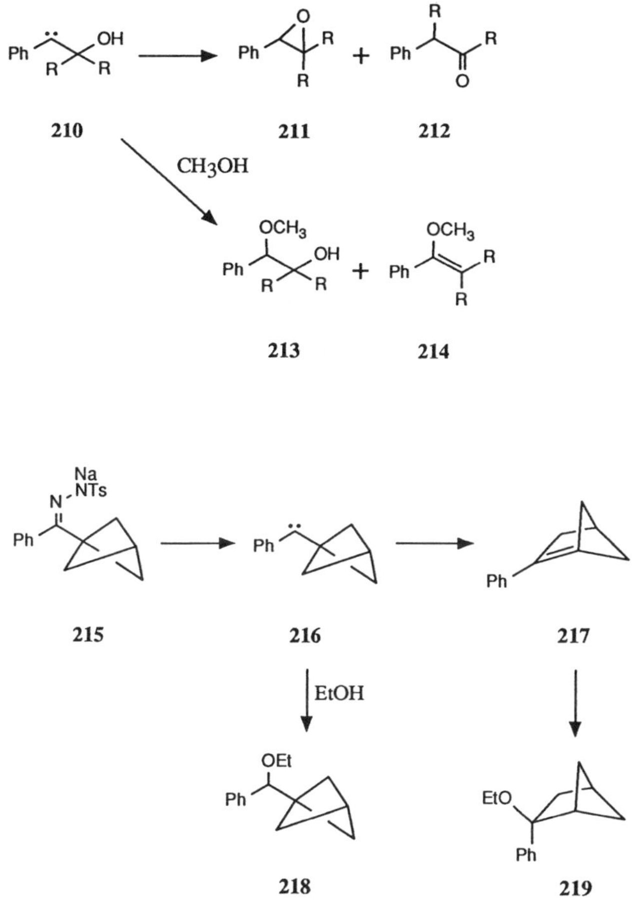

Scheme 28.

In summary, the propensity of arylcarbenes for O–H insertion reactions has been amply demonstrated by various techniques. We are tempted to attribute the almost universal reactivity of arylcarbenes toward O–H bonds to the stabilizing effect of aryl groups on the intermediates (mostly carbocations) of the insertion process.

D. Acylcarbenes

The photolysis of α-diazocarbonyl compounds (**220**) in alcohols can involve any of the following reactions: Wolff rearrangement to give ketenes (**226**) and eventually esters (**227**); insertion of the acylcarbene into C–H bonds (**221** → **223**) and

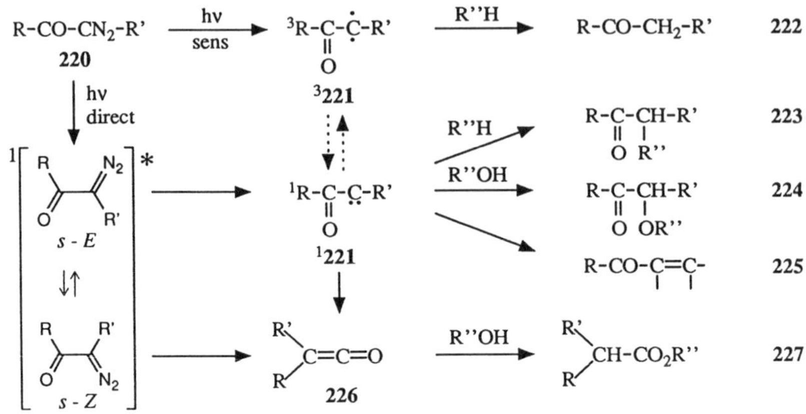

Scheme 29.

O–H bonds (**221** → **224**); abstraction of hydrogen by the acylcarbene (**221** → **222**); 1,2-shifts from R′ to the divalent carbon, with formation of enones (**225**) (Scheme 29). The Wolff rearrangement[1,130] proceeds, at least in part, concertedly with the extrusion of nitrogen from the excited singlet state of **220**. Diazoketones with locked *s-Z* conformation (e.g., R,R′ = $[CH_2]_{2-5}$) undergo the Wolff rearrangement almost exclusively whereas diazoketones with dominant *s-E* conformation (e.g., R = R′ = *t*-butyl) do not.[131] The migratory aptitudes in the photochemically induced Wolff rearrangement are in the following order: H > alkyl > aryl > SR > NR_2 > OR. Although sensitization favors the formation of **222**, the Wolff rearrangement is not completely quenched.[132] These observations indicate that singlet acylcarbenes, formed by ISC from the triplets, undergo the Wolff rearrangement competitively with intermolecular reactions.

As a consequence of facile intramolecular reorganization, insertion into O–H bonds is only a minor reaction of many acylcarbenes. (Alkoxycarbonyl)carbenes (**221**; R = OR″) stand out as notable exceptions. The O–H insertion reaction of (ethoxycarbonyl)carbene competes favorably with C–H insertion in the case of methanol[133] and 2-methyl-2-propanol[134] (OH: 1° CH = 17) but not in the case of 2-propanol (OH: 3° CH = 0.25).[132] Bis(alkoxycarbonyl)carbenes undergo O–H insertion as well as hydrogen abstraction, even in direct photolyses of dialkyl diazomalonates (**220**; R = OR″, R′ = $CO_2R″$).[135] The insertion of (alkoxycarbonyl)carbenes into O–H bonds is thought to proceed by way of the ylide mechanism. Supporting evidence comes from the photolysis of dimethyl diazoglutaconate (**228**) in methanol which afforded the products **229**, **230**, and **234** (≈ 1:1:1) whereas **232** was not observed (Scheme 30).[136] The ylide **233** is a plausible intermediate en route to **234**. Better yields of **234** were obtained from the Rh(II)-catalyzed decomposition of **228** in methanol.[137] Rhodium carbenoids proved to be very useful in

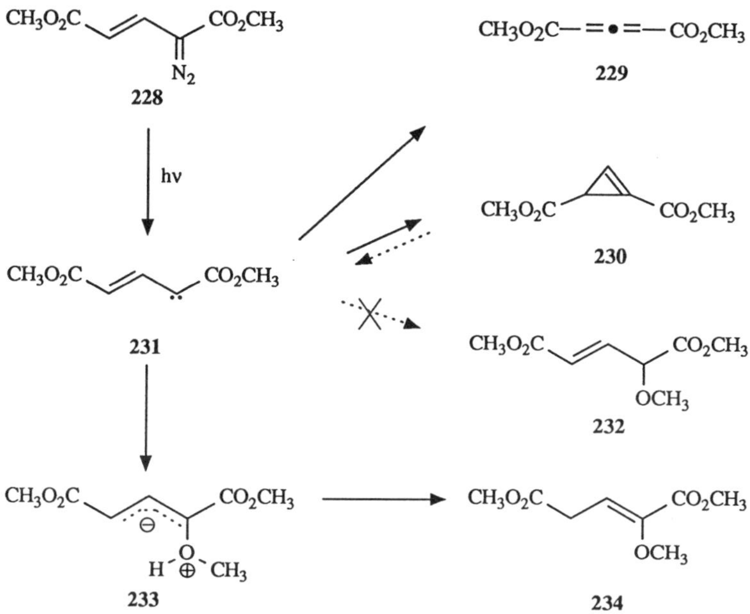

Scheme 30.

the synthesis of α-alkoxycarbonyl compounds, including six- to eight-membered rings, from diazo precursors.[138]

Interactions of the divalent carbon with aryl groups influence the reactivity of acylcarbenes substantially. The photolysis of $PhCN_2COCH_3$ in methanol proceeded with 35% O–H insertion,[139] as compared to virtually nil in the case of $CH_3CN_2COCH_3$. The triplet ground state of (methoxycarbonyl)(phenyl)carbene (**89a**; Scheme 13) was found to react rapidly with methanol ($k = 2.7 \times 10^8\ M^{-1}\ s^{-1}$).[140] According to recent LFP studies, the insertion of **89a** into O–H bonds appears to be channelled through an electrophilic singlet state (cf. Section IV. D).[63] In contrast, the 4-methoxy derivative **89b** gave rise to the carbocation **91**.[64] Not surprisingly, the 2-furyl group resembles anisyl in conferring nucleophilic reactivity on (alkoxycarbonyl)carbenes. The products **237** and **238**, obtained from the carbene **235** in a 3:1 ratio, are indicative of the delocalized carbocation **236** as an intermediate.[141] Analogous considerations apply to the allyl ether **240**, derived from the carbene **239** (Scheme 31).[136] The divergent behavior **239** and **231** (Scheme 30) demonstrates the dramatic effect of exchanging phenyl for methoxycarbonyl. Alternatively, nucleophilic reactivity can be attained by replacing alkoxycarbonyl with carboxylate. While (methoxycarbonyl)(4-nitrophenyl)carbene (**241**) underwent electrophilic addition to alkenes in preference to O–H insertion, extremely

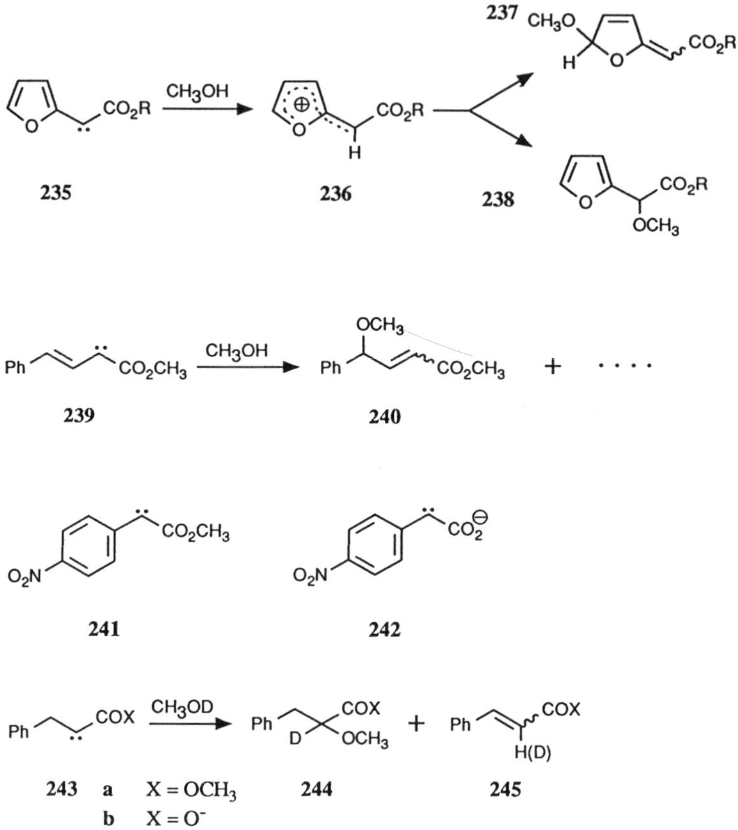

Scheme 31.

efficient capture of the analogous carboxylate **242** by methanol was observed.[142] When the benzylcarbenes **243a** and **243b** were compared, enhanced formation of the ether **244b** and incorporation of deuterium into the alkene **245b** pointed to a change in mechanism.[142b] Neighboring group participation of the carboxylate, with formation of α-lactones, is thought to be involved.

E. Halocarbenes

Dihalocarbenes are conveniently generated by base-induced α-elimination of HX from haloforms, often assisted by phase-transfer catalysis.[1] Haloform-base systems are known to convert alcohols into a variety of products, including alkyl formates, alkyl orthoformates, alkyl halides, and alkenes (Scheme 32). There is

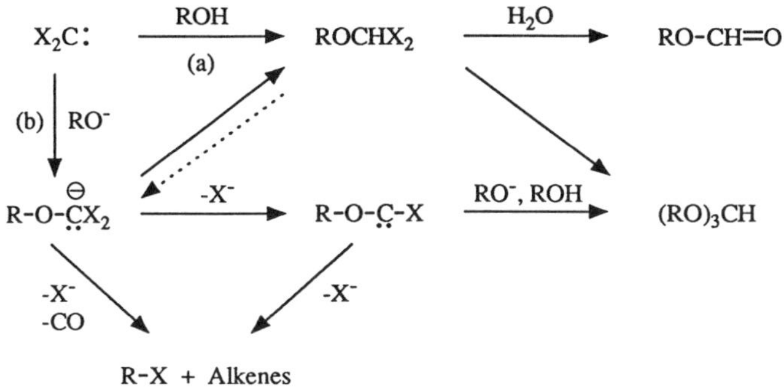

Scheme 32.

good reason to postulate the intervention of carbocations in the "deoxidation" reaction leading to alkyl halides and alkenes,[143] but conflicting evidence has also been presented.[144] The alkaline media do not allow to distinguish the insertion of dihalocarbenes into O–H bonds (path *a*) from electrophilic attack on alkoxides (path *b*). For further insight, the reaction of dihalocarbenes with alcohols should be studied in the absence of base, using appropriate propellanes as carbene precursors.[145]

Arylhalocarbenes (**247**), photochemically available from 3-aryl-3-halodiazirines (**246**),[85] are ideally suited for investigation by LFP techniques (Scheme 33). Strong transient absorptions in the UV region, with λ_{max} ranging from 310 nm (phenylchlorocarbene) to 355 nm (*p*-anisylchlorocarbene), have been assigned to the singlet ground states of the arylhalocarbenes. Although the kinetic studies focused on the addition reaction with alkenes,[146] some effort has been directed to the insertion of arylhalocarbenes into O–H bonds. The reaction course is normal in that

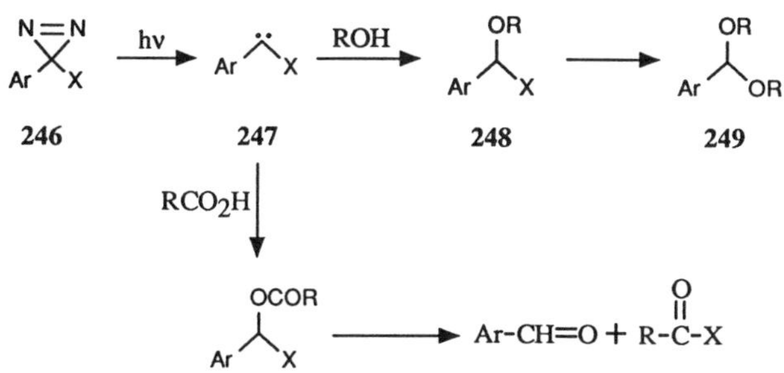

Scheme 33.

acetals (**249**) are formed by way of α-haloethers (**248**).[147] The reaction with methanol, however, is not first order in methanol concentration. Careful analysis of the reaction kinetics has shown that the arylhalocarbenes are more reactive towards methanol oligomers than to monomer.[147] For example, the rate constants derived for *p*-anisylchlorocarbene were *k* (oligomers) = $4.3 \times 10^9\ M^{-1}\ s^{-1}$ (per MeOH unit) and *k* (monomer) = $2 \times 10^7\ M^{-1}\ s^{-1}$ in isooctane at 25 °C. For 2-methyl-2-propanol, the converse behavior was observed, with the monomer being more reactive.[147] This was explained with the suggestion that the *t*-butyl groups protect the OH groups of the oligomer more than those of the monomer. Interestingly, 2,2-dimethyl-l-propanol is sufficiently unhindered to behave like methanol.[18b] Quenching of phenylchlorocarbene by 2,3-butanediol showed clean first-order kinetics.[18b] Presumably, intramolecular hydrogen bonding in dilute solution dominates the reaction of arylhalocarbenes with 1,2-diols to form 2-aryl-1,3-dioxolanes.[148] Linearity was also observed in the reactions of arylhalocarbenes with carboxylic acids that lead eventually to aromatic aldehydes and acyl chlorides (Scheme 33).[149] In these cases, dimers of the acids are the highest oligomers that form and the rate constants for these reactions were close to the diffusion-controlled limit ($k = 3.1 \times 10^9\ M^{-1}\ s^{-1}$ for phenylchlorocarbene and acetic acid). The heats of reaction of phenylhalocarbenes (X = F, Cl, Br) with methanol were determined by time-resolved photoacoustic calorimetry.[36] The reaction exothermicities are very similar (50 ± kcal/mol) and do not reflect the selectivity differences between the phenylhalocarbenes (X = F > Cl > Br). It has been suggested that this may be due to enhanced solvation of the less stable carbenes.

Benzylchlorocarbene (**251**) undergoes intramolecular rearrangement ($k_i = 6 \times 10^7$ s^{-1} at 25 °C, E_a = 4.8 kcal/mol)[150] in competition with intermolecular reactions. When 3-chloro-3-benzyldiazirine (**250**) was irradiated in the presence of methanol, the products were *E*- and *Z*-β-chlorostyrene (**252**), HCl, and phenylacetaldehyde dimethyl acetal (**253**) (Scheme 34).[151] Plots of the ratio **252:253** vs. $1/[MeOH]^2$

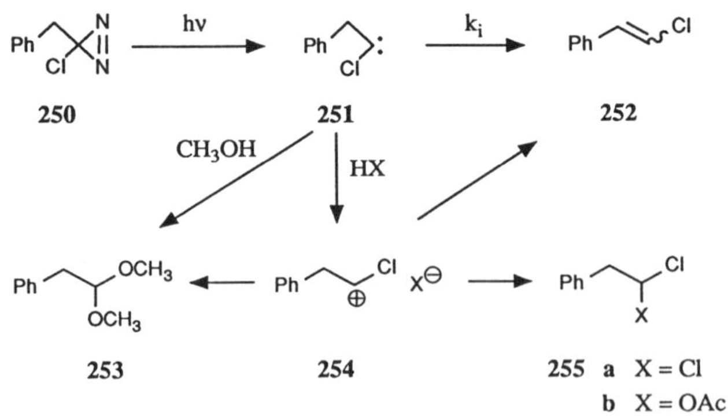

Scheme 34.

showed excellent linearity. The kinetic analysis suggests that reaction of **251** with two molecules of the methanol oligomer produces an intermediate that gives rise to **253** as well as to part of **252**. When the photolysis of **250** was carried out at various concentrations of hydrogen chloride, the products were **252** and 1,1-dichloro-2-phenylethane (**255a**).[152] In the presence of DCl, a mixture of deuterated and undeuterated **252** was formed in addition to deuterated **255a**. Similar results were obtained in the presence of acetic acid and DOAc.[57a] Absolute rates for the quenching of **251** by acids have recently been reported.[153] The rate constant for hydrogen chloride ($k = 4.7 \times 10^9\ M^{-1}\ s^{-1}$ at 25 °C) is close to the diffusion-controlled limit while that for acetic acid ($k = 1.0 \times 10^9\ M^{-1}\ s^{-1}$) is slightly smaller. Although chemical evidence (see above) points to the intervention of carbocations, no transient absorption of **254** was detected.

F. Oxacarbenes

Dimethoxycarbene (**258**), the prototype of nucleophilic carbenes, has been generated by pyrolysis of 7,7-dimethoxynorbornadiene derivatives (**256**)[154] and of hexamethoxycyclopropane (**257**).[155] In each case, **258** was efficiently trapped by methanol, with formation of trimethoxymethane (**261**) (Scheme 35). More recent, **258** was generated from 3,3-dimethoxydiazirine (**260**) and directly observed (λ_{max} = 255 nm) in matrix and in solution.[156] Photolysis of **260** in methanol afforded > 90% of **261**. In the range $0.08 \leq$ [MeOH] ≤ 1.54 the reaction was first order in [MeOH], with $k = 6.4 \times 10^6\ M^{-1}\ s^{-1}$.[157] Repetition of these experiments with MeOD revealed a kinetic isotope effect (KIE), $k_H/k_D = 3.3 \pm 0.5$, which is larger than any KIE so far encountered for the O–H insertion of singlet carbenes.[157] The reactivity of **258** with a variety of alcohols spanned four orders of magnitude, with *k* for carbene decay in 1 *M* ROH/MeCN ranging from $3.1 \times 10^4\ s^{-1}$ (EtOH) to $6.7 \times 10^8\ s^{-1}$ (HFIP).[158] A Brønsted correlation was obtained between log *k* and the pK_a of ROH. The appreciable value of α (–0.66) suggests a substantial degree of proton transfer in the transition state of O–H insertion. Laser flash photolysis monitoring of the reactions between **258** and the even more acidic quenchers AcOH and AcOD gave $k_H = 2.91 \times 10^9\ M^{-1}\ s^{-1}$ and $k_D = 2.84 \times 10^9\ M^{-1}\ s^{-1}$, respectively. The data indicate the absence of a significant KIE in this case, where *k* approaches the diffusion-controlled limit.[157]

Related observations were made with the more reactive (methoxy)(methyl)carbene (**263**), generated from the diazirine **262** by LFP.[159] The absorption of **263** was quenched by methanol to give curved plots of *k* versus MeOH concentration, indicative of simultaneous reactions with monomeric and oligomeric methanol. From these data, $k_H = 5.3 \times 10^9\ M^{-1}\ s^{-1}$ and $k_D = 4.8 \times 10^9\ M^{-1}\ s^{-1}$ could be extracted for reactions of **263** with oligomeric MeOH(D).

Cyclic oxacarbenes are accessible by photorearrangement of cycloalkanones.[160] In particular, cyclobutanones (**264**) undergo efficient α-cleavage from S_1 (n,π^*) to produce a singlet 1,4-acyl-alkyl diradical (**265**) (Scheme 36). The reactions gener-

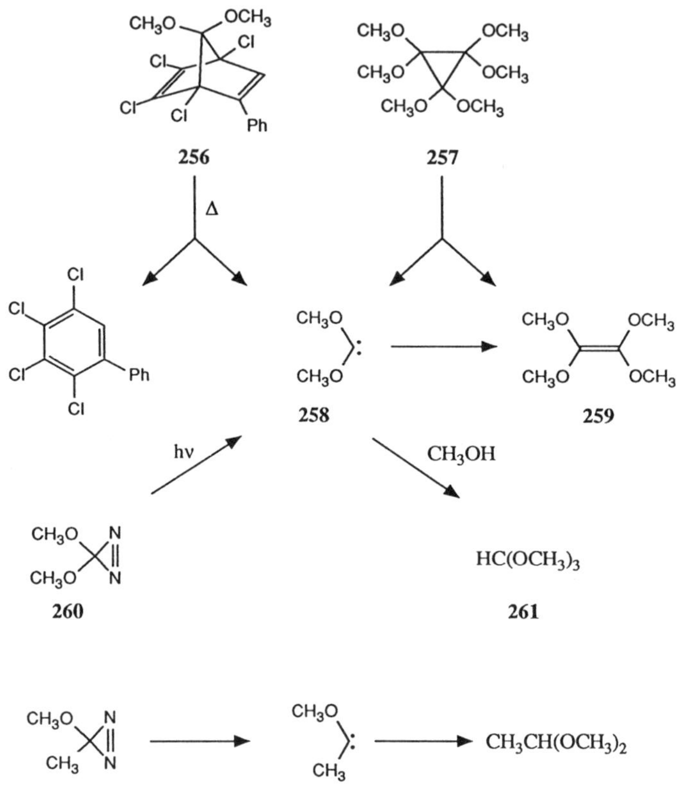

Scheme 35.

ally resulting from the diradical are (a) decarbonylation, (b) cycloelimination, and (c) oxacarbene formation. The oxacarbenes **266** have been trapped by alcohols to give the acetals **267**.[160] Matrix isolation and spectroscopic observation have also been reported.[161] Although retention of configuration at chiral α carbon atoms may argue for a concerted rearrangement (**264*** → **266**),[162] independent generation of the diradicals **265** has led to similar product patterns as were obtained in photolyses of **264**.[163] Preparative applications have begun to appear, including formation of the bicyclic acetals **269** from the hydroxyalkyl-substituted cyclobutanones **268**.[164] Higher cycloalkanones undergo α-cleavage predominantly from T_1 (n,π^*), and mainly yield disproportionation products. Ring expansion to give oxacarbenes and hence acetals has been observed with some conformationally biased cyclopentanones and cyclohexanones.[160]

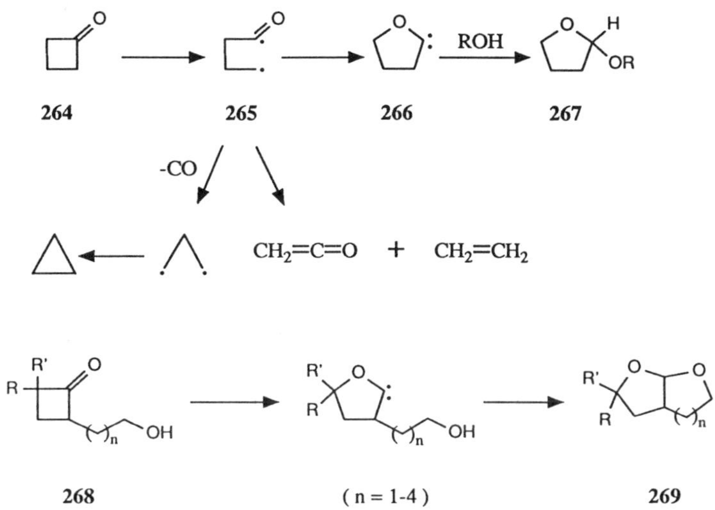

Scheme 36.

Glycosylidenes constitute a subgroup of oxacarbenes that is of special interest to the synthetic organic chemist.[165] The *gluco*-configurated diazirine **270**, as well as the *galacto*- and *manno*-configurated isomers, are highly reactive species, loosing nitrogen at room temperature or below (Scheme 37). Glycosylidenes (e.g., **271**), were found to insert efficiently into the O–H bonds of alcohols and phenols with the formation of glucosides (e.g., **272**). Although the distereoselectivity (α:β ratio) was only moderate in most cases, remarkable chemoselectivity in favor of the more acidic OH groups of polyhydric alcohols was noted. While **271** could not be observed directly, LFP of **270** in the presence of pyridine gave rise to the intense absorption of the pyridinium ylide **273**.[166] By monitoring the rates of formation of **273**, rate constants for the reaction of **271** with alcohols were obtained. From the kinetic results, one can favor a mechanism involving proton transfer from the alcohol to the carbene. The lower reactivity of **271**, as compared with **263**, may be due to an interaction of the neighboring benzyloxy group and the carbene center.

Apparent O–H insertion reactions of (methoxy)(phenyl)carbene (**277**) have been achieved with a variety of carbene precursors (Scheme 38). However, the azirine **274** gave only minor amounts (6%) of the acetal **278**,[167] and the intervention of diazonium ions has not been excluded in the decomposition of the tosylhydrazone **276**.[168] The diazirine **275** remains as a reliable source of **277** in preparative and LFP studies.[169] The kinetic behavior of **277** (λ_{max} = 290 nm) as a function of ROH concentration was similar to that reported for arylchlorocarbenes (Section V. E).

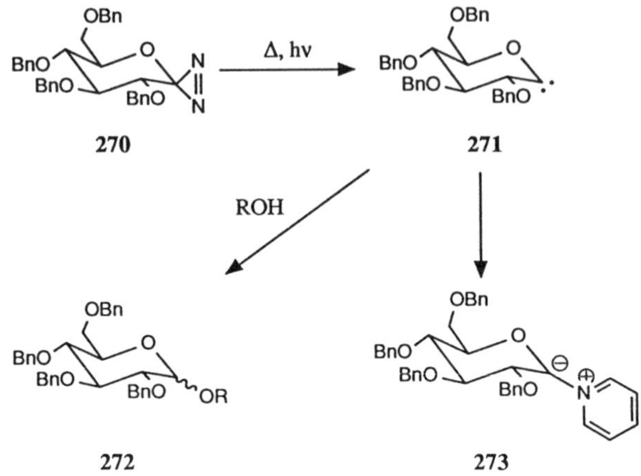

270 271 272 273

Scheme 37.

The reaction of **277** with oligomeric methanol in pentane was found to be diffusion-controlled ($k > 1 \times 10^{10}$ M^{-1} s^{-1}) whereas the reaction with monomeric 2-methyl-2-propanol had $k = 2.7 \times 10^8$ M^{-1} s^{-1}. Alternative approaches to aryl-substituted oxacarbenes include the photorearrangement of 9-alkoxytriptycenes (**279**),[170] benzocyclobutenediones (**100**; Scheme 15, Section IV. C),[67] and acylsilanes (**105**; Scheme 16, Section IV. C).[68–71] Time-resolved spectroscopy has confirmed the intervention of carbocations in the reaction of **105** with alcohols.[71] The application of LFP techniques to **280** was frustrated by the carbene's inefficient generation from **279**.[71]

Reference should be made here to the reactions of vinylogous oxacarbenes with alcohols (Scheme 39). Cyclopropenone acetals (**282**) undergo thermolysis at 70–80 °C with formation of the carbenes **283** which can be trapped by alcohols to afford the *ortho* esters **285**.[171] The carbenes **286** and **289**, generated from tosylhydrazone and diazo precursors, similarly accept the proton of alcohols at the divalent carbon. The delocalized carboxonium ions **287** and **290** thus produced are attacked by nucleophiles with substitution (**287** → **288**) and dealkylation (**290** → **291**), respectively.[172] The carbenes **294** are thought to arise in photolysis mixtures of the acetylenic ketones **292** and 2,3-dimethyl-2-butene.[173] The products **295**, obtained in the presence of alcohols, apparently exclude the intervention of delocalized carboxonium ions. It is not clear why the behavior of **294** should differ from that of **283**, **286**, and **289**.

The reactions of alkoxyhalocarbenes (**297**) with alcohols are normal in that the adducts **299**, detected by NMR in the case of X = F,[158] solvolyze rapidly to give orthoformates (Scheme 40). The reaction of the "invisible" (chloro)(methoxy)car-

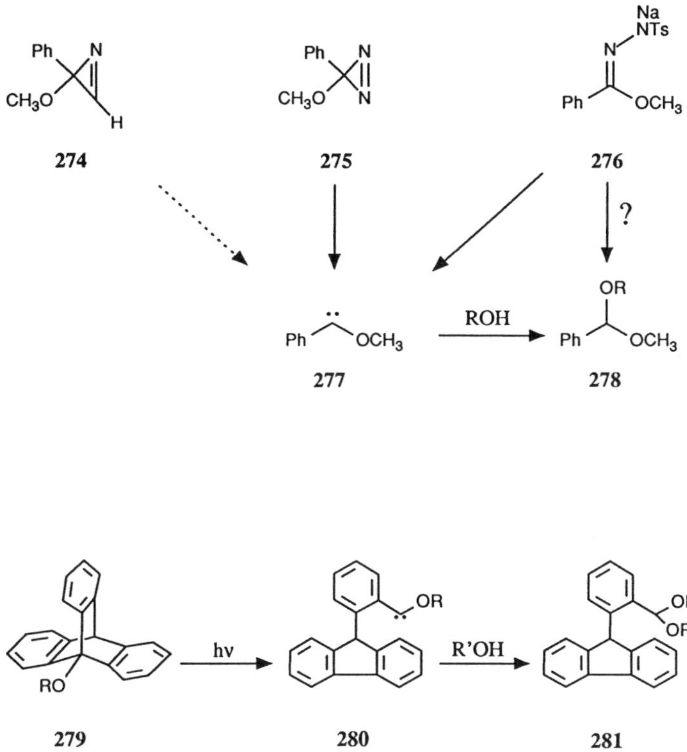

Scheme 38.

bene (**297b**) with methanol was monitored by time-resolved photoacoustic calorimetry,[36] the bimolecular rate constant being $k = 6.8 \times 10^4\ M^{-1}\ s^{-1}$. (Fluoro)(methoxy)carbene (**297a**) proved to be surprisingly "inert" to alcohols. EtOH, MeOH, and TFE failed to quench the carbene at 1 *M* ROH in acetonitrile.[158] Only with HFIP and HOAc was **297a** quenched competitively with its decay by dimerization and other pathways ($k = 9 \times 10^3\ s^{-1}$ and $2.0 \times 10^7\ s^{-1}$ for 1 *M* HFIP and HOAc, respectively). These data, as well as a substantial KIE ($k_H/k_D = 1.95$ for AcOH/D), point to proton transfer even in the case of **297a**. The sequence of carbene stabilization energies, $(MeO)_2C: > F(MeO)C: > Cl(MeO)C:$, is not reflected by the reaction rates with O–H bonds. $(MeO)_2C:$, the most stable carbene, is in fact the most reactive. These findings are at variance with concerted insertion into O–H bonds (**297** → **299**) but are readily explained in terms of intervening carbocations (**298**).

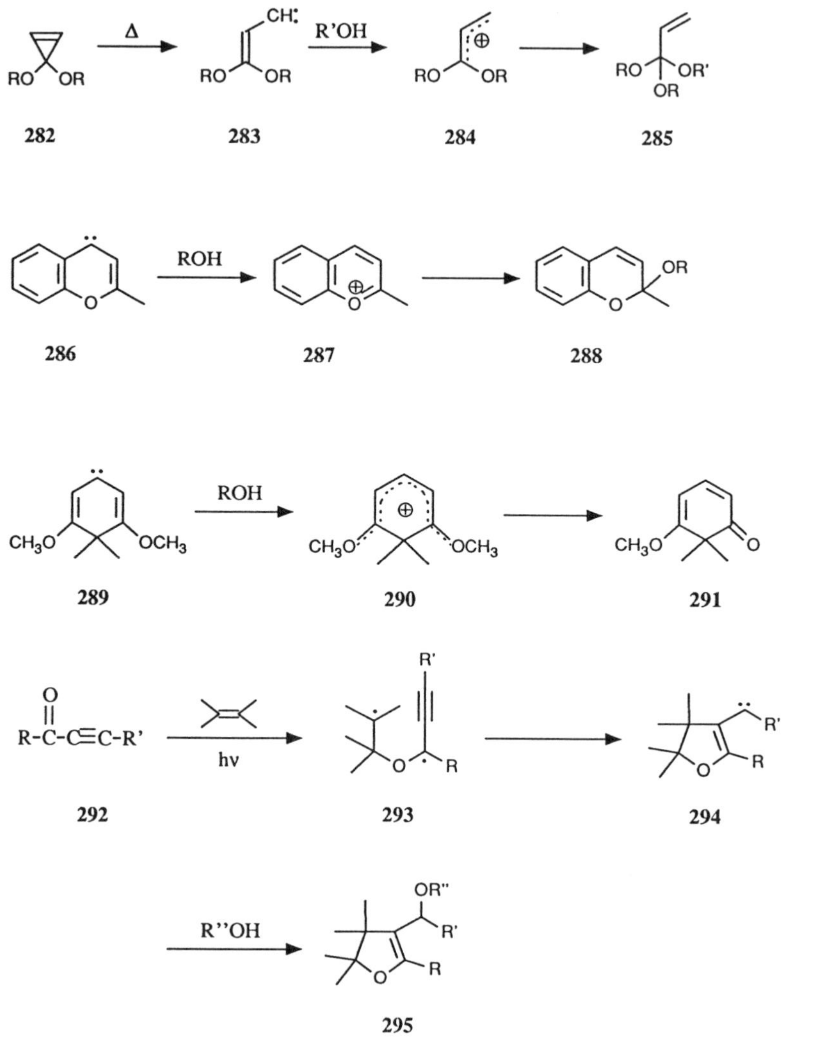

Scheme 39.

VI. CONCLUSION AND OUTLOOK

At the outset, we identified the reaction pathways by which carbenes can interact with O–H bonds. The mechanistic problems are by no means unique but are related to those encountered with C–H insertion and C=C addition reactions; however,

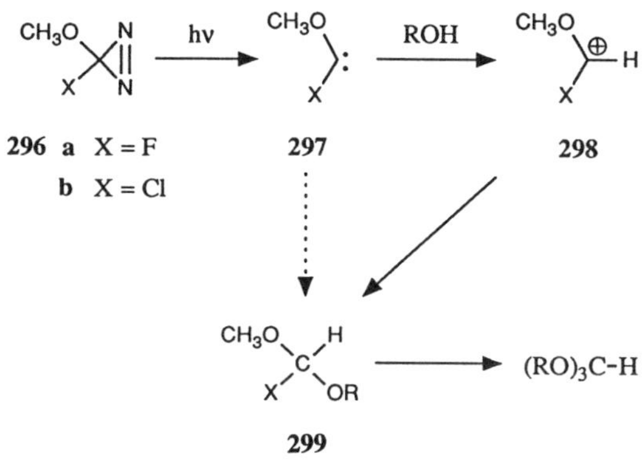

Scheme 40.

the insertion into O–H bonds is both simpler and more challenging, due to the lack of stereochemical criteria. We have seen how mechanistic insight can be gained in spite of such limitations. In the course of time, the work published in this field has advanced from accidental observation to purposefully conducted research. The application of laser flash photolysis to O–H insertion reactions has already provided a rich harvest, including the detection of carbocations as transient intermediates.

Nomenclature defines carbenium ions and carbenes as an acid–base pair. This formal relationship has now been amply realized by experiment—not only in exceptional cases but far beyond the limits that would have been predicted by conventional measures of carbene philicity. The mechanistic borderlines of O–H insertion reactions remain to be explored in more detail. In particular, the reactions of dialkylcarbenes with O–H bonds may provide a fresh look at some old problems of carbocation chemistry such as ion pairing, neighboring group participation, and charge delocalization.

Although much of the work reviewed here has been mechanistically oriented, the O–H insertion reaction of carbenes may be also viewed as a tool for making C–O bonds. Preparative applications, such as syntheses of glycosides, have begun to appear and show considerable promise for future development.

REFERENCES

1. For reviews, see: (a) Regitz, M., Ed; *Carbene(oide), Carbine*; Houben-Weyl, Thieme: Stuttgart, 1989; Vol. E19b; (b) Wentrup, C. *Reactive Molecules*; Wiley: New York, 1984; Chapter 4; (c) Moss, R. A., Jones, M., Jr., Eds; *Carbenes*; Wiley: New York, 1973, 1975; Vols. I and II; (d) Kirmse, W. *Carbene Chemistry*, 2nd ed.; Academic Press: New York, 1971; Chapter 7.

2. (a) Bethell, D.; Newall, A. R.; Stevens, G.; Whittaker, D. *J. Chem. Soc. B* **1969**, 749; (b) Bethell, D.; Newall, A. R.; Whittaker, D. *J. Chem. Soc. B* **1971**, 23.
3. Kirmse, W. *Justus Liebigs Ann. Chem.* **1963**, *666*, 9.
4. Hoffmann, R.; Zeiss, G. D.; Van Dine, G. W. *J. Am. Chem. Soc.* **1968**, *90*, 1485.
5. (a) Herzberg, G.; Shoosmith, J. *Nature* **1959**, *183*, 180; (b) Herzberg, G. *Proc. R. Soc. London, Ser. A* **1961**, *262*, 291; (c) Herzberg, G.; Johns, J. W. C. *Proc. R. Soc. London, Ser. A* **1966**, *295*, 107; (d) Herzberg, G.; Johns, J. W. C. *J. Chem. Phys.* **1971**, *54*, 2276.
6. Shavitt, I. *Tetrahedron* **1985**, *41*, 1531.
7. Leopold, D. G.; Murray, K. K.; Miller, A. E. S.; Lineberger, W. C. *J. Chem Phys.* **1984**, *83*, 4849.
8. Schuster, G. B. *Adv. Phys. Org. Chem.* **1986**, *22*, 311.
9. Field, K. W.; Schuster, G. B. *J. Org. Chem.* **1988**, *63*, 4000.
10. (a) Trozzolo, A. M.; Murray, R. W.; Wasserman, E. *J. Am. Chem. Soc.* **1962**, *84*, 4991; (b) Wasserman, E.; Trozzolo, A. M.; Yager, W. A.; Murray, R. W. *J. Chem. Phys.* **1964**, *40*, 2408; (c) Barash, L.; Wasserman, E.; Yager, W. A. *J. Am. Chem. Soc.* **1967**, *89*, 3931; (d) Moser, R. E.; Fritsch, J. M.; Matthews, C. N. *Chem. Commun.* **1967**, 770.
11. Gutsche, C. D.; Bachman, G. L., Coffee, R. S. *Tetrahedron* **1962**, *18*, 617.
12. Closs, G. L.; Moss, R. A. *J. Am. Chem. Soc.* **1964**, *86*, 4042.
13. Closs, G.; Hutchinson, C. A., Jr.; Kohler, B. E. *J. Chem. Phys.* **1966**, *44*, 413.
14. Trozzolo, A. M.; Gibbons, W. A. *J. Am. Chem. Soc.* **1967**, *89*, 239.
15. (a) Closs, G. L.; Rabinow, B. E. *J. Am. Chem. Soc.* **1976**, *98*, 8190; (b) Hadel, L. M.; Platz, M. S.; Scaiano, J. C. *J. Am. Chem. Soc.* **1984**, *106*, 283.
16. (a) Eisenthal, K. B.; Turro, N. J.; Aikawa, M.; Butcher, J. A., Jr.; DuPuy, C.; Hefferon, G.; Hetherington, W.; Korenowski, G. M.; McAuliffe, M. J. *J. Am. Chem. Soc.* **1980**, *102*, 6563; (b) Sitzmann, E. V.; Langan, J.; Eisenthal, K. B. *J. Am. Chem. Soc.* **1984**, *106*, 1868.
17. Eisenthal, K. B.; Turro, N. J.; Sitzmann, E. V.; Gould, I. R.; Hefferon, G.; Langan, J.; Cha, Y. *Tetrahedron* **1985**, *41*, 1543.
18. (a) Griller, D.; Nazran, A. S.; Scaiano, J. C. *J. Am. Chem. Soc.* **1984**, *106*, 198; (b) Griller, D.; Nazran, A. S.; Scaiano, J. C. *Tetrahedron* **1985**, *41*, 1525.
19. Lapin, S. C.; Schuster, G. B. *J. Am. Chem. Soc.* **1985**, *107*, 4243.
20. Chuang, C.; Lapin, S. C.; Schrock, A. K.; Schuster, G. B. *J. Am. Chem. Soc.* **1985**, *107*, 4238.
21. Griller, D.; Majewski, M.; McGimpsey, W. G.; Nazran, A. S.; Scaiano, J. C. *J. Org. Chem.* **1988**, *53*, 1550.
22. Hadel, L. M.; Maloney, V. M.; Platz, M. S.; McGimpsey, W. G.; Scaiano, J. C. *J. Phys. Chem.* **1986**, *90*, 2488.
23. (a) Nazran, A. S.; Griller, D. *J. Am. Chem. Soc.* **1984**, *106*, 543; (b) Gilbert, B. C.; Griller, D.; Nazran, A. S. *J. Org. Chem.* **1985**, *50*, 4783; (c) Gilbert, B. C.; Griller, D.; Nazran, A. S. *J. Org. Chem.* **1987**, *52*, 4460.
24. Alt, R.; Gould, I. R.; Staab, H. A.; Turro, N. J. *J. Am. Chem. Soc.* **1986**, *108*, 6911.
25. Jones M. B.; Platz, M. S. *Tetrahedron Lett.* **1990**, *31*, 953.
26. Dorigo, A. E.; Li, Y.; Houk, K. N. *J. Am. Chem. Soc.* **1989**, *111*, 6942.
27. Platz, M. S.; Maloney, V. M. In *Kinetics and Spectroscopy of Carbenes and Biradicals*; Platz, M. S., Ed.; Plenum Press; New York, 1990; pp 239–352.
28. Leyva, E.; Barcus, R. L.; Platz, M. S. *J. Am. Chem. Soc.* **1986**, *108*, 7786.
29. Moss, R. A. *Acc. Chem Res.* **1980**, *13*, 58.
30. Moss, R. A.; Mallon, C. B.; Ho, C.-T. *J. Am. Chem. Soc.* **1977**, *99*, 4105.
31. (a) Rondan, N. G.; Houk, K. N.; Moss, R. A. *J. Am. Chem. Soc.* **1980**, *102*, 1770; (b) Moss, R. A.; Perez, L. A.; Wloskowska, J.; Guo, W.; Krogh-Jespersen, K. *J. Org. Chem.* **1982**, *47*, 4177.
32. Moss, R. A.; Turro, N. J. In *Kinetics and Spectroscopy of Carbenes and Biradicals*; Platz, M. S., Ed.; Plenum Press: New York, 1990; pp 213–238.

33. (a) Harding, L. B.; Schlegel, H. B.; Krishsnan, R.; Pople, J. A. *J. Phys. Chem.* **1980**, *84*, 3394; (b) Pople, J. A.; Raghavachari, K.; Frisch, M. J.; Blinkley, J. S.; Schleyer, P. v. R. *J. Am. Chem. Soc.* **1983**, *105*, 6389.
34. Yates, B. F.; Bouma, W. J.; Radom, L. *J. Am. Chem. Soc.* **1987**, *109*, 2250.
35. Peters, K. S. In *Kinetics and Spectroscopy of Carbenes and Biradicals*; Platz, M. S., Ed.; Plenum Press: New York, 1990, pp 37–49.
36. LaVilla, J. A.; Goodman, J. L. *J. Am. Chem. Soc.* **1989**, *111*, 712, 6877; *Tetrahedron Lett.* **1990**, *31*, 6287.
37. Kirmse, W.; Loosen, K.; Sluma, H.-D. *J. Am. Chem. Soc.* **1981**, *103*, 5935.
38. (a) Jones, W. M.; Ennis, C. L. *J. Am. Chem. Soc.* **1969**, *91*, 6391; (b) Major, C.; Jones, W. M. *J. Org. Chem.* **1978**, *43*, 4498; (c) Duell, B. L.; Jones, W. M. *J. Org. Chem.* **1978**, *43*, 4901.
39. (a) Untch, K. International Symposium on the Chemistry of Nonbenzenoid Aromatic Compounds, Sendai, Japan, 1970; (b) Mayor, C.; Jones, W. M. *Tetrahedron Lett.* **1977**, 3855.
40. (a) Tyner, R. L.; Jones, W. M.; Öhrn, Y.; Sabin, J. R. *J. Am. Chem. Soc.* **1974**, *96*, 3765; (b) Dewar, M. S.; Landman, D. *J. Am. Chem. Soc.* **1977**, *99*, 6179; (c) Radom, L.; Schaefer, H. F., III; Vincent, M. A. *Nov. J. Chim.* **1980**, *4*, 411; (d) Waali, E. E. *J. Am. Chem. Soc.* **1981**, *103*, 3604; (e) Glidewell, C.; Lloyd, D. *J. Chem. Res. Synop.* **1983**, *178*; (f) Kassaee, M. Z.; Nimlos, M. R.; Downie, K. E.; Waali, E. E. *Tetrahedron* **1985**, *41*, 1579.
41. (a) West, P. R.; Chapman, O. L.; Le Roux, J. *J. Am. Chem. Soc.* **1982**, *104*, 1779; (b) Chapman, O. L.; Abelt, C. J. *J. Org. Chem.* **1987**, *52*, 1218; (c) McMahon, R. J.; Abelt, C. J.; Chapman, O. L.; Johnson, J. W.; Kreil, C. L.; Le Roux, J. P.; Mooring, A. M.; West, P. R. *J. Am. Chem. Soc.* **1987**, *109*, 2456.
42. Harris, J. W.; Jones, W. M. *J. Am. Chem. Soc.* **1982**, *104*, 7329.
43. Kirmse, W.; Sluma, H.-D. *J. Org. Chem.* **1988**, *53*, 763.
44. Balci, M.; Winchester, W. R.; Jones, W. M. *J. Org. Chem.* **1982**, *47*, 5180.
45. Holm, K. H.; Skattebøl, L. *J. Am. Chem. Soc.* **1977**, *99*, 5480.
46. Kirmse, W.; Jendralla, H. *Chem. Ber.* **1978**, *111*, 1873.
47. Diaz, A.; Brookhart, M.; Winstein, S. *J. Am. Chem. Soc.* **1966**, *88*, 3133.
48. For reviews, see: (a) Jones, W. M.; Brinker, U. H. In *Pericyclic Reactions*; Marchand, A. P.; Lehr, R. E., Eds.; Academic Press: New York, 1977; Vol. I, p 159; (b) Jones, W. M. In *Rearrangements in Ground and Excited States*; de Mayo, P. Ed.; Academic Press: New York, 1980; Vol. I, p 149; (c) Moss, R. A.; Jones, M., Jr. *Reactive Intermediates*; Wiley: New York, 1981; Vol. 2, p 113.
49. Kirmse, W.; Richarz, U. *Chem. Ber.* **1978**, *111*, 1883.
50. Kirmse, W.; Lelgemann, R.; Ritzer, E. Unpublished results.
51. Wuest, J. D.; Madonik, A. M.; Gordon, D. C. *J. Org. Chem.* **1977**, *42*, 2111.
52. Kirmse, W.; Spaleck, W. Unpublished results.
53. Kirmse, W.; Pham van Chiem; Henning P. G. *J. Am. Chem. Soc.* **1983**, *105*, 1695; *Tetrahedron* **1985**, *41*, 1441.
54. Friedrich, K.; Kirmse, W. Unpublished results.
55. Friedrich, K.; Jansen, U.; Kirmse, W. *Tetrahedron Lett.* **1985**, *26*, 193.
56. Kirmse, W.; Lelgemann, R.; Friedrich, K. *Chem. Ber.* **1991**, *124*, 1853.
57. (a) Liu, M. T. H.; Chichti, N. H.; Tencer, M.; Tomioka, H.; Izawa, Y. *Tetrahedron* **1984**, *40*, 887; (b) Tomioka, H.; Hayashi, N.; Suigura, T.; Izawa, Y. *J. Chem. Soc., Chem. Commun.* **1986**, 1364.
58. Kirmse, W.; Kund, K. *J. Org. Chem.* **1990**, *55*, 2325.
59. (a) Kirmse, W.; Kund, K.; Ritzer, E.; Dorigo, A. E.; Houk, K. N. *J. Am. Chem. Soc.* **1986**, *108*, 6045; (b) Dorigo, A. E.; Li, Y.; Houk, K. N. *J. Am. Chem. Soc.* **1989**, *111*, 6942.
60. (a) McClelland, R. A.; Kanagasabapathy, V. M.; Steenken, S. *J. Am. Chem. Soc.* **1988**, *110*, 6913; (b) McClelland, R. A.; Kanagasabapathy, V. M.; Banait, N. S.; Steenken, S. *J. Am. Chem. Soc.* **1989**, *111*, 3966; (c) Mecklenburg, S. L.; Hilinski, E. F. *J. Am. Chem Soc.* **1989**; *111*, 5471; (d) Alonso, E. O.; Johnston, L. J.; Scaiano, J. C.; Toscano, V. G. *J. Am Chem. Soc.* **1990**, *112*, 1270;

(e) McClelland, R. A.; Mathivanan, N.; Steenken, S. *J. Am. Chem. Soc.* **1990**, *112*, 4857; (f) Bartl, J.; Steenken, S.; Mayr, H.; McClelland, R. A. *J. Am. Chem. Soc.* **1990**; *112*, 6918; (g) McClelland, R. A.; Kanagasabapathy, V. M.; Banait, N. S.; Steenken, S. *J. Am. Chem. Soc.* **1991**, *113*, 1009.
61. Kirmse, W.; Kilian, J.; Steenken, S. *J. Am. Chem. Soc.* **1990**, *112*, 6399.
62. Chateauneuf, J. E. *J. Chem. Soc., Chem. Commun.* **1991**, 1437.
63. Chiang, Y.; Kresge, A. J.; Pruszynski, P.; Schepp, N. P.; Wirz, J. *Angew. Chem.* **1991**, *103*, 1391; *Angew. Chem.* (Int. Ed. Engl.) **1991**, *30*, 1356.
64. Schepp, N. P.; Wirz, J. Unpublished results.
65. (a) Closs, G. L.; Böll, W. A.; Heyn, H.; Dev, J. *J. Am. Chem. Soc.* **1968**, *90*, 173; (b) Pincock, J. A., Morchat, R.; Arnold, D. R. *J. Am. Chem. Soc.* **1973**, *95*, 7536; (c) Arnold, D. R.; Humphreys, R. W.; Leigh, W. J.; Palmer, G. E. *J. Am. Chem. Soc.* **1976**, *98*, 6225; (d) For applications in synthesis, see Ref. 1a, pp 691–705.
66. Kilian, J.; Kirmse, W.; Steenken, S. Unpublished results.
67. Boate, D. R.; Johnston, L. J.; Kwong, P. C.; Lee-Ruff, E.; Scaiano, J. C. *J. Am. Chem. Soc.* **1990**, *112*, 8858.
68. Bourque, R. A.; Davis, P. D.; Dalton, J. C. *J. Am. Chem. Soc.* **1981**, *103*, 697.
69. (a) Brook, A. G.; Duff, J. M. *J. Am. Chem. Soc.* **1967**, *89*, 454; (b) Duff, J.M.; Brook, A. G. *Can. J. Chem.* **1973**, *51*, 2869.
70. Perrin, H. M.; White, W. R.; Platz, M. S. *Tetrahedron Lett.* **1991**, *32*, 4443.
71. Guth, M.; Kirmse, W.; Steenken, S. Unpublished results.
72. (a) Meerwein, H.; Rathjen, H.; Werner, H. *Chem. Ber.* **1942**, *75*, 1610; (b) Meerwein, H.; Disselnkötter, H.; Rappen, F.; v. Rintelen, H.; van de Vloed, H. *Justus Liebigs Ann. Chem* **1957**, *604*, 151.
73. Palazzo, F. C. *Gazz. Chim. Ital.* **1949**, *79*, 13.
74. Ogoshi, H.; Takezaki, Y. *Bull. Inst. Chem. Res.* (Kyoto University) **1960**, *38*, 299; *Chem. Abstr.* **1961**, *55*, 20913.
75. Kerr, J. A.; O'Grady, B. V.; Trotman-Dickenson, A. F. *J. Chem. Soc. A* **1967**, 897.
76. Ho, S. Y.; Lin, H. B. *J. Chin. Chem. Soc.* (Taipeh) **1973**, *20*, 27.
77. Hatekayama, S.; Bandow, H.; Okuda, M.; Akimoto, H. *J. Phys. Chem.* **1981**, *85*, 2249.
78. Hack, W.; Wagner, H. G.; Wilms, A. *Ber. Bunsenges. Phys. Chem.* **1988**, *92*, 620.
79. Koch, M.; Temps, F.; Wagener, R.; Wagner, H. G. *Z. Naturforsch.* **1989**, *44a*, 195.
80. Oikawa, S.; Tsuda, M. *J. Am. Chem. Soc.* **1985**, *107*, 1940.
81. McGarrity, J. F. In *The Chemistry of Diazonium and Diazo Groups*; Patai, S., Ed.; Wiley: Chichester, 1978; Chapter 6.
82. (a) Kirmse, W. *Angew. Chem.* **1976**, *88*, 273; *Angew. Chem. Int. Ed. Eng.* **1976**, *15*, 251; (b) Kirmse, W.; Siegfried, R; *J. Am. Chem. Soc.* **1983**, *105*, 950; (c) Banert, K.; Kirmse, W.; Wroblowsky, H.-J. *Chem. Ber.* **1983**, *116*, 3591; (d) Kirmse, W.; Zellmer, V.; Goer, B. *J. Am. Chem. Soc.* **1986**, *108*, 4912; (e) Bentley, T. W.; Goer, B.; Kirmse, W. *J. Org. Chem.* **1988**, *53*, 3066.
83. Myers, D. R.; Senthilnathan, V. P.; Platz, M. S.; Jones, M., Jr. *J. Am. Chem. Soc.* **1986**, *108*, 4232.
84. Morgan, S.; Platz, M. S.; Jones, M., Jr.; Myers, D. R. *J. Org. Chem.* **1991**, *56*, 1351.
85. Liu, M. T. H.; Stevens, I. D. R. In *Chemistry of Diazirines*; Liu, M. T. H., Ed.; CRC Press: Boca Raton, Florida, 1987; Vol. I, Chapter 5.
86. Bradley, G. F.; Evans, W. B. L.; Stevens, I. D. R. *J. Chem. Soc., Perkin Trans. 2* **1977**, 1214.
87. Stevens, I. D. R.; Liu, M. T. H.; Soundarajan, N.; Paike, N. *Tetrahedron Lett.* **1989**, *30*, 481.; *J. Chem. Soc., Perkin Trans. 2* **1990**, 661.
88. Morgan, S.; Jackson, J. E.; Platz, M. S. *J. Am. Chem. Soc.* **1991**, *113*, 2782.
89. Kropp, P. J. *Org. Photochem.* **1979**, *4*, 1.
90. Kropp, P. J.; Tise, F. P. *J. Am. Chem. Soc.* **1981**, *103*, 7293.
91. Iwamura, H. M.; Yoshimura, K. *J. Am. Chem. Soc.* **1974**, *96*, 2652.

92. Ipaktschi, J. *Chem. Ber.* **1972**, *105*, 1989.
93. Glaser, M.; Kirmse, W. Unpublished results.
94. (a) Eaton, P. E.; Hoffmann, K.-L. *J. Am. Chem. Soc.* **1987**, *109*, 5285; (b) Eaton, P. E.; White, J. A. *J. Org. Chem.* **1990**, *55*, 1321.
95. (a) Chen, N.; Jones, M., Jr. *J. Phys. Org. Chem.* **1988**, *1*, 305; (b) Chen, N.; Jones, M., Jr. *Tetrahedron Lett.* **1989**, *30*, 6969; (c) White, W. R.; Platz, M. S.; Chen, N.; Jones, M., Jr. *J. Am. Chem. Soc.* **1990**, *112*, 7794; (d) Chen, N.; Jones, M., Jr.; White, W. R.; Platz, M. S. *J. Am. Chem. Soc.* **1991**, *113*, 4981.
96. For a comprehensive review, see Backes J.; Brinker, U. H. in Ref. 1a, pp 391–510.
97. (a) Kirmse, W.; Schütte, H. *Chem. Ber.* **1968**, *101*, 1674; (b) Kirmse, W.; Schnurr, O.; Jendralla, H. *Chem. Ber.* **1979**, *112*, 2120.
98. Kirmse, W.; Rode, J.; Rode, K. *Chem. Ber.* **1986**, *119*, 3672.
99. Kirmse, W.; Hellwig, G. *Chem. Ber.* **1982**, *115*, 2744.
100. Warner, P. M.; Chu, I.-S. *J. Am. Chem. Soc.* **1984**, *106*, 5366; *J. Org. Chem.* **1984**, *49*, 3666.
101. For reviews, see: (a) Griffin, G. W. *Angew. Chem.* **1971**, *83*, 604; *Angew. Chem.* (Int. Ed. Engl.) **1971**, *10*, 537; (b) Griffin, G. W.; Bertoniere, R. N. in Ref. 1c, pp 305–349; (c) Bertoniere, R. N.; Griffin, G. W. *Org. Photochemistry* **1973**, *3*, 115.
102. (a) Walsh, T. D.; Powers, D. R. *Tetrahedron Lett.* **1970**, 3855; (b) Shi, M.; Okamoto, Y.; Takamuku, S. *J. Org. Chem.* **1990**, *55*, 3821; (c) Shi, M.; Shouki, K.; Okamoto, Y.; Takamuku, S. *J. Chem. Soc., Perkin Trans. 1* **1990**, 2443.
103. Klett, M. W.; Johnson, R. P. *Tetrahedron Lett.* **1983**, *24*, 1107; *J. Am. Chem. Soc.* **1985**, *107*, 3963.
104. Iwamura, M.; Tukada, H.; Iwamura, H. *Tetrahedron Lett.* **1981**, *21*, 4865.
105. Barcus, R. L.; Hadel, L. M.; Johnston, L. J.; Platz, M. S.; Savino, T. G.; Scaiano, J. C. *J. Am. Chem. Soc.* **1986**, *108*, 3928.
106. Horn, K. A.; Chateauneuf, J. E. *Tetrahedron* **1985**, *41*, 1465.
107. Wright, B. B.; Platz, M. S. *J. Am. Chem. Soc.* **1984**, *106*, 4175.
108. Suguwara, T., Iwamura, H.; Hayashi, H.; Sekiguchi, A.; Ando, W.; Liu, M. T. H *Chem. Lett.* **1983**, 1257.
109. Abdel-Wahab, A. A.; Doss, S. H.; Dürr, H.; Turro, N. J.; Gould, I. R. *J. Org. Chem.* **1987**, *52*, 429.
110. Hadel, L. M.; Platz, M. S.; Wright, B. B.; Scaiano, J. C. *Chem. Phys. Lett.* **1984**, *105*, 539.
111. Grasse, P. B.; Brauer, B.-E.; Zupancic, J. J.; Kaufmann, K. J.; Schuster, G. B. *J. Am. Chem. Soc.* **1983**, *105*, 6833.
112. Zupancic, J. J.; Grasse, P. B.; Lapin, S. C.; Schuster, G. B. *Tetrahedron* **1985**, *41*, 1471.
113. Grasse, P. B.; Zupancic, J. J.; Lapin, S. C.; Hendrich, M. P., Schuster, G. B. *J. Org. Chem.* **1985**, *50*, 2352.
114. Lapin, S. C.; Brauer, B.-E.; Schuster, G. B. *J. Am. Chem. Soc.* **1984**, *106*, 2092.
115. Li, Y.-Z.; Schuster, G. B. *J. Org. Chem.* **1986**, *51*, 3730; *J. Org. Chem.* **1987**, *52*, 3975.
116. Astles, D. J.; Girard, M.; Griller, D.; Kolt, R. J. Wayner, D. D. M. *J. Org. Chem.* **1988**, *53*, 6053.
117. Kirmse, W.; Hömberger, G. *J. Am. Chem. Soc.* **1991**, *113*, 3925.
118. Tomioka, H.; Suzuki, S.; Izawa, Y. *J. Am. Chem. Soc.* **1982**, *104*, 1047.
119. Tomioka, H.; Tabayashi, K.; Ozaki, Y.; Izawa, Y. *Tetrahedron* **1985**, *41*, 1435.
120. Casal, H. L.; Werstiuk, N. H.; Scaiano, J. C. *J. Org. Chem.* **1984**, *49*, 5214.
121. Platz, M. S. In *Kinetics and Spectroscopy of Carbenes and Biradicals*; Platz, M. S., Ed.; Plenum Press: New York, 1990; pp 143–211.
122. O'Leary, M. A.; Wege, D. *Tetrahedron* **1981**, *37*, 801.
123. (a) Denny, D. B.; Klemchuk, P. P. *J. Am. Chem. Soc.* **1958**, *80*, 3289; (b) Palik, E. C.; Platz, M. S. *J. Org. Chem.* **1983**, *48*, 963.
124. (a) Zimmerman, H. E.; Aasen, S. M. *J. Am. Chem. Soc.* **1977**, *99*, 2342; (b) Padwa, A.; Blacklock, T. J.; Getman, D.; Hatanaka, N. *J. Am. Chem. Soc.* **1977**, *99*, 2344.

125. Padwa, A.; Blacklock, T. J. *J. Am. Chem. Soc.* **1977**, *99*, 2345.
126. Kirmse, W.; Kund, K. *J. Am. Chem. Soc.* **1989**, *111*, 1465.
127. Bailey, R. J.; Card, P. J.; Shechter, H. *J. Am. Chem. Soc.* **1983**, *105*, 6096.
128. Tomioka, H.; Nunome, Y. *J. Chem. Soc., Chem. Commun.* **1990**, 1243.
129. Wiberg, K.; Waddell, S. T. *J. Am. Chem. Soc.* **1990**, *112*, 2194.
130. For reviews, see: (a) Meier, H.; Zeller, K. P. *Angew. Chem.* **1975**, *87*, 52; *Angew. Chem.* (Int. Ed. Engl.) **1975**, *14*, 32; (b) Ando, W. In *The Chemistry of Diazonium and Diazo Groups*; Patai, S, Ed.; Wiley: New York, 1978; pp 458–476; (c) Regitz, M.; Maas, G. *Diazo Compounds*; Academic Press: Orlando, 1986; pp 185–195.
131. Kaplan, F.; Mitchell, M. L. *Tetrahedron Lett.* **1979**, 759.
132. Tomioka, H.; Okuno, H.; Izawa, Y. *J. Org. Chem.* **1980**, *45*, 5278.
133. DoMinh, T.; Strausz, O. P.; Gunning, H. E. *J. Am. Chem. Soc.* **1969**, *91*, 1261.
134. Wydila, J.; Thornton, E. R. *Tetrahedron Lett.* **1983**, *24*, 233.
135. (a) Jones, M., Jr.; Ando, W.; Hendrick, M. E.; Kulczycki, A., Jr.; Howley, P. M.; Hummel, K. F.; Malament, D. S. *J. Am. Chem. Soc.* **1972**, *94*, 7469; (b) Galucci, R. R.; Jones, M., Jr. *J. Org. Chem.* **1985**, *50*, 4404.
136. Kirmse, W.; Schmitz, C. Unpublished results.
137. Davies, H. M. L.; Smith, H. D.; Korkor, O. *Tetrahedron Lett.* **1987**, *28*, 1853.
138. (a) Heslin, J. C.; Moody, C. J.; Slawin, A. M. Z.; Williams, D. J. *Tetrahedron Lett.* **1986**, *27*, 1403; (b) Moody, C. J.; Taylor, R. J. *Tetrahedron Lett.* **1987**, *28*, 5351; *J. Chem. Soc., Perkin Trans. 1* **1989**, *721*; (c) Davies, M. J., Moody, C. J.; Taylor, R. J. *J. Chem. Soc., Perkin Trans. 1* **1991**, *1*, 9.
139. Tomioka, H.; Okuno, H.; Kondo, S.; Izawa, Y. *J. Am. Chem. Soc.* **1980**, *102*, 7123.
140. Fujiwara, Y.; Tanimoto, Y.; Hoh, M.; Hirai, K.; Tomioka, H. *J. Am. Chem. Soc.* **1987**, *109*, 1942.
141. Hoffman, R. V.; Shechter, H. *J. Org. Chem.* **1974**, *39*, 2939.
142. (a) Tomioka, H.; Tabayashi, K.; Izawa, Y. *Chem. Lett.* **1985**, 1103; (b) Tomioka, H.; Hirai, K., Tabayashi, K.; Murata, S.; Izawa, Y.; Inagaki, S.; Okajima, T. *J. Am. Chem. Soc.* **1990**, *112*, 7692.
143. Keating, J. T.; Skell, P. S. In *Carbonium Ions*; Olah, G. A.; Schleyer, P. v. R., Eds.; Wiley: New York, 1970; Vol. II, p 573.
144. Likhotvorik, I. R.; Jones, M., Jr.; Yurchenko, A. G.; Krasutsky, P. A. *Tetrahedron Lett.* **1989**, *30*, 5089.
145. (a) Hartwig, J. F.; Jones, M., Jr.; Moss, R. A.; Lawrynowicz, W. *Tetrahedron Lett.* **1986**, *27*, 5907; (b) Le, N. A.; Jones, M., Jr.; Bickelhaupt, F.; de Wolf, W. H. *J. Am. Chem. Soc.* **1989**, *111*, 8491.
146. Moss, R. A.; Turro, N. J. In *Kinetics and Spectroscopy of Carbenes and Biradicals*; Platz, M. S., Ed.; Plenum Press: New York, 1990; pp 213–238.
147. Griller, D.; Liu, M. T. H.; Scaiano, J. C. *J. Am. Chem. Soc.* **1982**, *104*, 5549.
148. Liu, M. T. H.; Kokosi, J. *Heterocycles* **1985**, 3049.
149. Griller, D.; Liu, M. T. H.; Montgomery, C. R.; Scaiano, J. C. *J. Org. Chem.* **1983**, *48*, 1359.
150. Liu, M. T. H.; Bonneau, R. *J. Am. Chem. Soc.* **1990**, *112*, 3915.
151. Liu, M. T. H.; Subramanian, R. *J. Chem. Soc., Chem. Commun.* **1984**, 1062; *J. Chem. Soc. Perkin Trans. 2* **1986**, 1233.
152. Liu, M. T. H.; Subramanian, R. *J. Org. Chem.* **1985**, *50*, 3218.
153. Liu, M. T. H.; Chateauneuf, J. E. *J. Chem. Soc., Chem. Commun.* **1991**, 1575.
154. Hoffmann, R. W. *Angew. Chem.* **1971**, *83*, 595; *Angew. Chem.* (Int. Ed. Engl.) **1971**, *10*, 529; *Acc. Chem. Res.* **1985**, *18*, 248.
155. Moss, R. A.; Cox, D. P. *Tetrahedron Lett.* **1985**, *26*, 1931.
156. Moss, R. A.; Wlostowski, M.; Shen, S.; Krogh-Jespersen, K.; Matro, A. *J. Am. Chem. Soc.* **1988**, *110*, 4443.
157. Moss, R. A.; Shen, S.; Wlostowski, M. *Tetrahedron Lett.* **1988**, *29*, 6417.

158. Du, X.-M.; Fan, H.; Goodman, J. L.; Kesselmayer, M. A.; Krogh-Jespersen, K.; LaVilla, J. A.; Moss, R. A.; Shen, S.; Sheridan, R. S. *J. Am. Chem. Soc.* **1990**, *112*, 1920.
159. Sheridan, R. S.; Moss, R. A.; Wilk, B. K.; Shen, S.; Wlostowski, M.; Kesselmayer, M. A.; Subramanian, R.; Kmiecik-Lawrynowicz, G.; Krogh-Jespersen, K. *J. Am. Chem. Soc.* **1988**, *110*, 7563.
160. For reviews, see: (a) Morton, D. R.; Turro, N. J. *Adv. Photochem.* **1974**, *9*, 197; (b) Yates, P.; Loutfy, R. O. *Acc. Chem. Res.* **1975**, *8*, 209; (c) Brinker, U. H.; Erdle, W. in Ref. 1a, pp 623–663.
161. Kesselmayer, M. A.; Sheridan, R. S. *J. Am. Chem. Soc.* **1987**, *109*, 5029.
162. Quinkert, G.; Jacobs, P. *Chem. Ber.* **1974**, *107*, 2473.
163. Miller, R. D.; Gölitz, P.; Janssen, J.; Lemmens, J. *J. Am. Chem. Soc.* **1984**, *106*, 7277.
164. (a) Pirrung, M. C. *Angew. Chem.* **1985**, *97*, 1073; *Angew. Chem.* (Int. Ed. Engl.) **1985**, *24*, 1043; (b) Pirrung, M. C.; Chang, V. K.; De Amicis, C. V. *J. Am. Chem. Soc.* **1989**, *111*, 5824.
165. (a) Briner, K.; Vasella, A. *Helv. Chim. Acta* **1989**, *72*, 1371; (b) Briner, K.; Vasella, A. *Helv. Chim. Acta* **1990**, *73*, 1764; (c) Vasella, A. *Pure Appl. Chem.* **1991**, *63*, 507; (d) Mangholz, S. E.; Vasella, A. *Helv. Chim. Acta* **1991**, *74*, 1808.
166. Vasella, A.; Briner, K.; Soundarajan, N.; Platz, M. S. *J. Org. Chem.* **1991**, *56*, 4741.
167. Padwa, A.; Rasmussen, J. K. *J. Am. Chem. Soc.* **1975**, *97*, 5912.
168. Crawford, R. J.; Raap, R. *Can. J. Chem.* **1965**, *43*, 126.
169. Moss, R. A.; Shen, S.; Hadel, L. M.; Kmiecik-Lawrynowicz, G.; Wlostowska, J.; Krogh-Jespersen, K. *J. Am. Chem. Soc.* **1987**, *109*, 4341.
170. Iwamura, H.; Tukada, H. *Tetrahedron Lett.* **1978**, *19*, 3451.
171. Butler, G. B.; Herring, K. H.; Lewis, P. L.; Sharpe, V. V.; Veazey, R. L. *J. Org. Chem.* **1977**, *42*, 679.
172. Hömberger, G.; Kirmse, W.; Lelgemann, R. *Chem. Ber.* **1991**, *124*, 1867.
173. Saba, S.; Wolff, S.; Schröder, C.; Margaretha, P.; Agosta, W. C. *J. Am. Chem. Soc.* **1983**, *105*, 6902.

ABSOLUTE KINETICS OF INTRAMOLECULAR ALKYLCARBENE REACTIONS

Robert A. Moss

Advances in Carbene Chemistry
Volume 1, pages 59–88.

ISBN: 1-55938-545-6

I. INTRODUCTION

Carbene reactions in solution are generally extremely rapid, so that flash photolytic techniques are normally required to obtain (absolute) kinetic data. Relatively stable triplet carbenes, such as diphenylcarbene, react with styrene and dienes on the microsecond time scale and can be followed by flash lamp methodology.[1] More reactive triplet (fluorenylidene)[2] and singlet (Ph–C–Cl)[3] carbenes, however, react in the submicrosecond time domain; their study requires (nanosecond) laser flash photolytic systems.

The advent of these installations, about 1980, stimulated many quantitative studies of the reactivity of singlet carbenes in solution. Attention initially focused on carbenes that exhibited reasonably strong UV spectra and on the intermolecular reactions of carbenes, notably addition to alkenes.[4]

More recently, Platz et al. have demonstrated that "invisible" carbenes (i.e., those that lack a useful UV absorption) react with, for example, pyridine, affording strongly absorbing ylides that could be monitored to yield kinetic data for the carbenes.[5] This innovation has made possible many quantitative studies of singlet intramolecular carbene reactions in solution, which is the subject of this chapter. Our attention herein will focus particularly on the experimental rate constants and activation parameters associated with elemental carbenic reactions such as 1,2-hydrogen and 1,2-carbon migrations. A table that compiles these data is included as an appendix to this chapter.

We note at the outset, however, that there are two presently unsolved questions that temper our conclusions; These are (1) the possible intervention of carbene/alkene complexes, and (2) the extent of carbene-mimetic reactions that occur via the excited states of the carbene precursors instead of in the carbenes themselves.

II. THE 1,2-HYDROGEN MIGRATION

The 1,2-hydrogen migration is the most common singlet, intramolecular carbenic reaction [cf. Eq. (1)].[6–8] Despite many published examples of this migration, absolute kinetic studies were lacking until very recently. It has been conventional wisdom that 1,2~H reactions occur so rapidly that their kinetics will be very difficult to obtain.[6,7] Theoretical calculations of the rearrangement barrier for methylcarbene (Me–C–H) to ethene vary from 0–27 kcal/mol, depending on the computational

(1)

method (a tabular summary appears in Ref. 9). The use of large basis sets and zero point corrections, however, leads to a calculated activation barrier of only 0.6 kcal/mol,[9] so that the Me–C–H rearrangement should indeed be very rapid.

On the other hand, substitutions at the "spectator" carbene substituent, R in Eq. (1), greatly affect the activation barrier. Even with R = Me, the 1,2~H rearrangement of Me–C–Me to propene is calculated to possess a barrier of ~5 kcal/mol,[10] which makes it reasonable to look for this carbene in solution by laser flash photolysis (LFP). Moreover, the substitution of lone-pair-carrying heteroatomic spectator substituents for R in Eq. (1) further increases the 1,2~H barrier. The calculated dependence of E_a on X for the reaction CH_3–C–X → CH_2=CHX is $E_a = 0.6$, 4.7, 11.5, 19, or 27 kcal/mol, when X = H, Me, Cl, F, or OMe, respectively.[9,10]

This computational data strongly encourages LFP searches for disubstituted alkylcarbenes in solution, and suggests that their 1,2~H rearrangements should be observable on the micro- or nanosecond time scale. A cautionary note, however, derives from the possibility that the product-forming 1,2~H shift might occur in an excited state of the carbene precursor, rather than in the carbene [cf. Eq. (2)]. To the extent that product CH_2=CHX is formed by the k^* excited state rearrangement, instead of the k_C carbenic rearrangement, the carbene will be more difficult to detect. We will return to this problem in Section IV.A, below.

How well do the experimental findings correspond to predictions from theory?

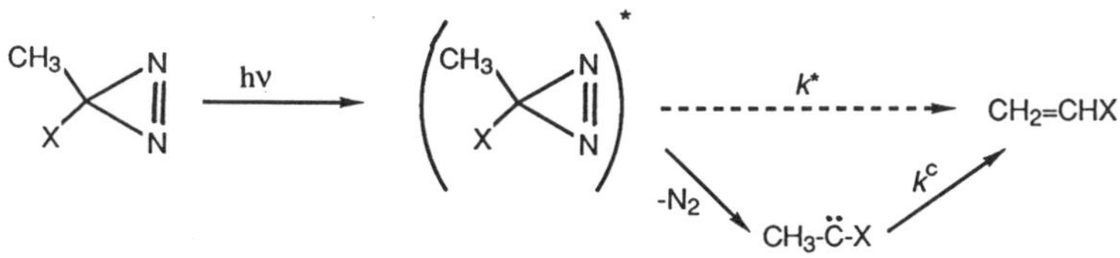

(2)

Prior to the recent LFP studies, there were indications of at least small barriers to 1,2~H rearrangements. For example, Su and Thornton reported product-based isotope and aromatic substituent effects for 1,2~H rearrangements of thermally generated 1-aryl-2-propylidenes [cf. Eq. (3)].[11] From the low intramolecular primary kinetic isotope effects ($k_H/k_D \approx 1.2$–1.6), which were highest when Ar = *p*-ClC_6H_4, and from the aromatic substituent effects on the internal/terminal alkene

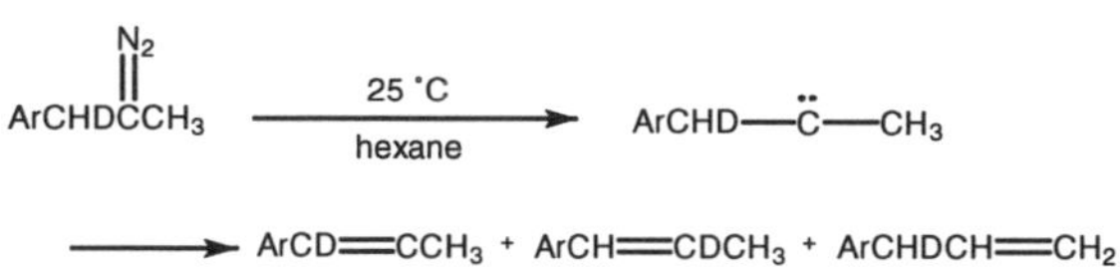

(3)

product ratio (Hammett $\rho \approx -1$), it was suggested that the reaction transition state involved an "early" hydride-like H shift that traversed a small, but real, barrier.

Similarly, the existence of a barrier could be inferred from the sensitivity of the *E*/*Z* alkene product ratio to the electronic character of R in the 1,2~H rearrangements of RCH_2–C–Cl or RCH_2–C–Ph.[12] Again, however, product distribution data does not address the possibility that the observed selectivities are decided in the carbene precursor and not in the carbene.

More recently, Stevens et al. estimated an upper limit to the barrier for 1,2~H shifts in dialkylcarbenes.[13] For example, 2-carbenepentanoic acid, generated thermolytically from the corresponding diazirine at 100 °C, gave the anticipated 1,2~H olefinic products, but *no* γ-methyl-γ-butyrolactone resulting from intramolecular capture of the carbene by the carboxylic acid group [cf. Eq. (4)].[13] From the known rate constant for the *intermolecular* reaction of Ph–C–Cl and acetic acid (1.8×10^9 $M^{-1}s^{-1}$), together with the estimated effective concentration of "COOH" in the carbenapentanoic acid (1.1 *M*), Stevens et al. estimated $k \approx 2 \times 10^9$ s^{-1} for the intramolecular carbene lactonization. Because the lactone is not formed, the 1,2~H shifts that supercede lactonization are estimated to have $k > 2 \times 10^{11}$ s^{-1}.

Further assumptions concerning the preexponential factor for 1,2~H migration at 100 °C, lead to an estimated barrier of ≤ 2 kcal/mol for the H shift. Clearly,

$$CH_3\text{—}\ddot{C}\text{—}CH_2CH_2COOH \longrightarrow CH_2\text{=}CHCH_2CH_2COOH + CH_3CH\text{=}CHCH_2COOH \qquad (4)$$

"indirect" experimental studies agree that there is a real, if small, barrier to 1,2~H shifts for dialkylcarbenes; this sets the stage for the time-resolved studies made possible by LFP methodology.

The parent dialkylcarbene, Me_2C, calculated to have a 1,2~H shift barrier of ~5 kcal/mol,[10] has recently been examined by LFP.[14] Laser flash photolysis of diazirine (**1**) at 351 nm afforded no transient absorbance, but, in the presence of pyridine, a transient absorbance at 364 nm was observed and attributed to the Me_2C/pyridinium ylide, **2**.[5,14a]

From the slope of the observed correlation between $1/\phi_Y$ and 1/[pyridine], (where ϕ_Y is the ylide absorbance), together with assumptions concerning the singlet–triplet equilibrium of Me_2C, and the rate constant for ylide formation ($k > 10^9$ M^{-1} s^{-1}), it was concluded that the lifetime of Me_2C in pentane solution was ~5 ns in the *absence* of pyridine. Me_2C therefore undergoes a rapid 1,2~H shift (> 2 ×

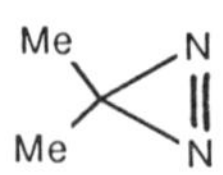

1

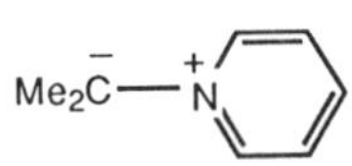

2

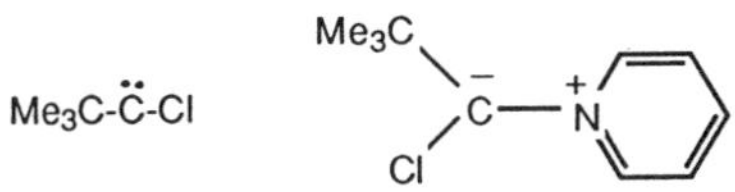

3 **4**

10^8 s^{-1}) to propene, and it also reacts with solvent. Nevertheless, the LFP experiment demonstrates that Me_2C has a finite lifetime in solution, consistent with a small barrier to the 1,2~H rearrangement. This conclusion is emphatically reinforced by the observation of a kinetic isotope effect of 3.2 (!) on the lifetime of Me_2C-d_6 versus Me_2C in pentane at 25 °C. The lifetime of Me_2C in solution, therefore, is largely controlled by the rate of the 1,2~H (or 1,2~D) rearrangement.

Dimethylcarbene is difficult to study, even by LFP methods, because of its low stability, high reactivity, and short lifetime. Heteroatom-substituted alkylcarbenes, which are stabilized by heteroatom lone pair donation, are generally better behaved under LFP conditions. Accordingly, their intramolecular rearrangements have been intensively examined. The ylide probe method is essential to following the reaction kinetics in many of these cases, and it will be considered first.

III. THE YLIDE METHOD

Although aryl substituted carbenes, such as Ph–C–Cl, can be directly observed in the UV spectrum and readily followed by LFP,[4b] alkyl and alkylhalocarbenes are often spectroscopically "invisible". To follow their kinetics, Jackson et al. introduced the ylide method.[5] *tert*-Butylchlorocarbene (**3**) cannot be directly observed by UV spectroscopy following LFP of its diazirine precursor, but, in pyridine/toluene, carbene **3** is trapped to form ylide **4**, λ_{max} ~375 nm. The observed pseudo first-order rate constants for the growth of ylide **4** are linear in [pyridine], with the correlation's slope affording $k_{abs} = 2.4 \times 10^9\ M^{-1}\ s^{-1}$ for the carbene/pyridine reaction. More important, the intercept at a pyridine concentration of zero gives k_0 (~$1.1 \times 10^7\ s^{-1}$), the rate constant for the disappearance of **3** in the absence of pyridine.

Rate constant k_0 represents the *sum* of all those processes that destroy **3** in toluene, including dimerization, reactions with solvent or impurities, reaction with the carbene precursor (i.e., with diazirine to form azine), *and* intramolecular reactions. Usually, we are most interested in the latter processes. In order to safely identify the experimental k_0 with intramolecular carbenic reactions, we must demonstrate by product analysis that the yield of the intramolecular products exceeds ~90%; dimer, azine, and solvent-derived products must be absent. Less desirably, the observed k_0 can be "corrected" for competitive intermolecular reactions.

The bimolecular dimerization and azine formation can be minimized by working at low carbene precursor concentrations. Also, intramolecular carbene rearrangements are usually faster than intermolecular reactions of the carbene with aprotic, saturated hydrocarbon solvents. Thus, it is often possible to set conditions under which the carbene affords “only” intramolecular products, and the pyridine ylide method gives a k_O that is representative of the rate constant for the carbene’s intramolecular reactions.

An additional benefit of the method is that the ylide can be used to monitor other *intermolecular* reactions of the carbene. Thus, if an olefinic trap is added to the pyridine/toluene solution in which **3** is generated by LFP, the olefin and pyridine will compete for the carbene. At a constant pyridine concentration, the observed pseudo first-order rate constant for ylide formation will appear to *increase* with increasing olefin concentration, according to Eq. 5. Thus, the second-order rate constant for the addition of **3** to, for example, *trans*-3-hexene, can be obtained from the slope of a plot of k_{obsd} for the formation of ylide **4** versus hexene concentration at constant pyridine concentration.[5] The value obtained is 4.3×10^7 M^{-1} s^{-1} in toluene at 25 °C. It is also possible to extract this rate constant by following the decrease in ylide absorbance as a function of olefin concentration.[5] For further discussion of the ylide method, the reader is referred to Chapter 3 of this volume.

$$k_{obsd} = k_O + k_{pyr}[\text{pyr}] + k_o[\text{olefin}] \quad (5)$$

A caution is in order. If there is a significant non-carbene pathway to an apparent carbene product [cf. Eq. (2)], then relative product yields cannot be simply used to partition the ylide-derived intramolecular rearrangement rate constant k_O into rate constants for competing intramolecular reactions. This problem will be discussed below.

Finally, we note that photoacoustic calorimetry measurements have furnished the heats of reaction for ylide formation.[15] For example, the reactions of Me–C–Cl or Ph–C–Cl with pyridine in heptane are accompanied by $\Delta H \approx -20$ to -21 kcal/mol, suggesting that the exothermically-formed ylides are unlikely to revert to carbene and pyridine under typical LFP conditions.

IV. REARRANGEMENTS OF ALKYLHALOCARBENES

A. General Survey

Bonneau et al. applied the ylide method to determinations of the absolute kinetics for the rearrangements of carbenes **3** and **5–11**.[16] LFP of diazirine precursors in pyridine–isooctane at 25 °C gave absorption spectra of ylides derived from the carbenes. Analyses of the ylide absorbances as a function of pyridine concentration, afforded extrapolated lifetimes for the carbenes in the absence of pyridine. Converted to rate constants for the disappearance of the carbenes, k ranged from $> 10^8$ s^{-1} for Et–C–Cl (**6**), Pr–C–Cl (**7**), *i*-Pr–C–Cl (**8**), and **10**, to a low value of $\sim 6 \times 10^5$ s^{-1} for Cy–C–Cl (**11**). Intermediate rate constants were determined for Me–C–

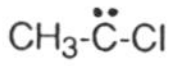
5

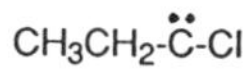
6

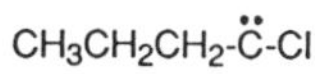
7

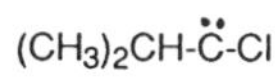
8

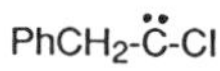
9

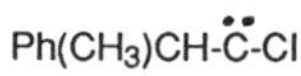
10

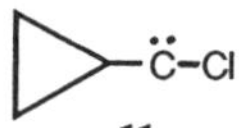
11

Cl (**5**) ($k \approx 3 \times 10^6$ s^{-1}), $PhCH_2$–C–Cl (**9**) ($k \approx 6 \times 10^7$ s^{-1}), and *t*-Bu–C–Cl (**3**) ($k \approx 1 \times 10^7$ s^{-1}).

By reference to the literature, these rate constants were associated with 1,2~H shifts for carbenes **5–10**, with a 1,2~Me migration for **3**, and with a 1,2~C ring expansion to chlorocyclobutene with Cy–C–Cl.[16] Accompanying product analyses, however, were not reported, and the 1,2~Me shift claimed for **3** is, in fact, only a minor reaction channel (see Section IV.H, below).

If we provisionally accept $k \approx 3 \times 10^6$ s^{-1} for the 1,2~H shift of Me–C–Cl, then alkylation at the methyl group (as in **6–8**) must enhance *k* by more than an order of magnitude. This is consistent with a hydride-like H shift, in which a developing positive charge at the migration origin can be stabilized by alkyl substituents. A similar conclusion is reached upon comparison of carbenes **10** and **9**, where the additional methyl substituent in **10** enhances 1,2~H migration. The rearrangements of carbenes **3**, **5**, **9**, and **11** will be considered in greater detail in Sections IV.B, IV.C, IV.D, and IV.H.

MINDO/3 calculations of activation parameters for the rearrangements of the geometry-optimized carbenes were also reported.[16] E_a values (kcal/mol) were Me–C–Cl (9.4), Et–C–Cl (6.0), *i*-Pr–C–Cl (2.5), $PhCH_2$–C–Cl (4.3) and **10** (2.3); $\Delta S^{\ddagger}$ (MNDO) was calculated to be about –3 to –4 eu for each carbene.[16] The calculated E_as reflect the observed rate constants reasonably well, except that Et–C–Cl appears out of place, relative to $PhCH_2$–C–Cl.

B. Chloromethylcarbene

The case of Me–C–Cl (**5**) has been studied in detail by Liu and Bonneau[17] and by LaVilla and Goodman.[18,19] The pyridine ylide method, with 3-chloro-3-methyldiazirine as the LFP precursor, gives $k = 3 \times 10^6$ s^{-1} for the 1.2~H shift that converts Me–C–Cl to vinyl chloride.[16,17] Pyridine ylide studies of the competitive trapping of Me–C–Cl by alkenes of varying electronic character indicated that the carbene is ambiphilic in its additions to olefins.[17]

The rearrangement of Me–C–Cl was also investigated by photoacoustic calorimetry, where separate, time-resolved heat evolutions could be observed for [diazirine → Me–C–Cl + N_2] and [Me–C–Cl → CH_2=CHCl].[18] From the latter process, ΔH_f (Me–C–Cl) could be obtained (50.7 ± 4.7 kcal/mol in heptane). A lifetime of 740 ± 80 ns was extracted for Me–C–Cl in heptane at 22 °C,[18] corresponding to $k = 1.3 \pm 0.12 \times 10^6\ s^{-1}$ for 1,2~H, in reasonable agreement with the previous value.[16,17] A photoacoustic calorimetric study of Me–C–Br gave $k \sim 5.6 \times 10^6\ s^{-1}$ for the 1,2~H shift (heptane, 22 °C).[19] Note the somewhat faster rearrangement of the less-stabilized Me–C–Br, relative to Me–C–Cl. It was also observed[18] that k (1,2~H) for Et–C–Cl or *i*-Pr–C–Cl was $> 10^8\ s^{-1}$, again in agreement with previous observations[16] (see Section IV.A).

Temperature dependent kinetic studies gave $E_a = 4.9 \pm 0.5$ kcal/mol and $\log A = 9.7 \pm 0.3\ s^{-1}$ (corresponding to $\Delta S^{\ddagger} = -16$ eu at 295 K) for the 1,2~H shift in Me–C–Cl.[18] Note that the observed activation energy is less than half of the *ab initio* calculated value of 11.5 kcal/mol[9] (see Section II). Moreover, the low preexponential factor for the unimolecular rearrangement of Me–C–Cl "suggests a highly ordered transition state"[18] and is not in accord with $\Delta S^{\ddagger} = -3.5$ eu calculated for this reaction by MNDO.[16] We will encounter other discrepancies between observed and calculated activation parameters in carbenic rearrangements.

Assuming a similar A factor as found for Me–C–Cl, and recalling that $k \geq 10^8\ s^{-1}$ for the rearrangements of Et–C–Cl and *i*-Pr–C–Cl,[16,18] E_a for the 1,2~H shifts would be ≤ 2.4 kcal/mol.[18] These lower activation energies reflect alkyl stabilization of the hydride shift transition states for RCH_2–C–Cl.

The photochemical generation of Me–C–Cl from its precursor diazirine, however, appears to be more complicated than we have thus far indicated.[19] There seems to be a second reactive intermediate, in addition to the carbene, that yields vinyl chloride. The identity of this intermediate is uncertain, with an excited state of the diazirine or a carbene/alkene complex the most frequently mentioned possibilities (see Section IV.C, below). The case of Me–C–Cl is representative and will be considered in detail. The discussion will be applicable to our consideration of benzylchlorocarbene in Section IV.C.

Photolytic generation of Me–C–Cl in heptane yields > 90% of vinyl chloride, but, in the presence of an olefin (e.g., $Me_2C{=}CMe_2$), both vinyl chloride (1,2~H product) and chloropentamethylcyclopropane (intermolecular addition product) are formed. If the carbene were the *only* product-forming intermediate [Eq. (6)], then the product ratio, cyclopropane/vinyl chloride (Cyc/VCl), would be linearly related to $Me_2C{=}CMe_2$ concentration, as in Eq. (7). Experimentally, however, the product ratio affords a strongly curved correlation with $Me_2C{=}CMe_2$ concentration that levels off at olefin concentrations above 0.05 *M*. Only about 66% of the carbene is trapped at this point; one-third of the initial diazirine ends up as vinyl chloride, even at twice the olefin concentration.[19] Similar results are obtained with Me–C–Br.[19]

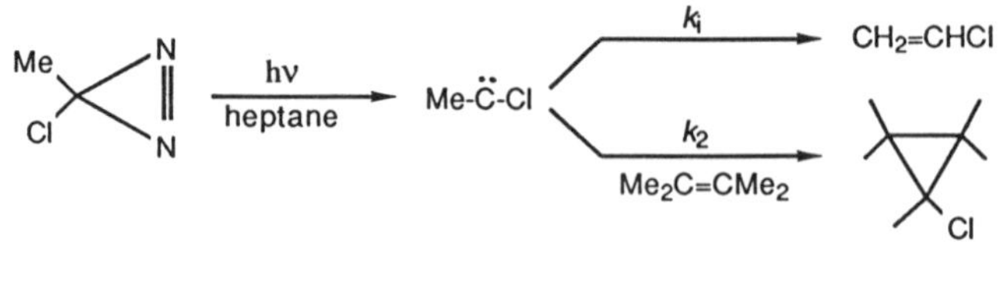

(6)

$$\frac{[\mathrm{Cyc}]}{[\mathrm{VCl}]} = \frac{k_2}{k_i[\mathrm{Me_2C{=}CMe_2}]} \tag{7}$$

On the other hand, the inverse product ratio, (VCl/Cyc) is linear with $1/[Me_2C{=}CMe_2]$, consistent with the mechanistic scheme shown in Eq. (8), and its kinetic expression, Eq. (9).[19] The key feature of Eq. (8) is that the excited diazirine not only gives the carbene, but that it also directly affords the 1,2~H shift product [cf. Eq. (2), above].

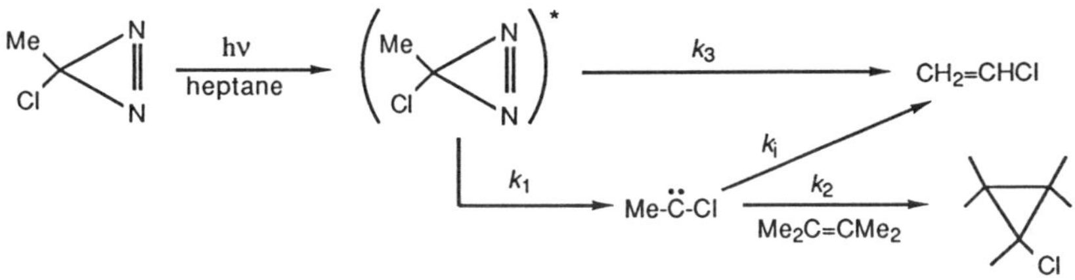

(8)

$$\frac{[\mathrm{VCl}]}{[\mathrm{Cyc}]} = \frac{k_3}{k_1} + \frac{[k_i(k_1 + k_3)]}{k_1 k_2[\mathrm{Me_2C{=}CMe_2}]} \tag{9}$$

Note that *any* second product-forming intermediate satisfies Eq. (9); it need not be the excited diazirine. This point was made by Warner, who formulated Eqs. (8) and (9) using an excited singlet carbene in place of the excited diazirine shown here.[20] One can also substitute an isomeric diazo compound for the excited diazirine.[21] And, as we will see below, Eq. (8) can be altered to include a carbene/alkene complex as the second intermediate.

Photoacoustic calorimetric and thermochemical studies also support the existence of two product-forming reactive intermediates in the photolysis of chloromethyldiazirine, with the minor pathway [k_3 in Eq. (8)] very fast (< 10 ns) and the major (carbenic) pathway (k_i) much slower ($\tau \approx 740$ ns; $k \approx 1.3 \times 10^6$ s^{-1}).[19]

The key feature of Eq. (8) is the partition of excited diazirine between vinyl chloride and carbene, represented by the rate constant ratio, k_3/k_1. According to Eq. (9), this ratio is the intercept, at $Me_2C{=}CMe_2$ concentration equal to zero, of the linear relation between [VCl]/[Cyc] and $1/[Me_2C{=}CMe_2]$. From the data, k_3/k_1 is 0.48, so that 67% of the VCl comes from Me–C–Cl and the remaining 33% stems directly from the excited diazirine.[19] The complications that emerge from a detailed consideration of Me–C–Cl are echoed in the chemistry of benzylchlorocarbene. Indeed these problems are general and must be anticipated in other studies of alkylcarbenes.

C. Benzylhalocarbenes

Much effort has been devoted to benzylchlorocarbene, **9**, largely by M. T. H. Liu and his associates, who suggest that this carbene's reactivity is complicated by the

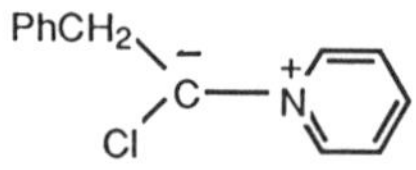

12

PhCH=CHCl

(*Z,E*)-**13**

intervention of carbene/alkene complexes.[22–25] More recently, the absolute rate constant for the 1,2~H rearrangement of $PhCH_2$–C–Cl to *Z*- and *E*-2-chlorostyrene has been obtained both by the pyridine ylide method[26–28] and by direct observation of the carbene.[27,28]

In the presence of pyridine, $PhCH_2$–C–Cl generated from an isooctane solution of the appropriate diazirine by ambient temperature LFP gives rise to ylide **12**.[26] Monitored at 379 nm, the absorbance of **12** is linear with pyridine concentration, and the correlation slope affords $k = 4.2 \pm 0.2 \times 10^9\ M^{-1}\ s^{-1}$ for ylide formation. The intercept at concentration of pyridine equal to zero gives $k = 6.7 \times 10^7\ s^{-1}$ for the sum of all destructive channels for the carbene in the absence of pyridine. In cyclohexane, the yield of the 1,2~H products, **13**, is 73% so that *k* (1,2~H) is taken as $4.9–6.7 \times 10^7\ s^{-1}$.[26] Repetitions of these experiments give *k* (1,2~H) = $6.0 \pm 0.5 \times 10^7\ s^{-1}$ [27] and $6.1 \pm 0.1 \times 10^7\ s^{-1}$ [28] in isooctane at 25 °C.

$PhCH_2$–C–Cl can also be observed directly upon LFP of its diazirine precursor in isooctane. The carbene is very short-lived ($\tau \approx 3–18$ ns), but its UV spectrum can be at least partially observed from 290–320 nm.[27] Decay of $PhCH_2$–C–Cl generates **13**, absorbing in the 275–305 nm region, with a rise time of ~18 ns. The rate constant for carbenic decay at ~300 nm and 24 °C is given as *k* (1,2~H) = $5.4 \pm 0.2 \times 10^7\ s^{-1}$, [27,28] in good agreement with the measurements made by the ylide method.

Activation parameters for the 1,2~H shift of $PhCH_2$–C–Cl were obtained by determining *k* as a function of temperature, using either the pyridine ylide or direct observation methods. The results were $E_a = 4.5 \pm 0.3$ kcal/mol, log $A = 11.1 \pm 0.2\ s^{-1}$ (ylide), and $E_a = 4.83 \pm 0.24$ kcal/mol, log $A = 11.3 \pm 0.2\ s^{-1}$ (direct).[28] These values should be considered more reliable than the previous estimate of $E_a = 6.4$ kcal/mol,[23] which involved measurements of the *relative* rate constant for $PhCH_2$–C–Cl addition to $Me_2C{=}CMe_2$ versus 1,2~H shift, and assumptions concerning the activation energy for the addition reaction.

Comparisons of *k* and E_a for the 1,2~H shifts of $PhCH_2$–C–Cl and Me–C–Cl (see Section IV.B.) reveal the assistance that phenyl substitution affords to the 1,2~H shift. The rate constant (5×10^7 to $6.7 \times 10^7\ s^{-1}$) is greater for $PhCH_2$–C–Cl than for Me–C–Cl (10^6 to $3 \times 10^6\ s^{-1}$), and the activation energy is slightly lower. Somewhat surprisingly, the rearrangement of $PhCH_2$–C–Cl is assisted by a more favorable *A* factor (log $A \approx 11.2$ vs. 9.7 for Me–C–Cl). We also note, in passing, that relative reactivity studies of $PhCH_2$–C–Cl addition reactions indicate that the carbene is an ambiphile,[29] as is Me–C–Cl.[17]

The effect of substitutents on the rate constants and activation parameters for the 1,2~H shifts of *p*-X–$PhCH_2$–C–Cl has also been determined.[30] The *p*-CF_3 and *p*-Cl substituted carbenes were studied by LFP using direct observational and pyridine probe methods; k (1,2~H) and Arrhenius parameters were as follows: p-CF_3 (24 °C) $k = 2.2 \times 10^7\ s^{-1}$, $E_a = 4.9 \pm 0.2$ kcal/mol, log $A = 10.9 \pm 0.1\ s^{-1}$; p-Cl (25 °C), $k = 4.2 \times 10^7\ s^{-1}$, $E_a = 4.5 \pm 0.1$ kcal/mol, log $A = 10.9 \pm 0.1\ s^{-1}$. *Estimates* of k (1,2~H) were also offered for the p-CH_3O ($1.7 \times 10^8\ s^{-1}$) and p-CH_3 ($1.2 \times 10^8\ s^{-1}$) substituted carbenes. These were derived from the relative rates of carbene addition to $Me_2C{=}CMe_2$ versus 1,2~H, by taking the absolute rate constants of the addition reactions as ~$5.8 \times 10^8\ M^{-1}\ s^{-1}$.

Therefore, at ~25 °C, the values of k (1,2~H) for p-X–$PhCH_2$–C–Cl as a function of X are (CH_3O) 1.7×10^8, (CH_3) 1.2×10^8, (H) ~6×10^7, (Cl) 4.2×10^7, and (CF_3) 2.2×10^7 (s^{-1}). The rate enhancing effect of electron-releasing substituents X on k is small, but real, leading to a Hammett ρ of ~ −1.0 (vs. σ_p), consistent with a hydride-like shift of H and the imposition of a small positive charge at the migration origin in the rearrangement transition state.[30]

Bonneau et al. have also studied benzylbromocarbene ($PhCH_2$–C–Br), reporting k(1,2~H) = $5.6 \pm 0.4 \times 10^7\ s^{-1}$ (25 °C, isooctane), as determined by the LFP-pyridine ylide method.[31a] Surprisingly, this rate constant is nearly identical to that of $PhCH_2$–C–Cl; the less-stabilized bromocarbene would be expected to rearrange more rapidly than $PhCH_2$–C–Cl, as is observed for Me–C–Br versus Me–C–Cl, where the 1,2~H shift is faster with the bromocarbene by ~2–4 times (cf. Section IV.B).

Estimates of E_a = 4.7[31b] or 6.6[31a] kcal/mol have been published for the 1,2~H shift of $PhCH_2$–C–Br. These involve indirect measurements and various assumptions, and appear too high when compared to the *measured* E_a of 4.5–4.8 kcal/mol for the rearrangement of $PhCH_2$–C–Cl.[28] From relative reactivity studies, $PhCH_2$–C–Br, like $PhCH_2$–C–Cl and Me–C–Cl, appears to be an ambiphile in its intermolecular addition to olefins.[31a]

As with Me–C–Cl, the competitive intramolecular 1,2~H shift and intermolecular addition reactions of $PhCH_2$–C–Cl (or $PhCH_2$–C–Br) cannot be kinetically fitted to a scheme of parallel reactions, as in Eq. (6). When an olefinic trap is present, the correlation of [Cyc]/[1,2~H product] versus [trap] is curved; it does not follow Eq. (7), and is inconsistent with the mechanism of Eq. (6). On the other hand, the inverse product ratio, [**13**]/[Cyc] is linear with 1/[trap], obeying a relation similar to Eq. (9), with VCl replaced by **13**.[22]

It was recognized that this required a second reactive intermediate, in addition to $PhCH_2$–C–Cl, and it was suggested[22] that this was likely to be a carbene/alkene (C/A) complex.[4b] Initially, the complex was pictured as the precursor of the cyclopropane adduct only,[22] but Warner showed that this was not in keeping with the observed kinetics.[20] Subsequent invocations of the C/A complex have therefore made it a precursor of both the 1,2~H-shift product (**13**) and the cyclopropane adduct [see Eqs. (10) and (11)].[23–25]

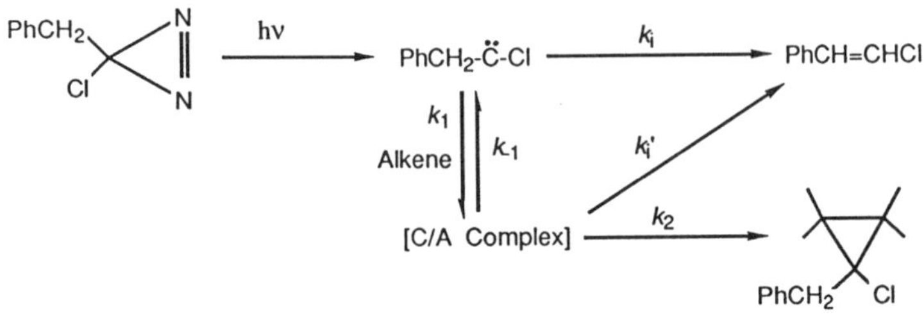

(10)

$$[\mathbf{13}]/[\mathrm{Cyc}] = k_i/k_2 + k_i/k_t\,[\mathrm{alkene}] \qquad (11)$$

The mechanism of Eq. (10), with application of the steady state assumption, leads to kinetic expression (11),[25] which satisfies the inverse dependence of the product ratio on [alkene], and is of the same overall form as Eq. (9) (see Section IV.B.). Indeed, as noted above, *any* second intermediate, properly situated in the mechanistic scheme, can account for the observed kinetics. These require two reactive intermediates; LFP demonstrates that one of them is the carbene, but neither kinetics nor spectroscopy identify the second species.

We will review the pros and cons of alternative mechanisms (8) and (10) below, in Section V. For now, we caution again that if Eq. (8), involving the excited diazirine as the second intermediate, is the correct representation, then the 1,2~H rearrangement products of Me–C–Cl, $PhCH_2$–C–Cl, and related carbenes, are not "pure" carbene products. The "trappability" of alkylcarbenes by alkenes may then be less a function of the carbenes' stabilities and more a reflection of the partitions of excited diazirine precursors between carbene formation or rearrangement with accompanying loss of nitrogen.

Note, too, that the LFP rate constants derived for 1,2~H rearrangements are not affected by these mechanistic complications as long as they come from observational studies on the carbene itself, or on a carbene/pyridine ylide. Only when "absolute" 1,2~H rate constants are estimated in part from *relative* product yields (e.g., 1,2~H vs. addition) must we be careful.

D. Cyclopropylhalocarbenes

As noted in Section IV.A, the 1,2~C ring expansion of chlorocyclopropylcarbene (**11**, Cy–C–Cl) to chlorocyclobutene [Eq. (12)], has been examined by the pyridine ylide LFP method, with k (1,2~C) reported to be ~$6 \times 10^5\ \mathrm{s}^{-1}$.[16] The carbene could

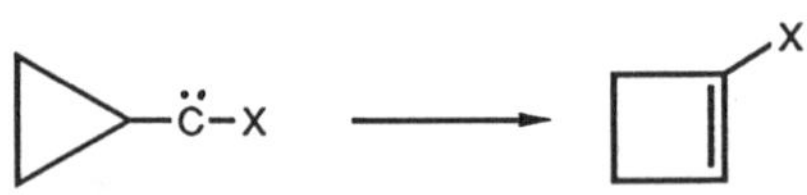

(12)

11, X = Cl **15**, X = Cl
14, X = F **16**, X = F

also be directly observed at 250 nm; the rate constant derived in this manner was ~3.8×10^5 s^{-1}, after correction for a competing dimerization of the carbene.[16]

As a prime example of a 1,2~C migration, carbene **11** has been subjected to further scrutiny,[32–34] and its fluoro analogue, **14** (Cy–C–F), has also been examined.[35] Details[32] of the initial study[16] of Cy–C–Cl reiterate k (1,2~C) = 3.8×10^5 s^{-1} (23 °C, isooctane) by direct observation of the carbene in a LFP experiment. Under these conditions, the value of k (dimerization) is estimated at ~6.7×10^9 M^{-1} s^{-1}.

Is a correction to the rate constant for dimerization necessary? The actual rate of dimerization depends on Cy–C–Cl concentration, and therefore on the laser power used in the photolysis of the diazirine precursor. In preparative studies, where the dimer was formed, 300-mJ, 15-ns pulses of a Nd:YAG laser were used at 355 nm and a frequency of 10 Hz; whereas, in kinetic runs, a single 5–30-mJ, 0.2-ns laser pulse was administered. Clearly, the preparative results are not transferable to the kinetic conditions, where the post-pulse Cy–C–Cl concentration will be much lower. Indeed, nuclear magnetic resonance (NMR) and gas chromatography (GC) product studies of Eq. (12) at both –18 and 21 °C, where the diazirine precursor was decomposed with 70–80-mJ, 14-ns, 351-nm laser pulses, showed the formation of cyclobutene **15** in > 90% yield, with no evidence for significant dimer or azine formation.[34]

Independent LFP studies of the Cy–C–Cl rearrangement gave k (1,2~C) = 9×10^5 s^{-1} (20 °C, isooctane) by direct observation of the carbene at 250 nm,[33] whereas the pyridine ylide method afforded k (1,2~C) = 1.5×10^6 s^{-1}, in reasonable agreement with the direct value, although higher than the previously reported pyridine ylide value.[16,32] Moreover, Cy–C–Cl was quenched by addition to $MeCH{=}CMe_2$ [k (addn) = $6.0 \pm 1.0 \times 10^6$ M^{-1} s^{-1} at 20 °C].[33] From the product ratio, [**15**]/[cyclopropane], the observed k (addn), and the concentration of olefin, k (1,2~C) was calculated to be $1.2 \pm 0.2 \times 10^6$ s^{-1}.[33] From these, and the previous results,[32] a k value of ~ 1×10^6 s^{-1} seems quite appropriate for the 1,2~C migration of Cy–C–Cl. The process is comparable in rate to the 1,2~H shift of Me–C–Cl, but it is more than an order of magnitude slower than the 1,2~H shifts in more "complicated" alkylchlorocarbenes, such as $PhCH_2$–C–Cl, Et–C–Cl, or *i*-Pr–C–Cl. We note, in passing, that rate constants for the intermolecular additions of Cy–C–Cl to a series of alkenes are generally about an order of magnitude less than the comparable rate constants of Ph–C–Cl. In analogy to Ph–C–Cl, however, Cy–C–Cl also behaves as an ambiphile in additions to alkenes.[33]

Activation parameters have been obtained for the rearrangement of Cy–C–Cl. By direct observation of the carbene, Liu and Bonneau obtained $E_a = 7.4$ kcal/mol and log $A = 11.1$ s^{-1},[32] corresponding to $\Delta G^\ddagger \approx 9.8$ kcal/mol, $\Delta H^\ddagger \approx 6.8$ kcal/mol, and $\Delta S^\ddagger \approx -10$ eu. In contrast, Moss et al. reported a considerably lower $E_a = 3.0 \pm 0.4$ kcal/mol, with log $A = 8.2 \pm 0.2$ s^{-1}.[34] Conversion to Eyring parameters gave $\Delta G^\ddagger \approx 9$ kcal/mol, similar to the previous value,[32] but with a low $\Delta H^\ddagger$ (~2–3 kcal/mol) and a very negative $\Delta S^\ddagger$ (~ –20 to –24 eu).[34]

The latter values were averages derived from Arrhenius parameters that were obtained from four different kinds of rate constant measurements: direct observations of the decay of Cy–C–Cl; pyridine ylide studies; quenching studies with 2-methyl-1-butene; and product based studies involving intermolecular addition versus 1,2~C shift, coupled with the measured rate constant for the addition reaction.[34] Not only are the rate constants and derived activation parameters in good mutual agreement, but the agreement between product-based and kinetically derived values suggests that there was little product formation from an excited diazirine precursor. The latter may competitively fragment to nitrogen, ethene, and chloroethyne,[33] but it does not significantly rearrange to chlorocyclobutene.

If we accept E_a ~3 kcal/mol, log A ~8 s^{-1}, $\Delta G^{\ddagger}$ ~ 9 kcal, $\Delta H^{\ddagger}$ ~3 kcal/mol, and $\Delta S^{\ddagger}$ ~ –20 to –24 eu as the most authoritative set of activation parameters for the 1,2~C shift of Cy–C–Cl,[34] we face a discrepancy between the experimental and *ab initio* calculated values. The latter (for *trans*-Cy–C–Cl) are $\Delta G^{\ddagger} = 9.0$ kcal/mol, $\Delta H^{\ddagger} = 8.2$ kcal/mol, and $\Delta S^{\ddagger} = -2.7$ eu.[33] There is good agreement between the experimental and calculated $\Delta G^{\ddagger}$, but the experimental $\Delta G^{\ddagger}$ is dominated by a large, negative $\Delta S^{\ddagger}$, whereas the calculated $\Delta G^{\ddagger}$ is composed mainly of $\Delta H^{\ddagger}$.

There is presently no definitive resolution to this conflict. The (parent) cyclopropylcarbene to cyclobutene rearrangement [Eq. (12), X = H] has been subjected to renewed theoretical study.[36–38] It was found that the more stable *trans*-Cy–C–H (**17**) undergoes ring expansion, whereas *cis*-Cy–C–H (**18**) is more likely to isomerize to **17** by bond rotation, or to fragment to ethene and ethyne. The 1,2~C shift of **17** is opposed by a barrier of at least 5 kcal/mol,[37,38] and the required, highly concerted atomic motions[36] may result in a large negative $\Delta S^{\ddagger}$.[38] The latter conclusions are probably also applicable to Cy–C–Cl, for which *ab initio* calculations[33] lead to similar results.

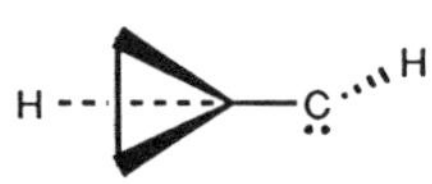

(*trans*) **17**

(*cis*) **18**

Other explanations have been offered to rationalize the experimental and computed values of $\Delta H^{\ddagger}$ and $\Delta S^{\ddagger}$ in the Cy–C–Cl rearrangement. Entropic control may be particularly important for reactions that cross relatively flat potential energy surfaces of the kind encountered in many carbene reactions.[34,39,40] The unusual motions required of the migrating group during rearrangement[36] translate into a "narrow" pass over the energy barrier, so that many energetically favorable collisions will be ineffectively translated into motion along the reaction coordinate and through the pass (i.e., the transmission coefficient will be $<< 1.0$).[34] A large negative entropy will therefore be observed, and the variational transition state of the reaction (which corresponds to the maximum on the $\Delta G^{\ddagger}$ profile) may not coincide with the calculated conventional minimum energy transition state.[39] Note that $\Delta S^{\ddagger} << 0$ is precedented in the 1,2~H shift of Me–C–Cl (–16 e.u.).[18]

Analogous phenomena are encountered in the rearrangement of Cy–C–F (**14**) to 1-fluorocyclobutene (**16**) [see Eq. (12)].[35] Photolysis of the appropriate diazirine affords **16** in > 90% yield, with no evidence of azine or carbene dimer formation. At low diazirine concentration, first order ylide formation is observed between Cy–C–F and pyridine, permitting application of the ylide kinetic methodology. This afforded k (1,2~C) = $1.4 \pm 0.2 \times 10^5\ s^{-1}$ (23 °C, isooctane) for Cy–C–F.[35] Variable temperature studies then generated activation parameters: $E_a = 4.2 \pm 0.5$ kcal/mol, $\log A = 8.3 \pm 0.3\ s^{-1}$; $\Delta G^{\ddagger} = 10.1$ kcal/mol, $\Delta H^{\ddagger} = 3.6$ kcal/mol, and $\Delta S^{\ddagger} \approx -22$ eu.[35]

Cy–C–F rearranges approximately seven times more slowly than Cy–C–Cl due to an activation energy barrier that is ~1 kcal/mol higher than that of the chlorocarbene.[34,35] Both Cy–C–F and Cy–C–Cl confront similar, unfavorable entropies of activation. The data are consistent with enhanced Cy–C–X carbene stabilization from better X → C electron donation when X = F, as opposed to X = Cl.[41,42] However, the magnitude of the change in E_a attending the F/Cl substitution, ~1 kcal/mol, is smaller than would be expected in analogy to the calculated E_a values for the 1,2~H shift of Me–C–X.[9] There, the F/Cl comparison affords ΔE_a ~7.5 kcal/mol. Similar differences between observed and calculated activation energies are observed with other alkylhalocarbenes.

E. Cyclobutylhalocarbenes

Cyclobutylchlorocarbene[43] (Cb–C–Cl, **19**) and cyclobutylfluorocarbene (Cb–C–F, **20**) have been photolytically generated from the diazirine precursors **21**. The products, **22** and **23**, arise by competitive 1,2~C (**22**) and 1,2~H (**23**) rearrangements of Cb–C–X; [see Eq. (13)]; yields are near quantitative, with no evidence of significant carbene dimer or azine formation.[35,44]

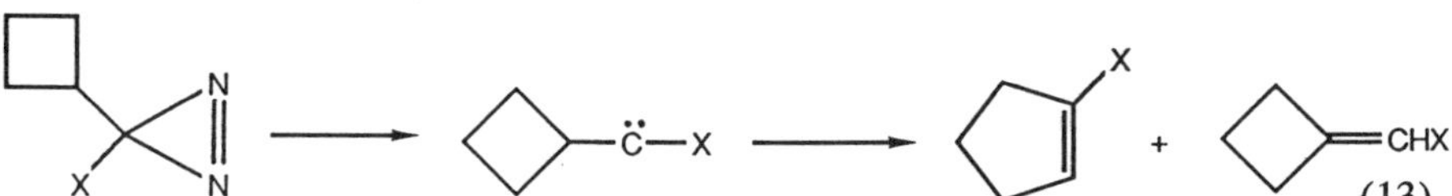

(13)

21 **19**, X = Cl **20**, X = F **22** **23**

LFP studies of Cb–C–Cl and Cb–C–F utilized the pyridine ylide technique and afforded rate constants that were sums of k (1,2~C) and k (1,2~H); these were 6.8 $\times 10^7\ s^{-1}$ (Cb–C–Cl, isooctane, 20 °C)[44] and $2.3 \times 10^6\ s^{-1}$ (Cb–C–F, isooctane, 23 °C).[35] Two conclusions follow from these results. First, as with the cyclopropylhalocarbenes, the greater overall reactivity of Cb–C–Cl versus Cb–C–F (a factor of ~30) is quite consistent with the expected carbene stabilizing effect of F versus Cl.[9,41,42] Second, the cyclobutylhalocarbenes are considerably more reactive than their cyclopropyl analogues (see Section IV.D). Rate constants for the carbene rearrangements are ($\times 10^{-6}$, s^{-1}): Cb–C–Cl, 68; Cy–C–Cl, 0.9; Cb–C–F, 2.3; Cy–C–F, 0.14. These trends support the importance of stabilizing electronic interactions operating between the cyclopropyl groups and the carbenic centers of Cy–C–X.[16,32–38,45]

The [**22**]/[**23**] product distributions can be used to partition the overall Cb–C–Cl and Cb–C–F rearrangement rate constants into specific 1,2~C and 1,2~H rate constants. For each cyclobutylcarbene, 1,2~C dominates at ambient temperature; the **22**/**23** ratios are Cb–C–Cl 2.17:1 (21 °C), and Cb–C–F 3.36:1 (23 °C).[35,44] Combined with the overall rearrangement rate constants (above), we obtain for Cb–C–Cl (s^{-1}), k (1,2~C) = 4.6×10^7 and k (1,2~H) = 2.1×10^7, whereas for Cb–C–F, the corresponding rate constants are 1.8×10^6 and 0.53×10^6. The internal comparisons here continue to manifest the F versus Cl carbene stabilizing effect. Comparisons of the 1,2~C rate constants for Cb–C–X with those of Cy–C–X (above) demonstrate the Cb > Cy ordering.

A cautionary note is in order. Competitive intermolecular addition/intramolecular rearrangement reactions have been carried out for Cb-C-Cl with several alkenes. In each case, the correlation of [cyclopropane adduct]/[**22** + **23**] versus alkene concentration was curved, whereas the inverse correlations, [**22** + **23**]/[adduct] versus 1/[alkene] were strictly linear ($r > 0.997$).[46] This suggests that the formation of **22** and **23** upon photolysis of **21** (X = Cl) cannot be simply attributed to the rearrangements of Cb–C–Cl, as written in Eq. (13), but that the more complicated mechanisms of Eqs. (8) and (9), or Eqs. (10) and (11), apply.

If, for example, the excited diazirine mechanism is adopted, then the data indicate that ~60% of **22** and **23** are formed from excited **21**-Cl, while ~40% of the products stem from Cb–C–Cl. Moreover, the **22**/**23** distribution would depend on the precursor, and this ratio for Cb–C–Cl alone would be higher than the overall observed product ratio. We must therefore be cautious in partitioning the ylide-based overall rate constant for Cb–C–Cl rearrangement (see above) into specific 1,2~C and 1,2~H rate constants.[44] The overall (1,2~C + 1,2~H) rate constant, however, reflects the behavior of the carbene alone, and is not affected by these complications.

Similar studies of the Cb–C–F system indicate that only about 12% of the products can form by a pathway that bypasses the carbene,[46] so that it is reasonable to interpret the photolysis of **21**-F in terms of Cb–C–F, as in Eq. (13), and to partition the overall ylide-derived rate constant into k (1,2~C) and k (1,2~H).

Activation parameters have been determined for Cb–C–F; values of E_a, log A, $\Delta G^{\ddagger}$, $\Delta H^{\ddagger}$, and $\Delta S^{\ddagger}$ are as follows: (1,2~C) 2.3 kcal/mol, 8.0 s^{-1}, 8.8 kcal/mol, 1.7 kcal/mol, and –24 eu; and (1,2~H) 3.8 kcal/mol, 8.5 s^{-1}, 9.6 kcal/mol, 3.2 kcal/mol, and –22 eu.[35] Comparisons of the activation parameters for the 1,2~C rearrangement of Cb–C–F with those for the analogous ring expansion of Cy–C–F (see Section IV.D) show that the Cb–C–F rearrangement is favored by $\Delta E_a \approx 2$ kcal/mol and $\Delta G^{\ddagger} \approx 1.3$ kcal/mol. Most of the difference is in the smaller enthalpic term of the Cb–C–F rearrangement ($\Delta\Delta H^\ddagger \approx 2$ kcal/mol), and can again be associated with the increased stability of the cyclopropyl versus the cyclobutyl carbene. Also, as previously, we see that the magnitudes of E_a or $\Delta H^\ddagger$ for the 1,2~C or 1,2~H Cb–C–F rearrangements are considerably smaller than anticipated for a fluorocarbene,[9] and that the $\Delta G^\ddagger$ barriers are again dominated by very unfavorable $\Delta S^\ddagger$ terms (cf. Section IV.D).

F. Halophenoxymethylcarbenes

The kinetics of the 1,2~H shifts of the halophenoxycarbenes (**24** and **25**) have been studied by the pyridine ylide LFP technique[35,47] [see Eq. (14)]. With the chlorocarbene **24**, only alkenes *Z*- and *E*-**26**-Cl were formed by 1,2~H shift in a combined yield > 90% (*Z*/*E* ≈ 4:1).[35,48] There were no significant complications due to dimer or azine formation.

24, X = Cl
25, X = F
26
(14)

At ~24 °C in isooctane, k (1,2~H) was 1.14 ; 0.06×10^8 s^{-1} for PhOCH$_2$–C–Cl.[47] The rate constant decreased with decreasing reaction temperature, falling to $6.0 \pm 0.4 \times 10^7$ at –10 °C,[47] and 3.6×10^7 s^{-1} at –32 °C.[35] An Arrhenius study at six temperatures gave $E_a = 2.8 \pm 0.6$ kcal/mol and log $A = 10.1$ s^{-1}.[47]

A pyridine ylide LFP study of the analogous fluorocarbene **25** gave k (1,2~H) = 1.3×10^7 s^{-1} at –32 °C.[35] This rearrangement produced > 90% of alkene *Z*-**26**-F. The preference for *Z*-alkene formation in the 1,2~H shifts of carbenes such as **24** and **25** has been previously discussed.[48]

At –32 °C, the rate constant for the 1,2~H shift of PhOCH$_2$–C–Cl exceeds that of PhOCH$_2$–C–F by approximately three times,[35] consistent with the previously noted R–C–Cl versus R–C–F trend. The difference here, however, is quite small, in part reflecting the influence of the phenoxy substituent's oxygen atom, which should promote the 1,2-hydride migration by stabilizing developing positive charge at the migration origin. The result is an early transition state, high rate constant, and reduced influence of the Cl or F "spectator" substituent.

The rate enhancing effect of the phenoxy substituent is apparent in a comparison of PhOCH$_2$–C–Cl and Me–C–Cl (see Section IV.B) where the rate constants for 1,2~H shifts at ~24 °C are 1.1×10^8 and 1×10^6 to 3×10^6 s^{-1}, respectively. The preexponential factors for these rearrangements are nearly equal, so that the differences in k (1,2~H) arise in ΔE_a, which is ~2 kcal/mol, favoring the rearrangement of PhOCH$_2$–C–Cl.

G. Haloneopentylcarbenes

The chloro- and fluorocarbenes **27** and **28** have been studied [cf. Eq. (15)].[35] The 1,2~H shifts afforded *Z*- and *E*-**29** in > 90% yields, with the *Z* isomers predominating in ratios of 12:1 for chlorocarbene **27**, and 2.5:1 (in $CDCl_3$) for fluorocarbene **28**.

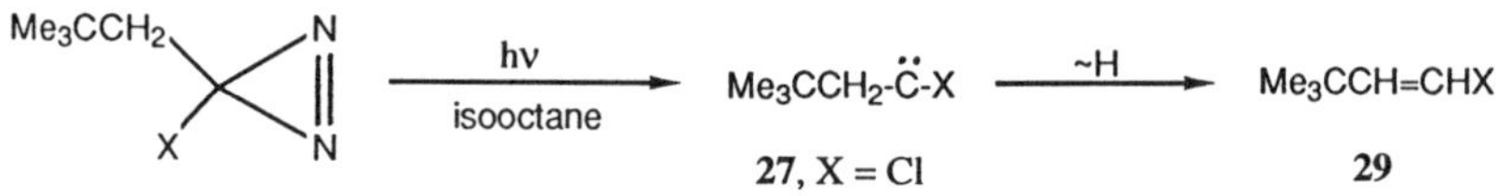

(15)

27, X = Cl
28, X = F
29

The pyridine ylide LFP method was used to obtain k (1,2~H) for each carbene at 20 °C, as well as activation parameters for **28**: k (1,2~H) = $1.4 \times 10^7\ s^{-1}$ for Me_3CCH_2–C–Cl, and $2.6 \times 10^6\ s^{-1}$ for Me_3CCH_2–C–F.[35] Activation parameters in the latter case were $E_a = 3.3$ kcal/mol, log $A = 8.9\ s^{-1}$, $\Delta G^\ddagger = 8.5$ kcal/mol, $\Delta H^\ddagger = 2.7$ kcal/mol, and $\Delta S^\ddagger \approx -20$ eu.[35]

As anticipated, there is a small (about fivefold) F versus Cl effect on k (1,2~H). The hydride shift of Me_3CCH_2–C–Cl is about 10 times faster than that of Me–C–Cl (Section IV.B), and about 10 times slower than hydride shifts in Et–C–Cl or Pr–C–Cl (see Section IV.A). These observations can be rationalized with reference to conventional electronic and steric effects.

Product distribution studies of the competitive, 1,2~H shifts of Me_3CCH_2–C–X versus the carbenes' additions to alkenes, as a function of alkene concentration, suggest that excited diazirines, carbene complexes, or other reactive intermediates contribute to the rearrangement products **29** [cf. Eq. (15)]. Application of Eq. (9) (see Section IV.B) to the data for Me_3CCH_2–C–Cl affords an intercept (k_3/k_1) that is consistent with ~50% of alkene formation arising from excited diazirine.[49] In the case of Me_3CCH_2–C–F, however, analogous data indicates that **29**-F comes almost exclusively (> 90%) from the carbene.[49]

(16)

28-d_1 ~**D** ~**H**

Remarkably, the product-based primary kinetic isotope effect associated with 1,2~H versus 1,2~D in carbene **28**-d_1, is 4.9 ± 0.25 at 22 °C in $CDCl_3$, or 6.4 ± 0.3 at –30 °C[49] [Eq. (16)]. There is little variation in k_H/k_D as calculated from either the *Z* or *E* ~D or ~H products. The isotope effect is also unchanged in pentane (5.3 ± 0.3 at 22 °C).

The extraordinarily high primary kinetic isotope effect is also apparent from k (1,2~D), determined by LFP ylide studies of Me_3CCD_2–C–F.[49] Here, we require extrapolation of k as a function of diazirine concentration to compensate for competitive azine formation.[50] We obtain k (1,2~D) = $4.0 \pm 0.3 \times 10^5\ s^{-1}$, from which comparison with k (1,2~H) = 2.6×10^6 s^{-1} for Me_3CCH_2-C-F (see above), affords k_H/k_D ~6.5.[49]

The temperature dependence of k_H/k_D is consistent with a classical isotope effect; there is no cogent evidence for tunneling in these hydride shifts. The unusual magnitude of k_H/k_D for fluoroneopentylcarbene is in keeping with the observed isotope effect of 3.2 on the lifetime of Me_2C versus Me_2C (d_6)[14b]; these isotope effects are much higher than the very small isotope effect (≤ 1.3) observed in the photoacoustic calorimetric determinations of k (1,2~H) and k (1,2~D) for CH_3–C–Cl and CD_3–C–Cl, respectively.[18] At this time, the origins of the large isotope effects with fluoroneopentylcarbene and dimethylcarbene remain unknown.

H. *tert*-Butylchlorocarbene

Carbene **3** afforded chlorodimethylcyclopropane (**30**) by γ-CH insertion, and trimethylchloroethene (**31**) by 1,2~Me shift [see Eq. (17)].[50,51] The disappearance of **3**, followed by the LFP-pyridinium ylide method,[5,16] gave $k_d \approx 1 \times 10^7$ s^{-1}. Of course, k_d is the sum of rate constants for all the reaction channels utilized by **3**. Under typical reaction conditions, the formation of **30** and **31** is competitive with the azine-forming intermolecular reaction of **3** with its diazirine precursor.[51] The ylide-derived rate constant is therefore too high, but it can be corrected by extrapolation.[50]

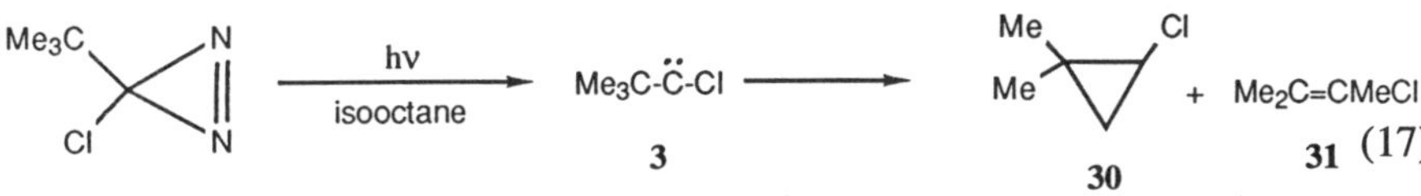

(17)

Rate constants for carbene disappearance are first obtained by the pyridine ylide method at a number of decreasing initial diazirine concentrations. Then, the observed (linear) correlation of k_d versus diazirine concentration is extrapolated to a diazirine concentration equal to zero, where the rate constant for azine formation can be neglected. The extrapolated k_d then represents the conversion of **3** to intramolecular products **30** and **31**, because the azine was the only other significant reaction product.[50]

The corrected k_d, $9.3 \pm 1.1 \times 10^5$ s^{-1} (isooctane, 20 °C), is about 10 times less than the uncorrected rate constant. At 20 °C, the **30/31** product ratio is 2.92:1, so that a straightforward partition of k_d gives k (CH insertion) = 6.9×10^5 s^{-1} and k (1,2~Me) = 2.4×10^5 s^{-1}. The rate constant partition ignores the possibility that **30** or **31** might be formed directly from an excited state diazirine, so that the specific rate constants should be treated with caution. Platz, however, has provided some evidence that excited state product formation may be of limited importance in Eq. (17) (cf. Section V, below).

Variable temperature studies on the **30/31** distribution indicate that E_a (insertion) – E_a (1,2~Me) = 3.4 ± 0.3 kcal/mol, whereas $\Delta S^\ddagger$ (insertion) – $\Delta S^\ddagger$ (1,2~Me) ≈ 13 eu. Methyl migration is therefore favored over γ-CH insertion by enthalpy, but the entropic term even more strongly favors insertion, which dominates at ambient temperatures.

Comparisons to other reactions indicate that the intramolecular reactions of carbene **3** are quite slow. The rate constant for the γ-insertion is roughly comparable to that of the 1,2~C ring expansion of Cy–C–Cl, whereas the 1,2~Me shift of **3** is slower than other measured R–C–Cl rearrangements.[50a]

Substitution of F for Cl in **3** gives *t*–Bu–C–F. Based on experimental results (see above) or theory,[9] we expect the intramolecular reactions of *t*–Bu–C–F to be slower than those of *t*–Bu–C–Cl. Indeed, "preparative" studies at ambient temperatures reveal mostly azine formation from the photolysis of *t*-butylfluorodiazirine, accompanied only by traces of the γ-insertion product.[46]

I. Alkoxyalkylcarbenes

The calculations of Evenseck and Houk indicate that E_a for the 1,2~H shift of CH_3–C–X to CH_2=CHX increases from 19 to 27 kcal/mol as X is changed from F to OMe.[9] The experimental activation energies for 1,2~H shifts in R–C–F appear to be much smaller than 19 kcal/mol (see above); however, alkylfluorocarbene rearrangements are already "slow", and the increase in E_a expected for R–C–OMe could "shut off" intramolecular reactions by making them slower than competitive intermolecular processes such as dimerization or azine formation. Under typical solution phase reaction conditions, intramolecular reactions become uncompetitive when their rate constants fall significantly below ~10^5 s^{-1}.

Me–C–Cl readily undergoes the 1,2~H rearrangement with $k \approx 1$–3×10^6 s^{-1} (see Section IV.B). Me–C–OMe, however, gives < 10% of the analogous product, methylvinyl ether, when generated from a diazirine precursor thermally or photolytically in pentane; the 1,2~H shift is suppressed, relative to azine formation.[52] Similarly, the 1,2~C ring expansions that occur with Cy–C–Cl or Cy–C–F (Section IV.D) are suppressed with cyclopropylmethoxycarbene (Cy–C–OMe). Generated by LFP from the diazirine, Cy-C-OMe reacts rapidly with methanol ($k \approx 2 \times 10^8$ M^{-1} s^{-1}), and adds as an ambiphile or nucleophile to alkenes, but it does not afford 1-methoxycyclobutene by ring expansion.[53] Indeed, the unsuccessful ring expansion, relative to the efficient methanolic trapping, permits an estimate of *k* (1,2~C) < 3 × 10^3 s^{-1} for Cy–C–OMe, representing a suppression of > 46, relative to Cy–C–F, where *k* (1,2~C) = 1.4 × 10^5 s^{-1}.[35]

Studies have also been made of alkoxyalkylcarbenes **32–34**.[54] Photolytic (0 °C) or thermolytic (25 °C) generation of **32** from the diazirine in pentane afforded many products, but significant 1,2~H shift to a vinyl ether did not occur,[54] even though these rearrangements occur readily in the chloro- or fluorocarbene analogues of **32** (see Section IV.G). Attempts to stimulate the reluctant hydride shift by substituting

$Me_3CCH_2\text{-}\ddot{C}\text{-}OMe$	$PhOCH_2\text{-}\ddot{C}\text{-}OMe$	$PhOCH_2\text{-}\ddot{C}\text{-}OCH_2CF_3$
32	**33**	**34**

a phenoxy group at the migration origin, as in carbene **33**, were unsuccessful in solution at ambient temperature; many products were formed in low yield.[54] Pyrolysis of the diazirine precursor at 95 °C, however, did give ~10% of *cis*-PhOCH═CHOMe, the anticipated 1,2~H product of **33**.[54]

With carbene **34**, the electron-donating, rearrangement-suppressing effect of the methoxy group of **33** should be mitigated by conversion to the trifluoroethoxy substituent. Solution phase thermal or photochemical generation of **34** from the appropriate diazirine (0–25 °C, isooctane) gave ~70% of the intermolecular azine product, with little or no 1,2~H product; however, thermal generation at 95 °C afforded ~50% of *cis*-$PhOCH{=}CHCH_2CF_3$.[54] These results indicate that 1,2~H shifts in R–C–OR′ are indeed opposed by higher activation energies than in comparable alkylhalocarbenes, but that the desired rearrangements can be elicited at elevated reaction temperatures.

The only example of an alkoxycarbene that appears to support significant 1,2~H rearrangement at ambient temperature is the glucose-derived glycosylidene carbene **35**.[55] Generated photochemically from the diazirine in MeCN, **35** affords a complex product mixture with the 1,2~H product **36** as "a major component."[55] It is unclear whether any of product **36** may have stemmed from an excited diazirine.

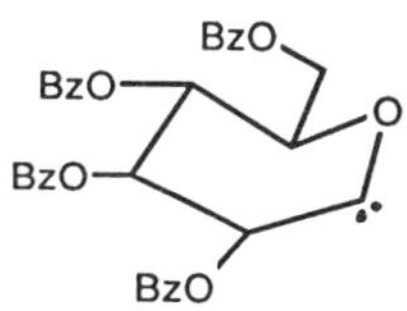

35 (Bz = benzyl)

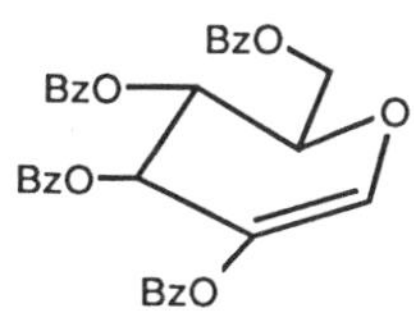

36

Laser flash photolysis–pyridine ylide experiments gave a lifetime of 772 ns for **35**, similar to that of Me–C–OMe.[52] The decay of **35**, $k \approx 1.3 \times 10^6\ s^{-1}$, represents the sum of unidentified reaction channels, in addition to 1,2~H.[55] [Note added in proof: In fact, **36** is formed in only 5% yield. The major products are azines. (Private communication from Professor A. Vasella, October 22, 1992)].

Carbene **35** also reacts as a nucleophile with alcohols,[55] phenols,[56] and electron-deficient alkenes.[57] These intermolecular properties are similar to those of Me–C–OMe,[52] although **35** reacts with MeOH and alkenes ~100 times more slowly.[55] The sluggish intermolecular reactivity of **35** could be due to steric or electronic factors originating at the large benzyloxy substituents.

J. α-Chloroalkylchlorocarbenes

Liu et al. examined carbenes **37** and **38**, where 1,2~Cl shifts are major reaction channels.[58,59] Carbene **37**, photochemically generated from the diazirine in isooctane, gave 90% of $PhMeC{=}CCl_2$, via a 1,2~Cl shift, and 10% of PhClC=CMeCl, so that the 1,2~Cl shift appears to be faster than either 1,2~Me or 1,2~Ph by a factor of ~9.

37 **38**

Carbene **37** added to $Me_2C{=}CMe_2$ competitively with the Cl shift, and the product ratio of (1,2~Cl)/(addition) correlated linearly with $1/[Me_2C{=}CMe_2]$. The data were interpreted in terms of Eqs. (10) and (11) (see Section IV.C).[58] A temperature dependence study of the product ratio, together with the assumption that E_a and A for the **37**/$Me_2C{=}CMe_2$ addition were identical to these parameters for the analogous addition of *p*-ClPh–C–Cl, led to *estimates* for the intramolecular kinetic parameters of carbene **37**: k (1,2~Cl) > 10^9 s^{-1}, E_a = 3.4 kcal/mol, and log A = 12.1 s^{-1}.

A further study of carbene **37** (λ_{max} 320 nm) by direct LFP observation at –20 °C to –70 °C has given a larger, more reasonable, extrapolated value of 3.5 ns for the lifetime of **37** at 21 °C. This corresponds to k (1,2~Cl = 2.8×10^8 s^{-1}). The activation parameters were now reported as $E_a = 3.4 \pm 0.14$ kcal/mol and log A = 11 s^{-1}.[60]

Photochemical generation of carbene **38** from the diazirine in isooctane afforded 95% of $PhClC{=}Cl_2$.[59] The product was taken as arising from a 1,2~Cl shift, rather than a 1,2~Ph shift, based on the results for carbene **37**, where ~Cl > ~Ph by at least 9:1.[58] Using the pyridine ylide LFP method, and correcting for concomitant azine formation, k (1,2~Cl) for **38** was estimated as 3.0×10^7 s^{-1} (isooctane, 25 °C), corresponding to a carbene lifetime of ~33 ns. Thus k (1,2~Cl) of carbene **38** is similar in magnitude to k (1,2~H) of $PhCH_2$–C–Cl (~ 5×10^7 to 7×10^7 s^{-1}).[27,28]

The Cl migration of carbene **37** appears to be about an order of magnitude faster than that of **38**, perhaps reflecting the influence of the α-Me for α-Cl structural change. A parallel acceleration of the 1,2~H migration attends the α-H to α-Me modification of $PhCH_2$–C–Cl to PhCHMe–C–Cl.[16]

K. Methylphenylcarbene

Most of this chapter has focused on carbenes that bear a heteroatomic spectator substituent, but here we will examine the kinetics of methylphenylcarbene, Me–C–

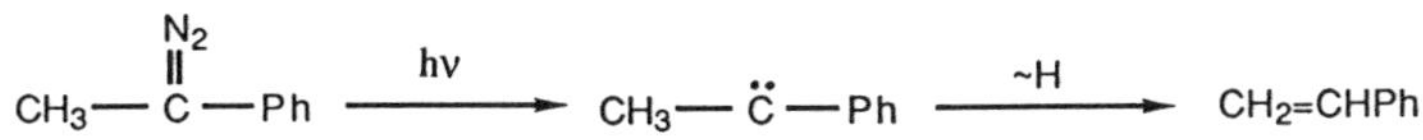

39 (18)

Ph (1-phenylethylidene, **39**), generated by photolysis of 1-phenyldiazoethane [cf. Eq. (18)].[61]

Me–C–X, where X is heteroatomic, is a ground state singlet; Me–C–Ph, however, is a ground state triplet.[61] The singlet-triplet separation is small ($\Delta G^\ddagger = 2.3$ kcal/mol in heptane), and spin state equilibration is rapid, relative to the reactions of either singlet or triplet Me–C–Ph.[61]

Triplet **39**, generated at very low temperature in a xenon matrix, rearranges to styrene upon warming to 65 K. The reaction, monitored by infrared (IR) spectroscopy, is presumed to occur via the thermally populated singlet Me–C–Ph, and exhibits $k = 2.9 \times 10^4$ s^{-1}. If $\Delta S^\ddagger \approx 0$ at 65 K, $\Delta G^\ddagger \approx \Delta H^\ddagger = 4.7$ kcal/mol for the *sum* of ΔH_{ST} (triplet → singlet) and ΔH (1,2~H).[62] If ΔH_{ST} is very small,[61] then the cryogenic experiment suggests that E_a for the 1,2~H shift of Me–C–Ph is ~5 kcal/mol, similar to the value calculated for the 1,2~H migration of Me–C–Me.[10]

The ambient temperature kinetics of Me–C–Ph were studied with LFP-pyridine ylide methodology.[61] Generated as in Eq. (18), Me–C–Ph formed an intensely absorbing ylide (λ_{max} 475 nm) with pyridine. Oxygen quenching experiments revealed that singlet–triplet equilibration of **39** occurred more rapidly than ylide formation, so that the singlet/triplet mixture was treated as a single kinetic unit.[61]

In pentane at ambient temperature, no more than 10% of styrene is formed in Eq. (18). The product data, combined with the value of k_d (for carbene disappearance) extrapolated from the pyridine-ylide experiments, and with $K = 0.02$ for the singlet–triplet equilibrium constant, led to an estimate of k (1,2~H) $\leq 6 \times 10^6$ s^{-1} for the rearrangement of singlet Me–C–Ph to styrene.[61] Assuming that $A = 10^{10}$ s^{-1} (similar to the 1,2~H shift of Me–C–Cl[18]), $E_a \geq 4.3$ kcal/mol, in good agreement with the earlier low temperature studies.[62]

An activation energy of 4–5 kcal/mol for the 1,2~H shift of Me–C–Ph is surprisingly high. It is similar to the barriers observed for H migration with Me–C–Cl or $PhCH_2$–C–Cl, and exceeds the barrier of Me_3CCH_2–C–F. Halogen spectator substituents are predicted to strongly enhance E_a (1,2~H) of R–C–X,[9] so that one is tempted to conclude that the Ph substituent is as potent in this regard as Cl or F. Note that although the 1,2~H shift of Me–C–Me is calculated to have $E_a \approx$ 5 kcal/mol,[10] it is shorter-lived than Me–C–Ph in pentane solution,[14] and presumably confronts a smaller barrier.

There is a strong solvent effect on k (1,2~H) for Me–C–Ph; the yield of styrene in Eq. (18) is enhanced approximately fivefold in MeCN compared to pentane, and styrene is the only volatile carbene product produced in the polar solvent. Analysis of LFP pyridine ylide data, corrected for the singlet–triplet carbene distribution ($K = 0.031$ in MeCN), leads to k (1,2~H) $= 2.0 \times 10^8$ s^{-1} for Me–C–Ph at 25 °C, equivalent to at least a 30-fold enhancement in k (1,2~H) attending the change to the polar solvent.[61] A parallel phenomenon is observed for k (1,2~H) of Me–C–Me in MeCN versus pentane.[14b]

The temperature dependence of the ylide-derived rearrangement rate constant in MeCN affords $E_a \approx 2.1$–2.3 kcal/mol and log $A = 8.3$–$10\ s^{-1}$.[61] Thus, altering the solvent from hydrocarbon to acetonitrile lowers E_a for the 1,2~H shift of Me–C–Ph by about 2 kcal/mol, consistent with a polar hydride-like transition state for the rearrangement.

Also consistent with the classical, hydride-like 1,2~H transition state is the kinetic behavior of ethylphenylcarbene, Et–C–Ph.[61] Studies analogous to those carried out with Me–C–Ph show that Et–C–Ph cleanly gives the 1,2~H shift products (*Z*- and *E*-1-phenylpropene) in dilute pentane solution. The ylide-derived k (1,2~H) $\approx 3.6 \times 10^8\ s^{-1}$ at 25 °C is at least 60 times larger than k (1,2~H) for Me–C–Ph under comparable conditions. Clearly, substitution of a methyl group at the migration origin strongly enhances k (1,2~H) versus Me–C–Ph. Similar effects are seen upon comparisons of Et–C–Me with Me–C–Me,[14b] Et–C–Cl with Me–C–Cl,[16,18] and PhCHMe–C–Cl with $PhCH_2$–C–Cl.[16,18] The activation energy for the 1,2~H shift of Et–C–Ph is reduced, relative to that of Me–C–Ph. Assuming again that log $A = 10\ s^{-1}$, we obtain $E_a \approx 2.0$ kcal/mol for Et–C–Ph versus $E_a \approx 4.3$ kcal/mol for Me–C–Ph.

V. CARBENE COMPLEXES VERSUS EXCITED DIAZIRINES

We saw above (Sections IV.B and IV.C) that the kinetic properties of Me–C–Cl and $PhCH_2$–C–Cl during competitive intermolecular and intramolecular reactions were not in accord with a single reactive intermediate (i.e., carbene only). To satisfy the observed inverse dependence of the intra/inter product ratio on 1/[trap], more complicated mechanisms were postulated. The most popular contenders for the role of the additional reactive intermediate were a carbene–alkene complex [see Eqs. (10) and (11), Section IV.C] or an excited diazirine [see Eqs. (8) and (9), Section IV.B]. It is not yet possible to definitively choose between these alternatives, but it is appropriate to briefly review the available experimental evidence.

Carbene–alkene (C–A) complexes were initially postulated to be reactive intermediates in addition reactions in order to account for the negative activation energies that were observed.[4b,63] Houk et al., however, calculated that C–A complexes should be unimportant in reactions of carbenes that were more reactive than CCl_2 (e.g., R–C–Cl).[40,64] Negative activation energies could instead be understood in terms of free energies of activation that were strongly dominated by unfavorable $T\Delta S^{\ddagger}$ terms, despite reaction enthalpies that smoothly and continuously decreased during the carbene–alkene addition reactions (i.e., $\Delta H^{\ddagger} < 0$).[40,64] Houk's reaction model accounted for the observed negative activation energies without invoking any reaction intermediate other than the carbene.

Later, it was reported that the intermolecular insertion of Ph–C–Cl into the Sn–H bond of n-Bu_3SnH, where an intermediate carbene complex appeared unlikely, exhibited a small negative activation energy (–0.13 ± 0.04 kcal/mol).[65] This observation lends weight to Houk's contention that carbene complexes are not obligatory intermediates in carbene–alkene additions. A similar conclusion followed from extensive LFP studies of Ph–C–Cl additions to alkenes.[66]

If C–A complexes are not important in the intermolecular addition reactions of reactive carbenes, are they necessary to a proper understanding of *intramolecular* reactions? The C–A complex mechanism readily accounts for the kinetic form of competitive carbene rearrangements and alkene additions [cf. Section IV.C, Eqs. (10) and (11)].[23–25,31] It also rationalizes several more subtle features of these reactions; for example, that the *Z*/*E* ratio of the β-halostyrene 1,2~H rearrangement product of $PhCH_2$–C–X depends on both the concentration and identity of an added alkene trap. This is easily understood in terms of Eq. (10); as long as the stereochemical outcomes of the 1,2~H shifts of the carbene and the C–A complex differ, the contribution of the latter, and therefore the overall styrene stereochemistry, will be affected by the efficiency of the k_2 step.

Additionally, the analytical form of Eq. (10), that is, Eq. (11), predicts that the y-intercept of the product ratio versus 1/[alkene] should depend on the identity of the alkene. This, too, is observed,[31] although the differences in these extrapolated intercepts are not large.

Despite its facility in accounting for the kinetic data, there are reservations attached to the C–A complex mechanism. There is no direct evidence for the existence of these complexes under typical reaction conditions. Their stability is suspect;[40,64] their association is likely to be very "loose",[31] in the limit, they are weakly bound proximity or contact pairs in solvent cages.[4b,63] Finally, if no alkene is present, no C–A complex can form, and the putative complexes become irrelevant to the carbene's intramolecular chemistry.

The principal alternative to C–A complexes is the excited diazirine, which is postulated to directly afford "carbene rearrangement products" in competition with the carbene [cf. Section IV.B]. Again, there are pros and cons to this alternative too. Contrary to the excited diazirine mechanism, as outlined in Eqs. (8) and (9), is the observation that the y-intercept of the product ratio versus 1/[alkene] correlation depends on the identity of the alkene[31]; this is consistent with the C–A complex mechanism. In the excited diazirine scenario, however, the y-intercept represents the partition of the excited state between nitrogen loss and direct rearrangement to product, and simple nitrogen loss to give the carbene; the presence or identity of an added carbene trap should not matter. On the other hand, as noted above, the differential dependence of the y-intercept on the alkene's identity is small, and may be distorted by extrapolation.

It has also been observed that the product ratios for the competitive intra- and intermolecular reactions of $ArCH_2$–C–Cl conform to the same Arrhenius correlation whether the carbenes are generated thermally or photochemically from their

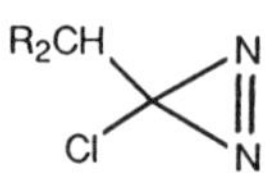

40

$R_2CH\text{-}\ddot{C}\text{-}Cl$

41

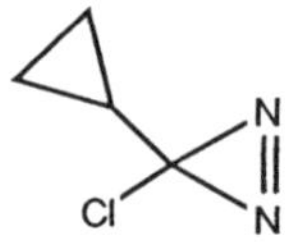

42

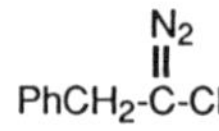

43

diazirine precursors.[25] This is consistent with the C–A complex mechanism, but not, in the absence of further assumptions, with the excited diazirine mechanism, where the excited state is only accessible photochemically.

The observed dependence of the *E*/*Z* intramolecular product ratios on an added trap is compatible with the excited diazirine mechanism as long as the *E*/*Z* product ratios from the excited diazirine or the carbene differ, which seems likely.

Important evidence supporting the excited diazirine mechanism comes from studies of the structural dependence of the carbene yield in the photolytic decompositions of diazirines **40**.[67] LFP of **40** affords carbene **41**, which, in the presence of pyridine gives the appropriate ylide. When the pyridine concentration > 0.5 *M*, the ylide signal intensity (ϕ_Y) is maximized, indicating that all of the carbene has been diverted to ylide. The values of ϕ_Y^{max} are found to be directly proportional to the bond dissociation energies of the α-CH bonds of **40**. Thus, the efficiency of ylide (or carbene) generation upon LFP of, for example, chloromethyldiazirine (**40**, R = H) is about twice that of chloroisopropyldiazirine (**40**, R = Me), where the α-CH bond is weaker. The ylide signal from the LFP of *t*-butylchlorodiazirine, which lacks an α-CH, is about 10 times stronger than the signal intensity from **40**, R = Me.

These observations are well explained by a mechanism in which the α-hydrogen of excited **40** undergoes 1,2~H shift in concert with N_2 loss, and in competition with carbene formation. The stronger the α-CH bond, the less 1,2~H shift, and the more efficient is the formation of carbene and, ultimately, ylide.[67] Simultaneously, it is assumed that the excited state of *t*-butylchlorodiazirine gives mostly carbene, with little rearrangement.

Chlorocyclopropyldiazirine (**42**) appears to form Cy–C–Cl upon LFP about as efficiently as *t*-butylchlorodiazirine gives *t*-Bu-C-Cl, so that it is suggested that chlorocyclobutene, the 1,2~C product of Cy–C–Cl, is formed only from the carbene, and not from excited diazirine **42**.[67] This may well be correct, but it is premature to generalize; there appears to be evidence compatible with a 1,2~C shift in the excited state of chlorocyclobutyldiazirine (**21**) (cf. Section IV.E).

White and Platz observe the formation of diazoalkane **43** when benzylchlorodiazirine is photolyzed at 350 nm in a glassy 3-methylpentane matrix at –175 °C.[67] This suggests that not only the excited diazirine, but also its isomerization product **43** could be intermediates in the chemistry ascribed to $PhCH_2$–C–Cl (see Section IV.C).

Finally, Platz and his associates have observed fluorescence from excited dialkyldiazirines (but not alkylhalodiazirines).[14b] The fluorescence intensity increased with decreasing temperature, suggesting the existence of a barrier to nitrogen loss from the excited diazirine. It is therefore reasonable and economical to identify the excited diazirine as a key intermediate that can afford either carbene or alkene upon nitrogen extrusion. The excited diazirine mechanism is particularly efficient in rationalizing the photochemistry of dimethyldiazirine.[14b]

Although we cannot definitively decide between the C–A complex and excited diazirine mechanisms with the evidence now at hand, we incline toward the latter as perhaps less *ad hoc*. More important, we must consider it whenever product analysis is combined with absolute kinetic data, if we want to derive accurate path-specific rate constants.

VI. CONCLUSION

The absolute kinetics of the 1,2~H and 1,2~C rearrangements of many alkylhalocarbenes are readily ascertained by laser flash photolysis. Intramolecular reactions with rate constants ranging from 10^5–10^8 s^{-1} are particularly well handled by the currently available methodology. The activation energies measured for the rearrangements are low (~ 2–6 kcal/mol), and the companion preexponential factors or entropies of activation are often surprisingly unfavorable (e.g., log $A \approx 8$ s^{-1} or $\Delta S^{\ddagger} \approx -20$ eu). Although the structural dependence of the 1,2~H shift, for example, is nicely compatible with the classical hydride-like transition state formulation, it is clear a deeper understanding of these "simple" reactions will require thorough computational exploration of their energy surfaces. In addition, the intervention of excited state carbene precursors as contributory sources of "carbene" products is an open and expanding concern.

APPENDIX

Table 1. Kinetics of Carbene Rearrangements

R (*in R–C–X*)	*X*	*Rearr. type*	*k* (s^{-1})	*Temp.* (°C)[a]	E_a (*kcal/mol*)	*log A* (s^{-1})	*Ref.*
Me	Me	1,2~H	~5×10^7	25			14a,b
Me	Cl	1,2~H	3×10^6	25			16, 17
			1.3×10^6	22	4.9	9.7	18
Me	Br	1,2~H	5–6×10^6	22			19
Et	Cl	1,2~H	$> 10^8$	25	2.4	(9.7)	16, 18
n-Pr	Cl	1,2~H	$> 10^8$	25			16
i-Pr	Cl	1,2~H	$> 10^8$	25	< 2.4	(9.7)	16, 18
$PhCH_2$	Cl	1,2~H	6×10^7	25			16, 27, 28
			$4.9\text{-}6.7 \times 10^7$	25			26
			5.4×10^7	24	4.5–4.8	11.1–11.3	27, 28

(*continued*)

Table 1. *(continued)*

R (*in R-C-X*)	*X*	*Rearr. type*	*k* (s^{-1})	*Temp.* (°C)[a]	E_a (*kcal/mol*)	*log A* (s^{-1})	*Ref.*
p-CF_3PhCH_2	Cl	1,2~H	2.2×10^{7}	24	4.9	10.9	30
p-$ClPhCH_2$	Cl	1,2~H	4.2×10^{7}	25	4.5	10.9	30
p-$MePhCH_2$	Cl	1,2~H	1.2×10^{8} [b]	24			30
p-$MeOPhCH_2$	Cl	1,2~H	1.7×10^{8} [b]	24			30
$PhCH_2$	Br	1,2~H	5.6×10^{7}	25	4.7–6.6[b]		31a,b
Ph(Me)CH	Cl	1,2~H	$> 10^{8}$	25			16
Cy[c]	Cl	1,2~C	6×10^{5}	25			16
			3.8×10^{5}	23	7.4	11.1	32
Cy[c]	Cl	1,2~C	$0.9–1.5 \times 10^{6}$	20	3.0	8.2	33, 34
Cy[c]	F	1,2~C	1.4×10^{5}	23	4.2	8.3	35
Cb[d]	Cl	1,2~C	4.6×10^{7}	20			44
		1,2~H	2.1×10^{7}	20			44
Cb[d]	F	1,2~C	1.8×10^{6}	23	2.3	8.0	35
		1,2~H	5.3×10^{5}	23	3.8	8.5	35
$PhOCH_2$	Cl	1,2~H	1.1×10^{8}	24	2.8	10.1	47
			3.6×10^{7}	–32			35
$PhOCH_2$	F	1,2~H	1.3×10^{7}	–32			35
Me_3CCH_2	Cl	1,2~H	1.4×10^{7}	20			35
Me_3CCH_2	F	1,2~H	2.6×10^{6}	20	3.3	8.9	35
t-Bu	Cl	[e]	1×10^{7}	25			16
		γ-CH ins[f]	6.9×10^{5}	20			50
		1,2~Me[f]	2.4×10^{5}	20			50
PhMeClC	Cl	1,2~Cl	2.8×10^{8}	21			60
$PhCl_2C$	Cl	1,2~Cl	3.0×10^{7}	25			59
Me	Ph	1,2~H	$< 6 \times 10^{6}$	25	≥ 4.3	10^{8}[g]	
			2×10^{8} [h]	25	2.1–2.3[h]	8.3–10[h]	61
Et	Ph	1,2~H	3.6×10^{8}	25	2.0	10^{8}[g]	61

Notes: [a]Unstated or ambient temperature conditions are listed as 25 °C. Unless otherwise stated, solvents were hydrocarbons.

[b]This value is an estimate.

[c]Cyclopropyl.

[d]Cyclobutyl.

[e]Sum of intermolecular and intramolecular processes.

[f]Corrected for intermolecular reaction. See, however, ref. 50a.

[g]Assumed value.

[h]In acetonitrile.

ACKNOWLEDGMENT

We are grateful to the National Science Foundation for support of our work described in this chapter.

REFERENCES

1. Closs, G. L.; Rabinow, B. E. *J. Am. Chem. Soc.* **1976**, *98*, 8190.
2. Zupancic, J. J.; Schuster, G. B. *J. Am. Chem. Soc.* **1980**, *102*, 5958.
3. Turro, N. J.; Butcher, J. A., Jr.; Moss, R. A.; Guo, W.; Munjal, R. C.; Fedorynski, M. *J. Am. Chem. Soc.* **1980**, *102*, 7576.
4. (a) Moss, R. A. *Acc. Chem. Res.* **1989**, *22*, 15. (b) Moss, R. A.; Turro, N. J. In *Kinetics and Spectroscopy of Carbenes and Biradicals*; Platz, M. S., Ed.; Plenum, New York, 1990; pp 213–238.
5. Jackson, J. E.; Soundararajan, N.; Platz, M. S.; Liu, M. T. H. *J. Am. Chem. Soc.* **1988**, *110*, 5595.
6. Baron, W. J.; DeCamp, M. R.; Hendrick, M. E.; Jones, M., Jr.; Levin, R. H.; Sohn, M. B. In *Carbenes*; Jones, M., Jr.; Moss, R. A.; Eds.; Wiley, New York, 1973; Vol. 1, pp 1–151.
7. Kirmse, W. *Carbene Chemistry*, 2nd ed.; Academic Press, New York, 1971.
8. Regitz, M., Ed. *Carbene (oide), Carbine*; Houben-Weyl, 4th ed., Vol. E19b, G. Thieme Verlag, Stuttgart, 1989.
9. Evenseck, J. D.; Houk, K. N. *J. Phys. Chem.* **1990**, *94*, 5518.
10. Evenseck, J. D.; Houk, K. N. *J. Am. Chem. Soc.* **1990**, *112*, 9148.
11. Su, D. T. T.; Thornton, E. R. *J. Am. Chem. Soc.* **1978**, *100*, 1872.
12. Tomioka, H.; Sugiura, T.; Masumoto, Y.; Izawa, Y.; Inagaki, S.; Iwase, K. *J.C.S. Chem. Commun.* **1986**, 693.
13. Stevens, I. D. R.; Liu, M. T. H.; Soundararajan, N.; Paike, N. *Tetrahedron Lett.* **1989**, *30*, 481.
14. (a) Modarelli, D. A.; Platz, M. S. *J. Am. Chem. Soc.* **1991**, *113*, 8985. (b) Modarelli, D. A.; Morgan, S.; Platz, M. S. *J. Am. Chem. Soc.* **1992**, *114*, 7034.
15. LaVilla, J. A.; Goodman, J. L. *Tetrahedron Lett.* **1990**, *31*, 6287.
16. Bonneau, R.; Liu, M. T. H.; Rayez, M. T. *J. Am. Chem. Soc.* **1989**, *111*, 5973.
17. Liu, M. T. H.; Bonneau, R. *J. Am. Chem. Soc.* **1989**, *111*, 6873.
18. LaVilla, J. A.; Goodman, J. L. *J. Am. Chem. Soc.* **1989**, *111*, 6877.
19. LaVilla, J. A.; Goodman, J. L. *Tetrahedron Lett.* **1990**, *31*, 5109.
20. Warner, P. M. *Tetrahedron Lett.* **1984**, *25*, 4211.
21. Avila, M. J.; Becerra, R.; Figuera, J. M.; Rodriguez, J. C.; Tobar, A.; Martinez-Utrilla, R. *J. Phys. Chem.* **1985**, *89*, 5489.
22. Tomioka, H.; Hayashi, N.; Izawa, Y.; Liu, M. T. H. *J. Am. Chem. Soc.* **1984**, *106*, 454.
23. Liu, M. T. H. *J.C.S. Chem. Commun.* **1985**, 982.
24. Liu, M. T. H.; Subramanian, R. *Tetrahedron Lett.* **1985**, *26*, 3071.
25. Liu, M. T. H.; Soundararajan, N.; Paike, N.; Subramanian, R. *J. Org. Chem.* **1987**, *52*, 4223.
26. Jackson, J. E.; Soundararajan, N.; White, W.; Liu, M. T. H.; Bonneau, R.; Platz, M. S. *J. Am. Chem. Soc.* **1989**, *111*, 6874.
27. Bonneau, R.; Liu, M. T. H.; Subramanian, R.; Linkletter, B.; Stevens, I. D. R. *Laser Chem.* **1989**, *10*, 267.
28. Liu, M. T. H.; Bonneau, R. *J. Am. Chem. Soc.* **1990**, *112*, 3915.
29. Liu, M. T. H.; Suresh, R. V.; Soundararajan, N.; Vessey, E. G. *J.C.S. Chem. Commun.* **1989**, 12.
30. Liu, M. T. H.; Bonneau, R. *J. Am. Chem. Soc.* **1992**, *114* (in press).
31. (a) Bonneau, R.; Liu, M. T. H.; Suresh, R. V. *J. Phys. Chem.* **1989**, *99*, 4802. (b) Liu, M. T. H.; Subramanian, R. *J. Phys. Chem.* **1986**, *90*, 75.
32. Liu, M. T. H.; Bonneau, R. *J. Am. Chem. Soc.* **1989**, *93*, 7298.
33. Ho, G-J.; Krogh-Jespersen, K.; Moss, R. A.; Shen, S.; Sheridan, R. S.; Subramanian, R. *J. Am. Chem. Soc.* **1989**, *111*, 6875.
34. Moss, R. A.; Ho, G-J.; Shen, S.; Krogh-Jespersen, K. *J. Am. Chem. Soc.* **1990**, *112*, 1638.
35. Moss, R. A.; Ho, G-J.; Liu, W. *J. Am. Chem. Soc.* **1992**, *114*, 959.
36. Wang, B.; Deng, C. *Tetrahedron* **1988**, *44*, 7335.
37. Shevlin, P. B.; McKee, M. L. *J. Am. Chem. Soc.* **1989**, *111*, 519.

38. Chou, J-H; McKee, M. L.; De Felippis, J.; Squillacote, M. *J. Org. Chem.* **1990**, *55*, 3291.
39. Blake, J. F.; Wierschke, S. G.; Jorgensen, W. L. *J. Am. Chem. Soc.* **1989**, *111*, 1919.
40. Houk, K. N.; Rondan, N. G.; Mareda, J. *Tetrahedron* **1985**, *41*, 1555.
41. Rondan, N. G.; Houk, K. N.; Moss, R. A. *J. Am. Chem. Soc.* **1980**, *102*, 1770.
42. Moss, R. A. *Acc. Chem. Res.* **1980**, *13*, 58.
43. Moss, R. A.; Fantina, M. E.; Munjal, R. C. *Tetrahedron Lett.* **1979**, *20*, 1277.
44. Moss, R. A.; Ho, G-J. *Tetrahedron Lett.* **1990**, *31*, 1225.
45. Moss, R. A.; Fantina, M. E. *J. Am. Chem. Soc.* **1978**, *100*, 6788. Moss, R. A.; Vezza, M.; Guo, W.; Munjal, R. C.; Houk, K. N.; Rondan, N. G. *J. Am. Chem. Soc.* **1979**, *101*, 5088.
46. Moss, R. A.; Ho, G-J. *J. Phys. Org. Chem.* **1993**, *6*, 126.
47. Chateauneuf, J. E.; Liu, M. T. H. *J. Org. Chem.* **1991**, *56*, 5942.
48. Tomioka, H.; Sugiura, T.; Masumoto, Y.; Izawa Y.; Inagaki, S.; Iwase, K. *J. Chem. Soc. Chem. Commun.* **1986**, 693.
49. Moss, R. A.; Ho, G-J.; Liu, W.; Sierakowski, C. *Tetrahedron Lett.* **1993**, *34*, 927.
50. Moss, R. A.; Ho, G-J. *J. Am. Chem. Soc.* **1990**, *112*, 5642.
50a. Moss, R. A.; Liu, E. *J. Chem. Soc. Chem. Commun.* **1993**, 1597.
51. Moss, R. A.; Munjal, R. C. *J.C.S. Chem. Commun.* **1978**, 775.
52. Sheridan, R. S.; Moss, R. A.; Wilk, B. K.; Shen, S.; Wlostowski, M.; Kesselmayer, M. A.; Subramanian, R.; Kmiecik-Lawrynowicz, G.; Krogh-Jespersen, K. *J. Am. Chem. Soc.* **1988**, *110*, 7563.
53. Moss, R. A.; Jang, E. G.; Fan, H.; Wlostowski, M.; Krogh-Jespersen, K. *J. Phys. Org. Chem.* **1992**, *5*, 104.
54. Moss, R. A.; Liu, W.; Ge, C-S. *J. Phys. Org. Chem.* **1993**, *6*, 376.
55. Vasella, A.; Briner, K.; Soundararajan, N.; Platz, M. S. *J. Org. Chem.* **1991**, *56*, 4741.
56. Briner, K.; Vasella, A. *Helv. Chim. Acta* **1990**, *73*, 1764.
57. Vasella, A.; Waldraff, C. A. A. *Helv. Chim. Acta* **1991**, *74*, 585.
58. Liu, M. T. H.; Murray, S. K.; Zhu, J. *J. C. S. Chem. Commun.* **1990**, 1650.
59. Liu, M. T. H.; Bonneau, R. *J. Org. Chem.* **1992**, *57*, 2483.
60. Liu, M. T. H.; Chateauneuf, J. E. *J. Phys. Org. Chem.* **1992**, *5*, 285.
61. Sugiyama, M. J.; Celebi, S.; Platz, M. S. *J. Am. Chem. Soc.* **1992**, *114*, 966.
62. McMahon, R. J.; Chapman, O. L. *J. Am. Chem. Soc.* **1987**, *109*, 683.
63. Turro, N. J.; Lehr, G. F.; Butcher, J. A., Jr.; Moss, R. A.; Guo, W. *J. Am. Chem. Soc.* **1982**, *104*, 1754.
64. Houk, K. N.; Rondan, N. G.; Mareda, J. *J. Am. Chem. Soc.* **1984**, *106*, 4291.
65. Jackson, J. E.; Soundararajan, N.; Platz, M. S.; Doyle M. P.; Liu, M. T. H. *Tetrahedron Lett.* **1989**, *30*, 1335.
66. Soundararajan, N.; Platz, M. S.; Jackson, J. E.; Doyle, M. P.; Oon, S. M.; Liu, M. T. H.; Anand, S. M. *J. Am. Chem. Soc.* **1988**, *110*, 7143.
67. White, W. R., III; Platz, M. S. *J. Org. Chem.* **1992**, *57*, 2841.

LASER FLASH PHOTOLYSIS STUDIES OF YLIDE-FORMING REACTIONS OF CARBENES

James E. Jackson and Matthew S. Platz

Advances in Carbene Chemistry
Volume 1, pages 89–160.

ISBN: 1-55938-545-6

I. INTRODUCTION

The influence of flash photolysis on carbene chemistry in solution dates back to 1976.[1] In that year appeared the landmark paper of Closs and Rabinow on the detection of diphenylcarbene. Closs and Rabinow used a conventional flash lamp with a time resolution of several microseconds to obtain the first absolute rate

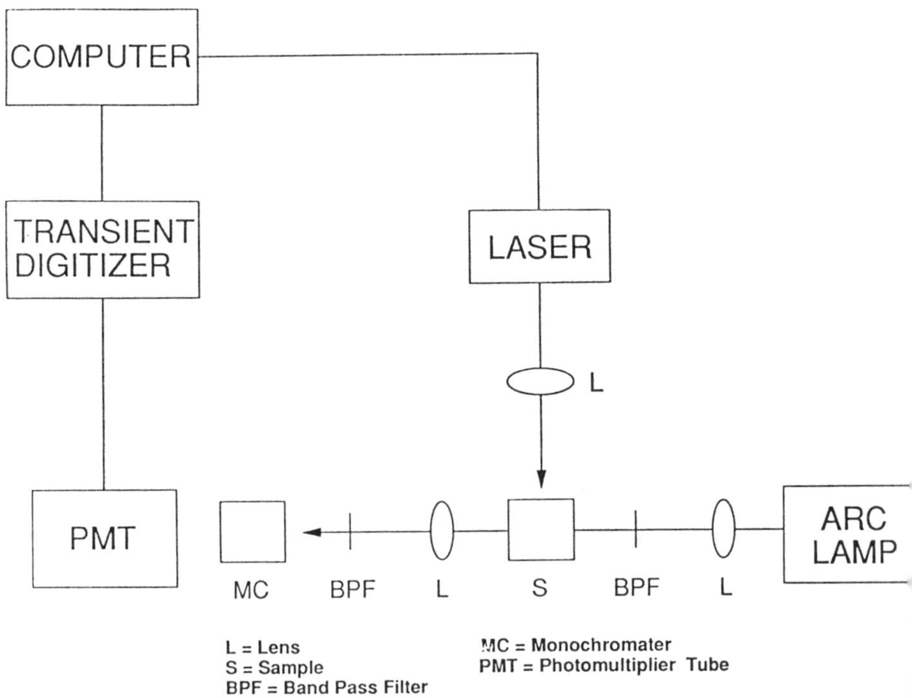

Figure 1. A simplified schematic of a laser flash photolysis apparatus.

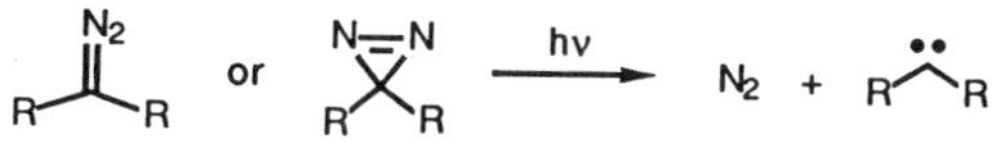

Scheme 1.

constant of a carbene reaction in solution. Subsequent to this study, high powered nano- and picosecond pulsed lasers were introduced, and they have tremendously enhanced the time resolution and sensitivity of the flash photolysis technique (for reviews see Ref. 2). Laser flash photolysis (LFP) has been increasingly applied to the study of carbene processes in the 1980s (for reviews see Ref. 3). In fact, LFP has by now become a standard tool in the study of reactive intermediates.

An LFP experiment is shown schematically in Figure 1. At the heart of the experiment is the intersection, within the sample cell, of two light beams of different color. The first beam, from the high-powered, short-pulse-width laser, is monochromatic and initiates the photochemistry that converts a precursor (usually a diazo compound or diazirine; see Scheme 1) to a carbene in 10^{-8} s or less.[4] The second beam is continuous and is typically provided by a xenon arc. The xenon arc lamp is frequently pulsed to increase its brightness by 20- or 30-fold for a few milliseconds.[5] This beam is not monochromatic, but a reasonably narrow band pass can be achieved by filtering the output of the arc lamp usually before and/or after traversing the sample cell. The xenon arc beam (the "monitoring beam") is directed through the sample and through a monochromator, and is focused into a photomultiplier tube (PMT) detector.[6] The PMT produces an electrical signal that is proportional to the incident light intensity, and that has a minimum time resolution of 2–10 ns. The time dependent output of the detector is typically captured by a transient digitizer and transferred to a computer for storage and processing.

The LFP experiment works best when the absorption maxima of the diazo compound or diazirine (λ^{D}_{max}), the carbene (λ^{C}_{max}) and the carbene reaction products (λ^{P}_{max}) are in widely separated regions of the ultraviolet and visible (UV–VIS) regions of the spectrum; unfortunately, this condition is not often met in the study of carbenes. For example, the absorption maxima of diphenyldiazomethane (DPDM) and diphenylcarbene (DPC) are virtually identical.[1] These absorption maxima are also very close to that of benzhydryl radical (**1**),[7] a common reaction product when DPC is generated in hydrocarbon solvents (see Eq. 1).

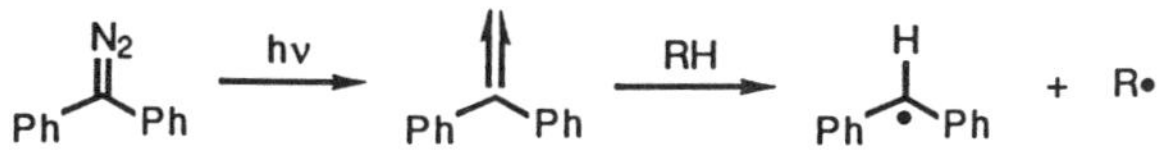

(1)

DPDM ^{3}DPC **1**

Several problems result from the overlapping of spectra. The LFP experiment monitors changes in optical density of the samples which are induced by the laser flash. In the DPDM example, the transient signal may incorporate five events: the

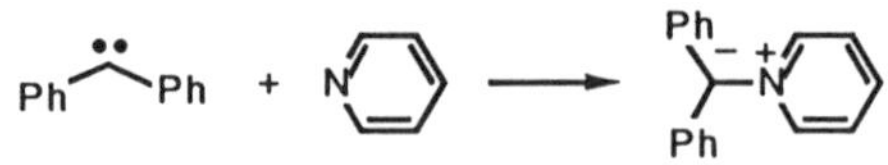

DPC 2

Scheme 2.

bleaching of the absorption of DPDM, the formation and decay of ^{3}DPC, and the formation and decay of the benzhydryl radical. The exact type of signal observed depends on the extinction coefficients of the three absorbing species at the monitoring wavelength. The shape of the transient signal may change drastically over only a few nanometers, reflecting variations in the relative extinction coefficients of the three absorbing species. Of these five processes, usually the only one of interest is the dynamics of the conversion of DPC to **1**. Despite the problems of the overlapping of absorptions, Hadel was able to obtain the absolute rate constants for reaction (1) with various hydrocarbons.[8]

In addition to the problems of spectral deconvolution, two other issues surface when the absorptions of the diazo precursor and carbene transient overlap. One wishes the sample to absorb most of the light of the laser pulse efficiently and thus produce a relatively large (10^{-4}–10^{-5} *M*) local concentration of carbene. This requires a reasonable absorbance (often as high as ~0.5–1.0) of the carbene precursor at the laser excitation wavelength. Because of spectral overlap, however, the absorption of the unphotolyzed sample at the monitoring wavelength will also then be quite high, which in turn means that it will be difficult to pass the monitoring light through the sample and into the detector. With strongly absorbing samples, nonuniform deposition of energy in the sample leads to inhomogeneous concentrations of the transient species and may also produce undesirable photoacoustic effects.[4] Finally, when the laser beam and monitoring beams are of nearly the same wavelength, scattered light from the much more intense laser beam will find its way to the detector. The scattered laser light can then seriously obscure the transient absorption signals.

These problems have frequently been overcome by trapping the transient carbenes with various Lewis bases to form ylides. As we will discuss in more detail in Section III, diphenylcarbene reacts with pyridine to form ylide **2** (see Scheme 2).[9]

Because ylide **2** absorbs strongly in the visible (λ_{max} = 500 nm, Figure 2),[10] one can excite DPDM at 308, 337 or 351 nm and monitor the ylide at a wavelength free of complicating absorptions of precursor, carbene, or other carbene reaction products such as benzhydryl radical. Since the rate constant for ylide growth is equal to that for carbene decay (*vide infra*), carbene kinetics are easily extracted from ylide rate data. Furthermore, ylides such as **2** typically have lifetimes that are orders of magnitude longer than their parent carbenes, so the growth curves of these ylides are not distorted by their subsequent decay (Figure 3). Therefore, ylides such as **2**

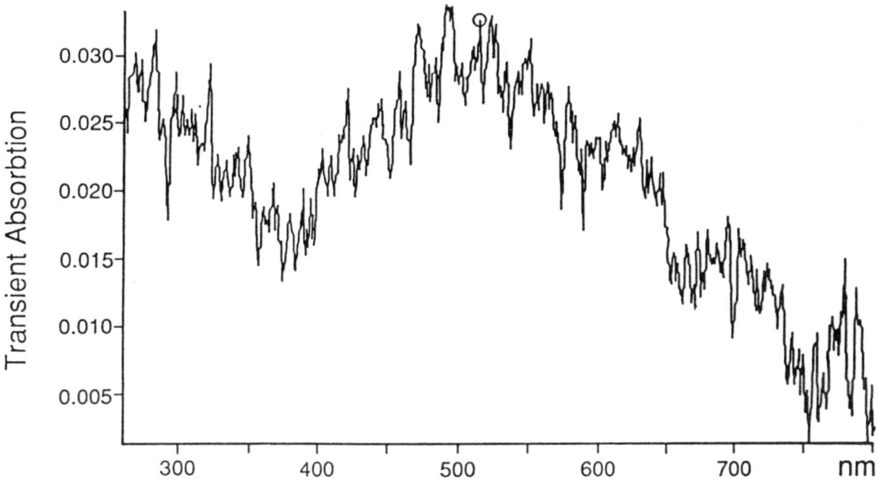

Figure 2. The transient absorption spectrum of ylide **2** derived from the reaction of diphenylcarbene with pyridine, in neat pyridine solution at room temperature.[9]

are ideal probes of the dynamics of the carbenes from which they are derived. The study of carbenes by LFP has been reviewed previously[3] and is not the aim of this chapter. It is our intention only to illustrate how ylides have been used to probe the dynamics of carbenes, frequently with more success than can be achieved by direct study of a carbene.

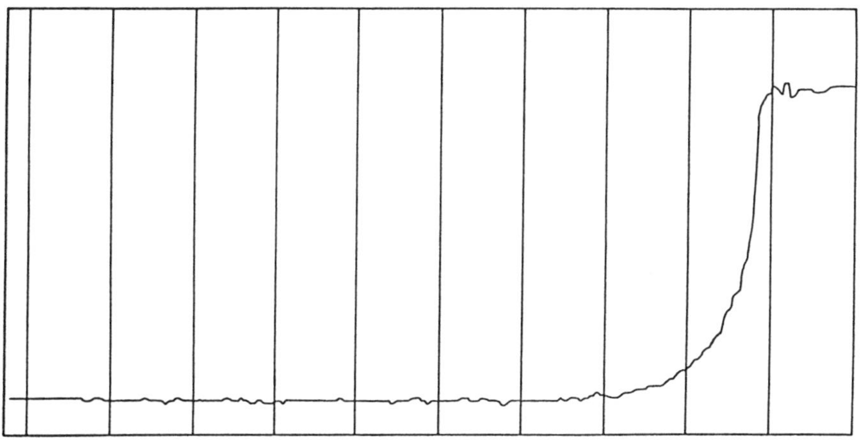

Figure 3. The growth of the transient absorption of the diphenylcarbene–pyridine ylide **2** with time, in neat pyridine solution at room temperature (horizontal unit is μs; the range of the vertical axis is absorbance units).

II. CARBENES WITH WELL-SEPARATED SINGLET GROUND STATES

The parent carbene, methylene, has been rigorously studied by chemical trapping, high resolution spectroscopy, and by theory (see Scheme 3).[11] There is universal agreement that the ground state of methylene is a triplet with a bond angle of 135°. The lowest singlet state has a filled in-plane sigma orbital, a bond angle of 102°, and is about 9 kcal/mol higher in energy than the triplet state.

Simple aromatic derivatives of methylene, such as phenylcarbene, 1- and 2-naphthylcarbene, and diphenylcarbene, also have triplet ground states.[12] Carbenes substituted with electron-withdrawing groups are known by low temperature electron paramagnetic resonance (EPR) spectroscopy to have triplet ground states (see Scheme 4).[12]

A singlet carbene ground state results from the placement of a heteroatom on the carbene carbon (Scheme 4). A heteroatom bearing a lone pair of electrons can

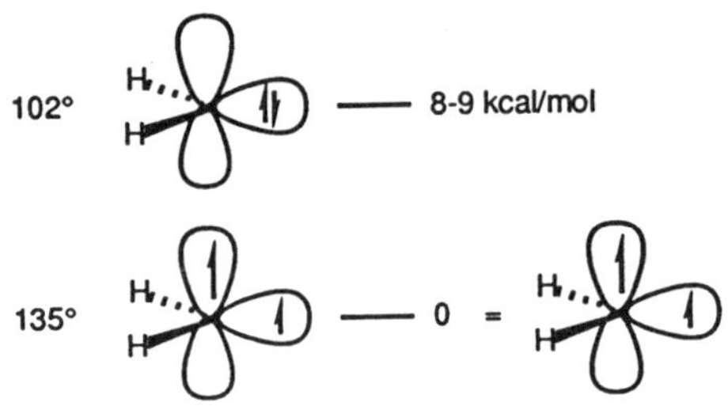

Scheme 3.

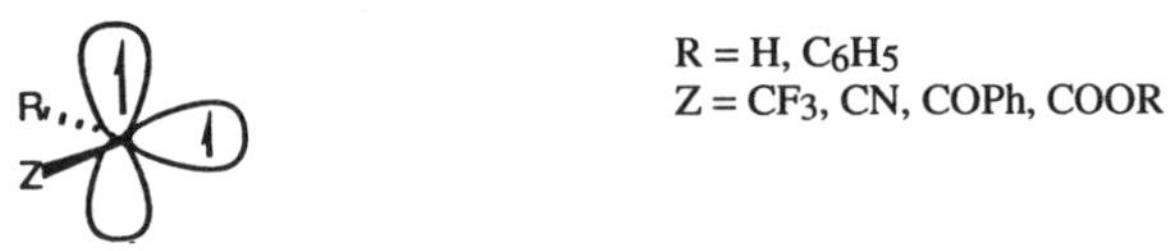

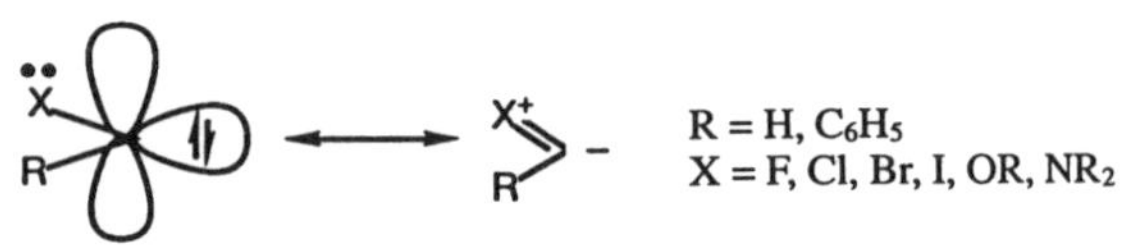

Scheme 4.

Example 1.

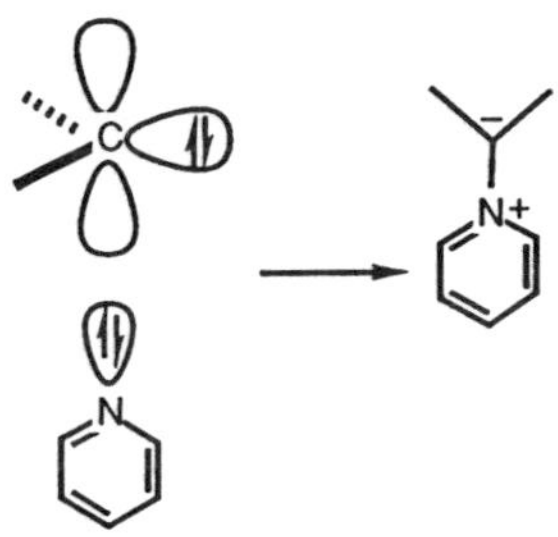

Scheme 5.

stabilize the singlet state by overlap with the empty *p* orbital,[13,14] which widens the energetic gap between the sp^2 and *p* (now π^*) orbitals.

It is also believed that incorporation of a carbene within a small ring, with resulting constriction of the angle at the carbene carbon, can produce a singlet ground state.[14] In this case, the relaxed triplet geometry (cf. 135° in $^3CH_2$) is not accessible to the carbene (see Example 1).

The reaction of a singlet carbene with a substrate such as pyridine is easily visualized. It is simply a Lewis acid–base condensation between the empty 2*p* orbital of the carbene and the nonbonding pair of electrons of the heterocycle (Scheme 5).

The reactions of singlet carbenes with substrates are expected, and have been found, to be exceedingly rapid (*vide infra*) (absolute rate constant, *k*, typically 10^8–10^9 $M^{-1}s^{-1}$). The importance of ylidic species in carbene reactions has long been appreciated by mechanistic chemists,[15] far predating their detection in laser flash photolysis experiments. As mentioned previously, the value of ylides as dynamical probes in LFP studies of carbenes results from their long-wavelength, high-extinction absorption and long lifetimes.

A. Chlorophenylcarbene

Chlorophenylcarbene (CPC) is easily generated from chlorophenyldiazirine (CPD) and has been studied more extensively than any other ground state singlet carbene.[13b,16] The only reported ylide it forms that has been detected by LFP is derived from pyridine (Scheme 6).[17]

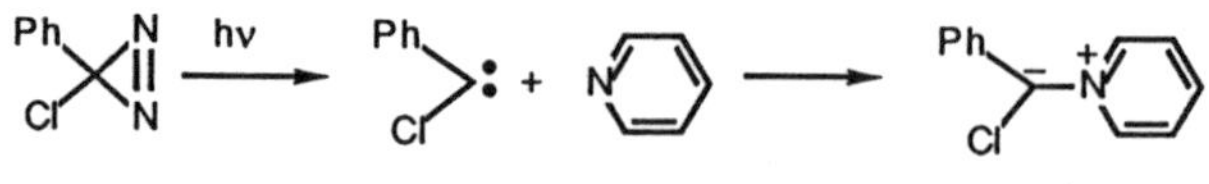

Scheme 6.

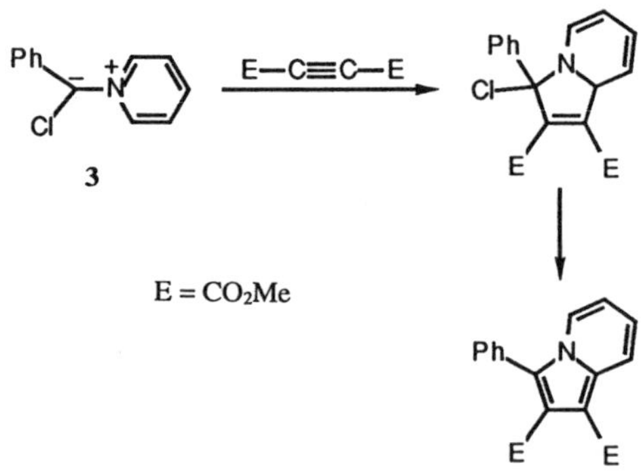

Scheme 7.

The absolute rate constant of this reaction shows little change with solvent variation (Table 1).[9] Pyridine ylide **3** can be intercepted with dimethyl acetylenedicarboxylate ($k = 1.6 \times 10^6\ M^{-1}\ s^{-1}$) to give heterocycle **4** in ~30% yield (Scheme 7).

Table 1. The Absolute Rate Constant of the Reaction of Chlorophenylcarbene with Pyridine as a Function of Solvent[9]

Solvent	ε^a	$k\ (M^{-1}\ s^{-1})$
Hexane	1.89	$(7.6 \pm 0.9) \times 10^8$
Toluene	2.38	$(3.7 \pm 0.4) \times 10^8$
Ethyl acetate	6.02	$(2.9 \pm 0.4) \times 10^8$
α,α,α-Trifluorotoluene	9.18	$(4.2 \pm 0.3) \times 10^8$
o-Dichlorobenzene	9.93	$(5.4 \pm 0.6) \times 10^8$
Methylethylketone	18.5	$(4.7 \pm 1.1) \times 10^8$
Propionitrile	27	$(3.6 \pm 0.4) \times 10^8$
Acetonitrile	36.2	$(3.9 \pm 0.4) \times 10^8$
Sulfolane	44	$(2.9 \pm 0.3) \times 10^8$

Note: [a]Dielectric constant

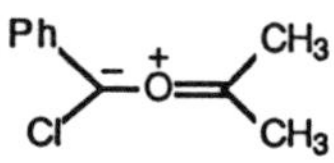

5

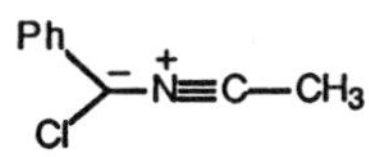

6

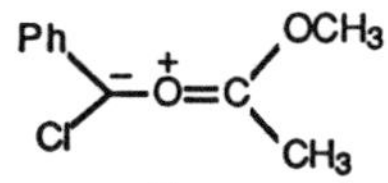

7

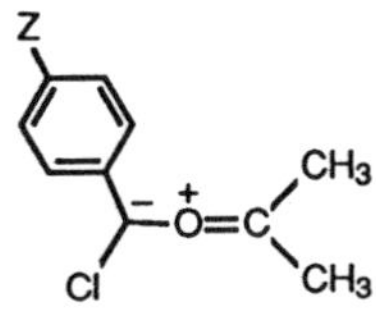

8

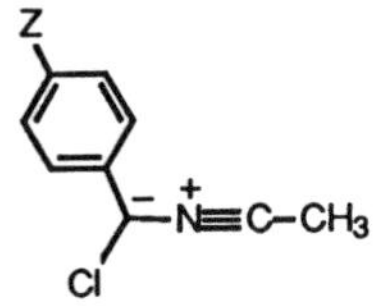

9

Z = NO_2, Cl, CF_3

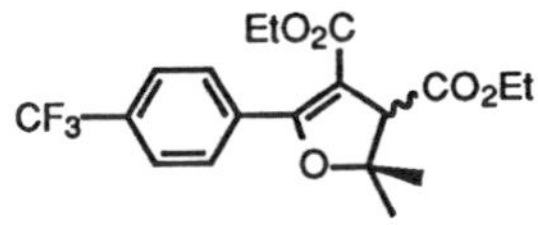

10

CPC does not react with either acetone, acetonitrile, or ethyl acetate to form ylides **5–7**; however, *p*-nitro, *p*-chloro, and *p*-trifluoromethyl substituted CPCs do react to form ylides **8** and **9** in acetone and acetonitrile.[18]

The ketone ylide **8** (Z = CF_3) is trapped by diethylfumarate (DEF) with a rate constant of $5.5 \times 10^8\ M^{-1}\ s^{-1}$ to ultimately yield **10**.

Chlorophenyldiazirine CPD has an absorption maximum at 353 nm in toluene and a minimum in its absorption spectrum between 310 and 320 nm.[19] Thus, it is a straightforward matter to excite the diazirine at 351 nm and to monitor the decay of CPC at 318 nm where it absorbs very strongly. This allowed Moss and Turro,[13b,16] followed later by our group,[20] to study the kinetics of CPC with olefins. Similarly, Griller and Scaiano studied the kinetics of reactions of CPC with alcohols and acetic acid by direct observation of the carbene.[21]

Pyridine ylide **3** has proven useful in probing the dynamics of CPC in one instance. Jones has discovered that CPC will abstract chlorine atoms from suitable donors to produce an α,α-dichlorobenzyl (DCB) radical (Scheme 8).[22]

The severe overlap in the absorption spectra of CPC and DCB complicated direct measurement of the absolute rate constant of the chlorine atom transfer reaction. It

CPC $\lambda_{max} = 318$ DCB $\lambda_{max} = 324$ nm

Scheme 8.

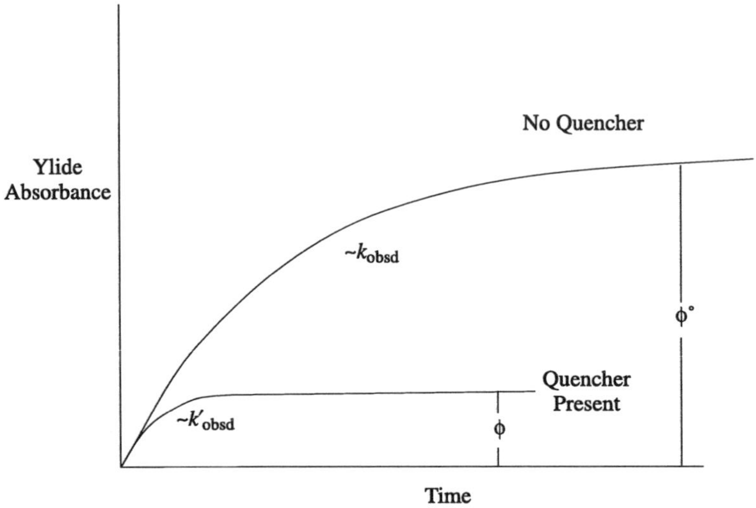

Figure 4. The effect of a chlorophenyl carbene (CPC) quencher [e.g., (CCl_3CN)] on the rate of formation of the CPC–pyridine ylide **3**.

was possible, however, to obtain this information by monitoring the growth of ylide **3** as a function of chlorine donor concentration as illustrated in Figure 4. Laser flash photolysis of chlorophenyldiazirine in the presence of dilute pyridine produces ylide **3**, whose formation can be resolved and the pseudo first-order observed rate constant of formation, k_{obsd}, calculated. As shown in Eq. (2), k_{obsd} can be dissected as

$$k_{obsd} = k_o + k_{pyr}[\text{pyr}] \tag{2}$$

where k_{pyr} is the absolute rate constant of reaction of CPC with pyridine,[23] and k_o is the sum of all other first-order and pseudo first-order decay processes involving

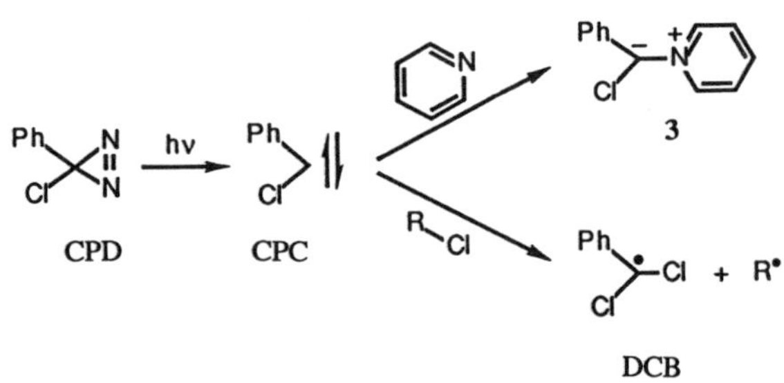

Scheme 9.

CPC. When the flash photolysis is repeated in the presence of RCl and pyridine, CPC partitions between reaction with the two quenchers (Scheme 9)*.

The presence of RCl both reduces the yield of ylide **3** (ϕ smaller than $\phi°$) and increases its observed pseudo first-order rate constant of formation ($k'_{obsd} > k_{obsd}$). In the presence of RCl the pseudo first-order rate constant for formation of ylide **3** is given by Eq. (3).

$$k'_{obsd} = k_o + k_{pyr}[pyr] + k_{RCl}[RCl] \tag{3}$$

Based on kinetic analyses originally outlined by Scaiano et al.,[23] one can obtain the rate constant of interest, k_{RCl}, by measuring k'_{obsd} at several concentrations of RCl and constant pyridine concentration. A plot of k'_{obsd} versus concentration of RCl will be linear (Figure 5) with a slope of k_{RCl} and an intercept at [RCl] = 0 of $k'_{obsd} = k_{obsd}$ ($= k_o + k_{pyr}[pyr]$).

It is also possible to obtain k_{RCl} from the Stern–Volmer equation (Eq. 4) by

$$\frac{\phi°}{\phi} = 1 + \frac{k_{RCl}}{k_{obsd}}[RCl] \tag{4}$$

monitoring the decrease in the yield of ylide (ϕ) as a function of RCl concentration (Figure 6).

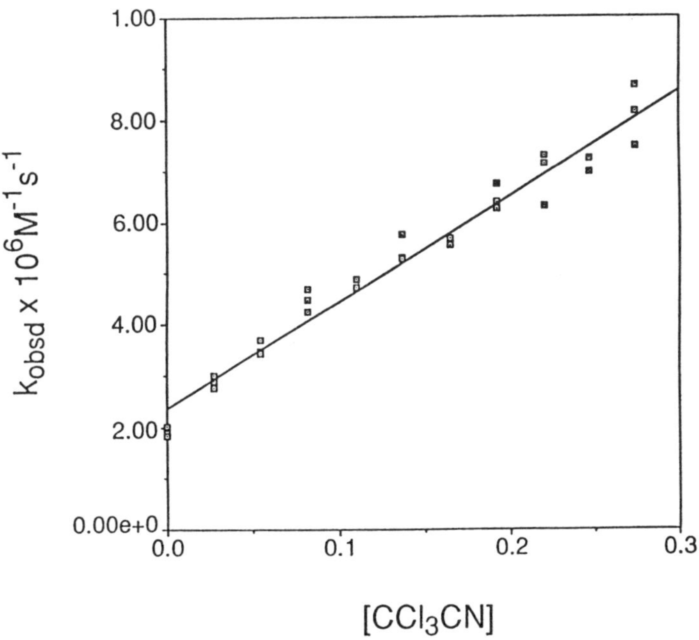

Figure 5. A plot of k'_{obsd} versus $[CCl_3CN]$. The slope is the bimolecular rate constant of reaction of CPC with CCl_3CN (k_{RCl}), and the intercept is k_{obsd}.

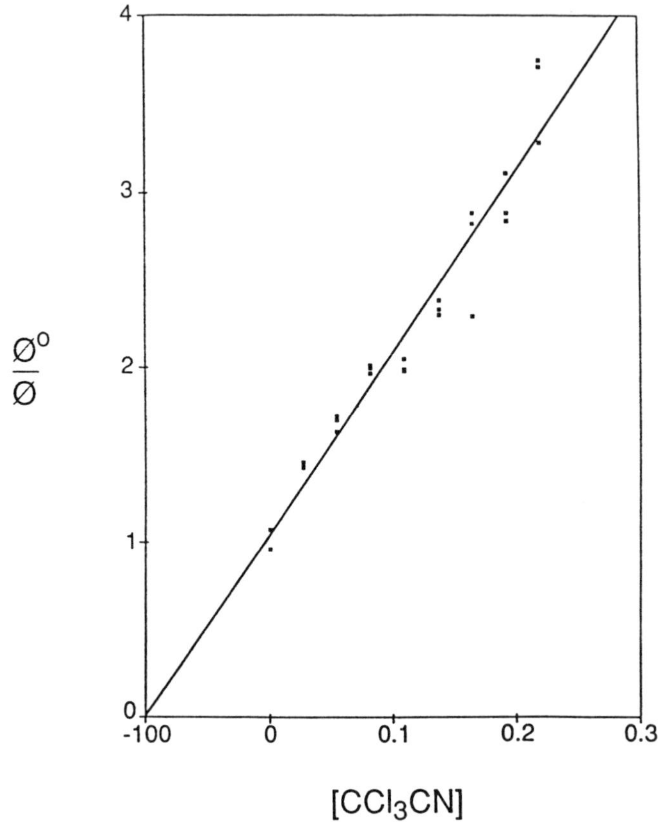

Figure 6. A Stern–Volmer plot of the effect of CCl_3CN on the yield of CPC–pyridine ylide **3**. The slope is $k_{RCl}\tau$ where $\tau = (1/k_{obsd})$.

The absolute rate constants for various chlorine atom transfers are listed in Table 2. Two trends are obvious; (1) that increasing solvent polarity increases the reaction rate (Figure 7), and (2) that increasing the ease of reduction of the alkyl halide increases the reaction rate. The entropy of activation of the reaction of CPC with $CCl_2(CN)_2$ is significantly more positive than the reaction of CPC with either CCl_3CN or alkenes.[13b,16] The data point to a polar atom transfer transition state. This reaction may also be viewed as an inner sphere electron transfer.[24] The transition state occurs quite early and, at least in the case of $CCl_2(CN)_2$, is much looser than in the reaction of CPC with alkenes, suggesting little carbon–chlorine bond formation.

Table 2. Rate Constants and Activation Parameters for the Reaction of CPC with Halogen Donors as Determined with the Pyridine Ylide Probe[9,22]

Substrate	*Solvent*	k_{obsd} $(M^{-1} s^{-1})$	E_a *(kcal/mol)*[b]	*log A*[b]
CCl_4	CH_3CN	$(3.829 \pm 1.506) \times 10^4$		
Cl_3CCN	Toluene	$(2.060 \pm 0.141) \times 10^7$	1.5 ± 0.3	8.4 ± 0.3
Cl_3CCN	CH_3CN	$(1.059 \pm 0.064) \times 10^9$	–0.5 ± 0.3	8.6 ± 0.2
$Cl_2C(CN)_2$	Toluene	$(8.450 \pm 1.420) \times 10^9$	1.4 ± 0.3	11.2 ± 0.2
$Cl_2C(CN)_2$	CH_3CN	$(2.602 \pm 0.334) \times 10^{10}$	1.2 ± 0.6	11.3 ± 0.5
Cl_3CBr	Toluene	$(7.543 \pm 0.006) \times 10^9$		
Cl_3CBr	CH_3CN	$(2.648 \pm 0.001) \times 10^{10}$		
NCS	Toluene	$(1.152 \pm 0.076) \times 10^{10}$		
NCS	CH_3CN	$(1.668 \pm 0.001) \times 10^{10}$		
CH_3CH_2I	Hexane	$(3.663 \pm 0.390) \times 10^6$		
CF_3CH_2I	Hexane	$(2.064 \pm 0.091) \times 10^8$		
CCl_3CN[a]	Toluene	$(1.142 \pm 0.214) \times 10^8$		

Notes: [a]Methoxyphenylcarbene.
[b]Compare to $E_a \cong$ 1–5 kcal/mol; log A $\cong$ 7–8 for alkene substrates.

The activation energy for the reaction of CPC with pyridine is 1.3 kcal/mol[25], which is very similar to that observed in the reaction of CPC with 1-hexene and ethyl acrylate.[13b,16,20] The activation energy for the reaction of CPC with tetramethylethylene is actually negative.[13b,16] To account for these observations it has been proposed that this reaction involves a reversibly formed complex (**11**) (Scheme 10).[13b,16]

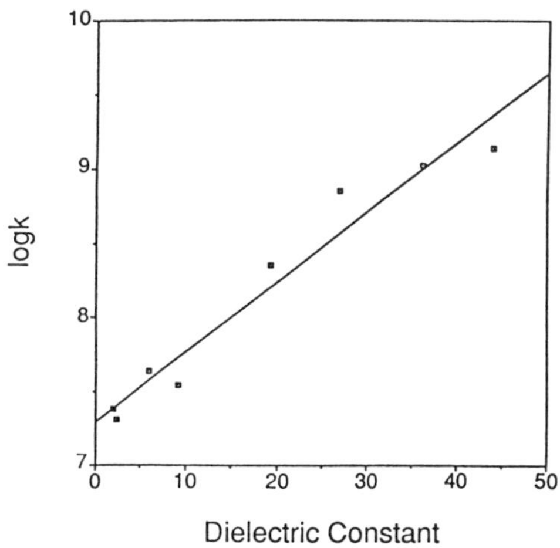

Figure 7. The effect of solvent polarity on the rate of reaction of CPC with CCl_3CN.

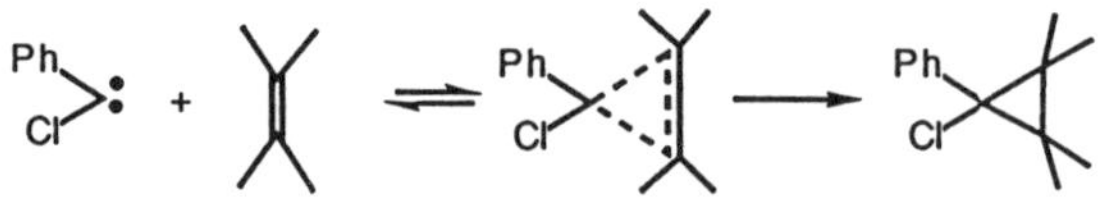

CPC 11

Scheme 10.

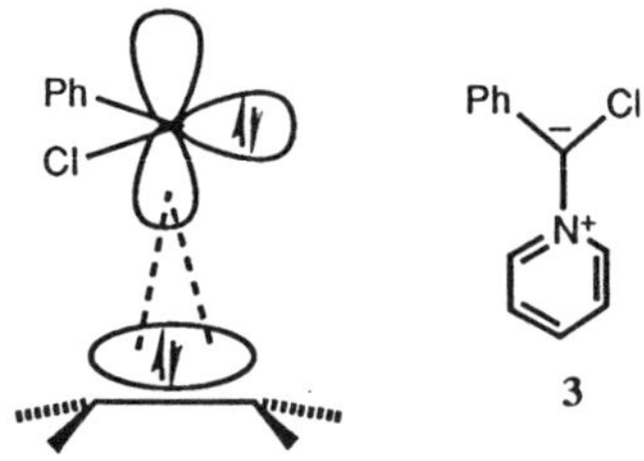

11 3

Example 2.

It seemed reasonable to us that putative complex **11** would be bound by charge-transfer interactions not unlike those found in ylide **3** (Example 2).

Thus, we regarded **3** as a model of an *irreversibly* formed complex; the kinetics give no indication of **3** reverting to carbene plus pyridine. The fact that the ylide- and cyclopropane-forming reactions have comparable activation energies would seem to imply a common concerted reaction mechanism and to diminish the need to postulate a *reversibly* formed complex.[25]

In fact, the activation energy for the reaction of CPC with tri-*n*-butylstannane is slightly negative, and it is difficult to visualize a reversibly formed intermediate in this reaction (Scheme 11).[25]

In our opinion, the negative activation energies observed in carbene reactions are a consequence of the entropic barriers described by Houk and co-workers.[26]

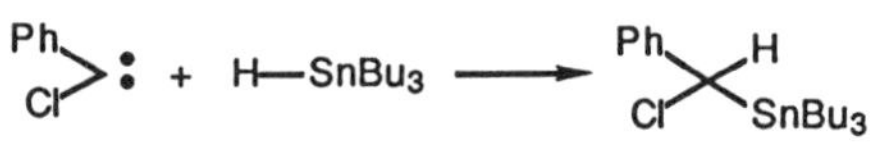

CPC

Scheme 11.

B. Alkylchlorocarbenes

Alkylchlorocarbenes do not absorb strongly above 300 nm[27] and as a consequence are not readily studied by LFP with UV–vis detection. It is in systems of this type, where the carbene is "invisible," that ylide-forming reactions are most valuable as probes of carbene dynamics.

Laser flash photolysis of *tert*-butylchlorodiazirine (**12**) in toluene at 25 °C produces no transient absorption above 300 nm.[17] When the experiment is repeated in the presence of pyridine a new transient is present (Scheme 12).

The transient is attributed to ylide **13** because its rate of formation is first order in a solution of pyridine [$k_{pyr} = (2.4 \times 10^9\ M^{-1}\ s^{-1})$] and the fact that carbene **14**, the precursor to putative ylide **13**, reacts with quenchers such as butylvinyl ether and *trans*-3-hexene with absolute rate constants (k_Q) comparable to that of CPC with these substrates.[17]

The lifetime of carbene **14** in toluene in the absence of pyridine can be deduced from the intercept of the pyridine quenching plot (see Eq. 3) and was found to be 80 ns. This corresponds to a k_{obsd} of $1.25 \times 10^7\ s^{-1}$, which is an upper limit for k_I, the total rate constant for both of the intramolecular decay channels, methyl migration and cyclization, and all intermolecular decay channels as well.

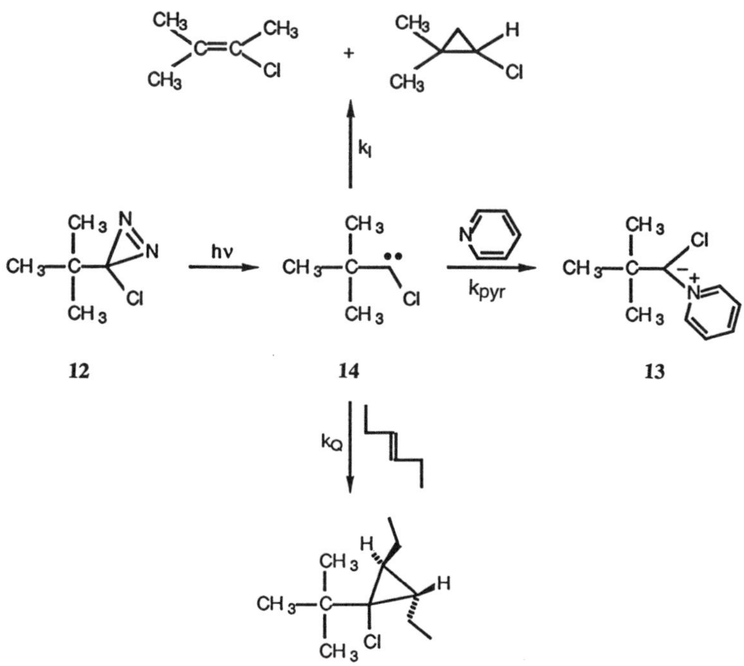

Scheme 12.

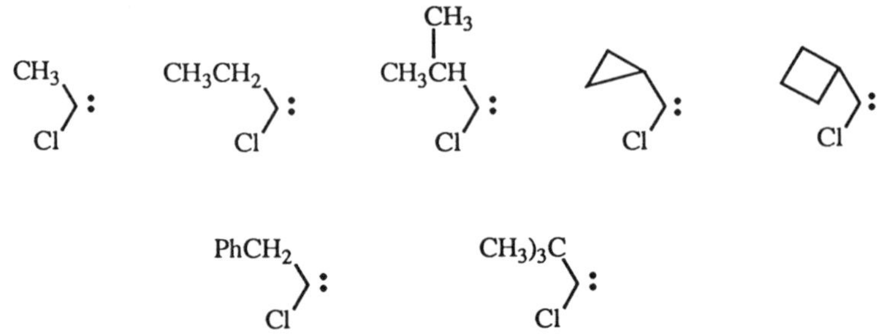

Example 3.

In the 1960s it was discovered that different mixtures of stable products arise from pyrolysis or photolysis of alkyldiazirines can lead to a different mixture of stable products.[28b] This led several groups to consider the possibility that the photolysis of alkyldiazirines produces either diazirine excited states or carbene excited states as product-forming species. Thus, one can worry whether ylide **13** is formed by capture of the excited state of diazirine **12** or by reaction with an excited state of carbene **14**. However, the fact that the lifetime of the precursor to the ylide is 80 ns rules out these possibilities. In solution at ambient temperature, an excited state lifetime of 80 ns is too long to be associated with a nonfluorescent excited state of either the diazirine or the carbene. In fact, Moss has demonstrated that the lifetime of the precursor of ylide **13** (carbene **14**) is controlled by reaction with the diazirine precursor **12**. Moss[29] has deduced by extrapolation that the lifetime of carbene **14** (the precursor of ylide **13**) is longer than 1 μs when the concentration of **12** equals zero. Thus, the data demonstrate that pyridine reacts with the ground state of the carbene and not with an excited state of the precursor.[29]

Subsequent to the 1988 report of Jackson,[17] the pyridine ylide probe method has been successfully applied to a large number of alkylchlorocarbenes[30] (see Example 3) and to the reactions of dichlorocarbene itself (Scheme 13).[31]

The pyridine probe method has been extremely useful in measuring the absolute rates of intramolecular rearrangement of alkylhalocarbenes.[30] This subject is reviewed in detail by Moss in Chapter 2 of this volume. The only other technique that has been applied to the direct measurement of alkylchlorocarbene dynamics is

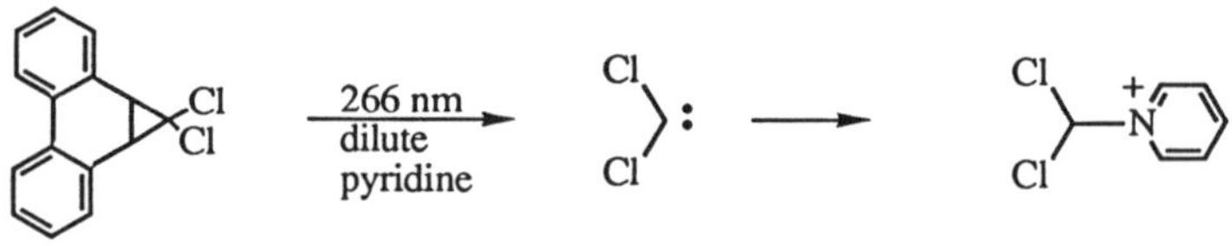

Scheme 13.

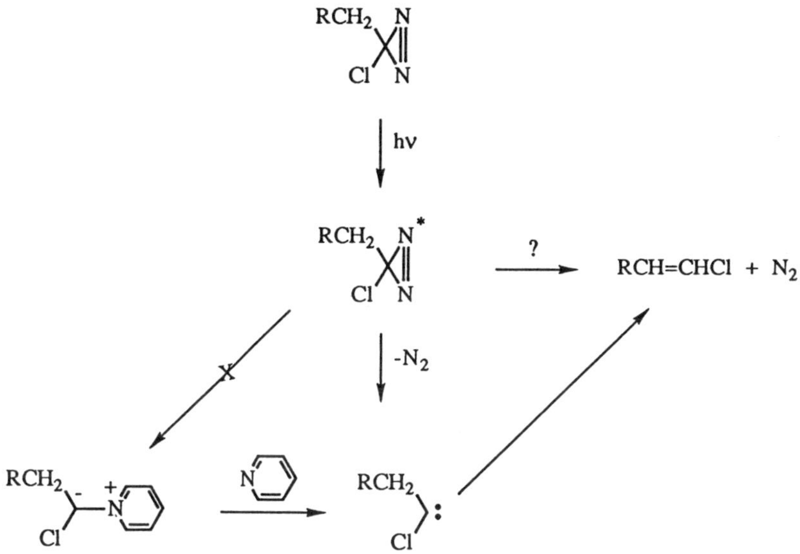

Scheme 14.

photoacoustic colorimetry (PAC), which does not rely on a UV–vis-active chromophore.[32]

Although pyridine ylides are not formed by the reaction of diazirine excited states with pyridine, it is still possible that some "carbene" rearrangement products can be formed in this manner (Scheme 14).

The pyridine probe method has given new insight into this problem. As mentioned previously, LFP of a diazirine in the presence of pyridine produces an ylide whose optical yield ($\phi°$, see Figure 4) and rate constant of formation (k_{obsd}) can be resolved. It was earlier demonstrated that upon increasing the concentration of pyridine both $\phi°$ (no carbene quencher present) and k_{obsd} increase; it is possible, however, to saturate $\phi°$. As the concentration of pyridine increases eventually *every* carbene that is produced in a laser pulse is captured by pyridine to produce a pyridine ylide. At saturating pyridine concentrations every carbene is diverted to ylide before it can either rearrange or react with solvent. A further increase in pyridine concentration leads to no further increase in $\phi°$ above the saturating concentration of pyridine ($\phi° = \phi°_{sat}$).

White and Platz have measured $\phi°_{sat}$ for several alkylchlorodiazirines.[33] Chlorocyclopropyl diazirine and *tert*-butylchlorodiazirine give very large values of $\phi°_{sat}$ whereas diazirines with a weak α-C–H bond give low values of $\phi°_{sat}$ (see Example 4). Significantly, there is an inverse isotope effect on the yield of $\phi°_{sat}$ with CD_3CCl and CH_3CCl. The conclusion seems inescapable that the α-carbon hydrogen atom migrates in the excited state of the diazirine to form alkene, in competition with

Carbene	ϕ°_{sat} *(relative)*
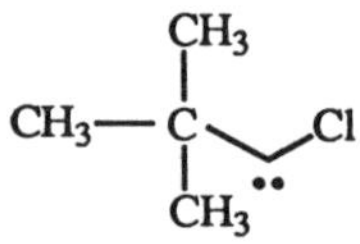	1
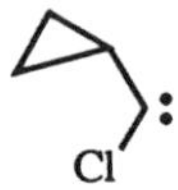	1
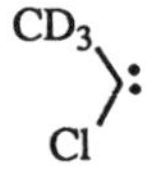	0.563
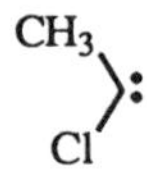	0.518
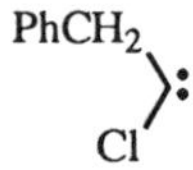	0.375
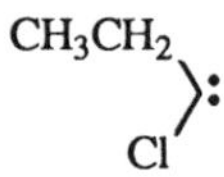	0.285
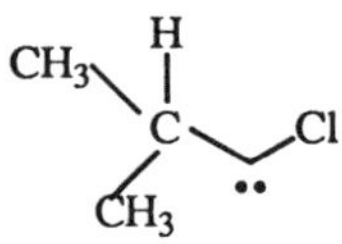	0.024

Example 4.

carbene formation. Diazirines which lack an α-C–H bond (e.g., *tert*-C_4H_9CCl), or which have a particularly strong α-C–H bond (e.g., chlorocyclopropyl), decompose to form the carbene very efficiently. There is no evidence for α-carbon migration in the excited states of simple *unstrained* alkylchlorodiazirines.

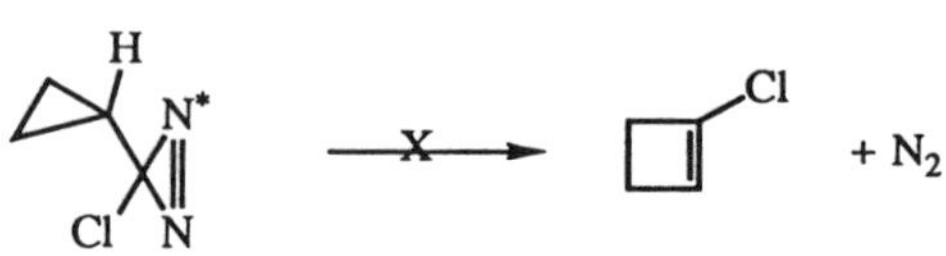

Scheme 15.

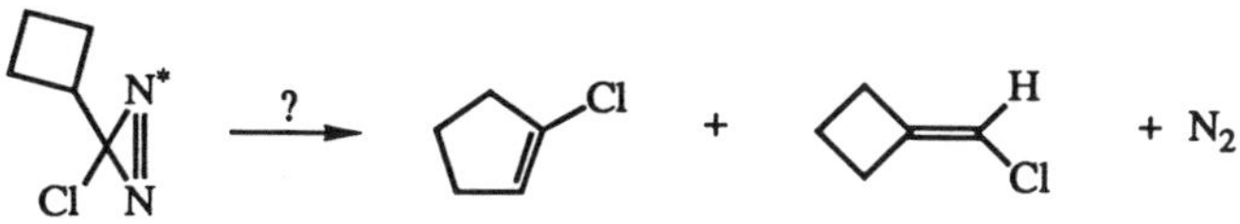

Scheme 16.

Scheme 17.

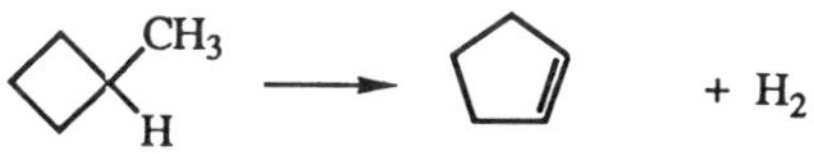

Scheme 18.

The rate data of White and Platz[33] are consistent with those of Moss where there is overlap.[34] Moss has concluded that there is no carbon migration in the excited state of cyclopropylchlorodiazirine (Scheme 15), but that migration may proceed in the excited state of cyclobutylchlorodiazirine (Scheme 16),[35] although other interpretations are possible. This is surprising at first glance because one imagines that strain relief in the cyclopropyl system is at least as great as in the cyclobutyl system. However, the isomerization of methylcyclopropane to cyclobutene (Scheme 17) leads to a small *increase* in strain energy[36] whereas the isomerization of methylcyclobutane leads to a large *decrease* in strain energy (Scheme 18). Since there is more strain-release driving force in rearranging the excited state cyclobutylchlorodiazirine than in cyclopropylchlorodiazirine, the lack of rearrangement in the excited state of the latter carbene can be rationalized.

The benzylchlorodiazirine system **15** has been extensively studied.[30,37] Laser flash photolysis of benzylchlorodiazirine **15** in toluene at 25 °C does not produce a detectable transient absorption above 300 nm. Laser flash photolysis in the presence of dilute pyridine produces the transient absorption spectrum of ylide **17** (Scheme 19).[38]

The absolute rate constants of reaction of benzylchlorocarbene **16** with pyridine, tetramethylethylene, butylvinyl ether, and *trans*-3-hexene are $4.2 \pm 0.2 \times 10^9\ M^{-1}\ s^{-1}$; $1.1 \times 10^8\ M^{-1}\ s^{-1}$; $7.2 \times 10^7\ M^{-1}\ s^{-1}$; and $6.8 \times 10^7\ M^{-1}\ s^{-1}$, respectively. The lifetime of **16** at 25 °C in toluene is 30 ns. This corresponds to a value of k_{obsd} in toluene of 4.9×10^7 to $6.7 \times 10^7\ M^{-1}\ s^{-1}$, which again is the upper limit of k_I, the total rate of 1,2-hydrogen migration in the carbene to form chlorostyrenes **18*E*** and **18*Z***.

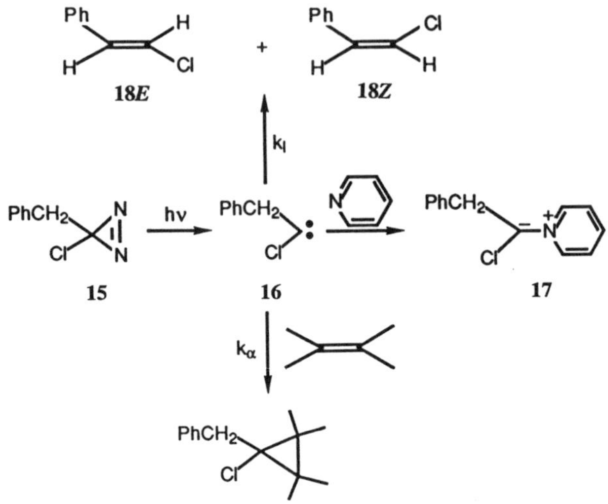

Scheme 19.

This system does remain somewhat controversial. Liu has discovered that photolysis of diazirine **15** in the presence of dilute tetramethylethylene or diethylfumarate produces the expected cyclopropane **20** along with the **18***Z* and **18***E* chlorostyrenes.[37] The *Z*/*E* ratio in the chlorostyrenes, however, is influenced by the presence of tetramethylethylene or diethylfumarate, which demonstrates that there are at least two chlorostyrene-forming intermediates present in the reaction. Liu and co-workers prefer to associate the second intermediate with the carbene–olefin complex **19** (Scheme 20).[30] Warner postulates that an excited state of the carbene may be the other styrene forming intermediate.[39] White and Platz[33] have proposed

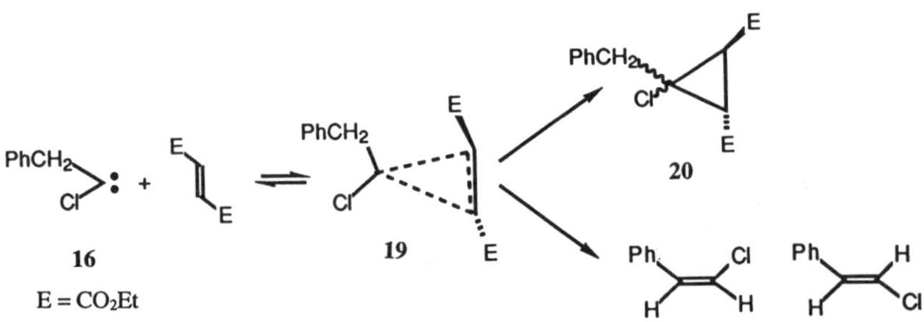

Scheme 20.

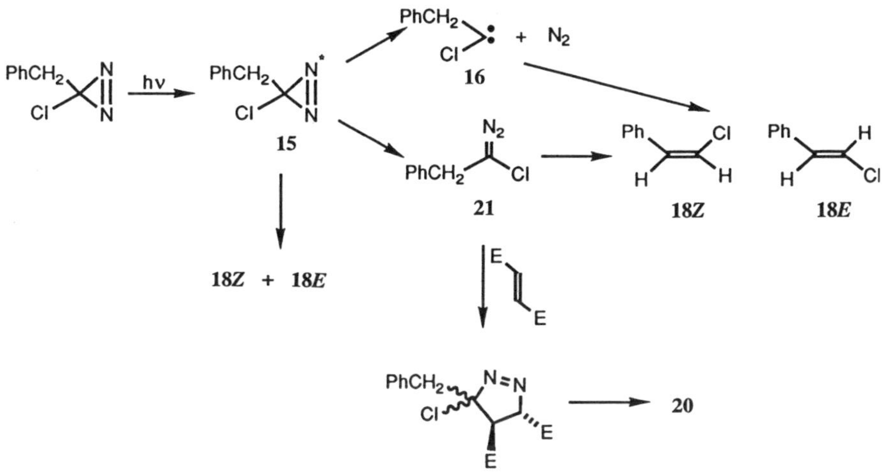

Scheme 21.

that upon photolysis, diazirine **15** may partition between loss of nitrogen and isomerization to form the diazo compound **21**, and undergo hydrogen migration in concert with nitrogen extrusion (Scheme 21). Obviously, the **18***Z*/**18***E* ratio produced from **15*** may differ from that produced by carbene **16**. Furthermore, the diazo compound can react with fumarate and eventually form cyclopropane **20**, or rearrange to its own characteristic *Z*/*E* ratio of styrenes by ionic pathways. Finally, an equilibrium between pre-*E* and pre-*Z* conformers of carbene **16**, which may have different intermolecular reactivities, cannot be ruled out.

It should be noted that a second, parallel pathway forming chlorostyrenes does not affect the interpretation of the LFP data previously presented,[30,38] although it does impact the product studies of Liu.[37]

C. Alkoxycarbenes

Three alkoxycarbenes have been trapped with pyridine to form ylides. Vasella et al.[40] have trapped a glucose-derived carbene (Scheme 22), Perrin and Platz[41] have

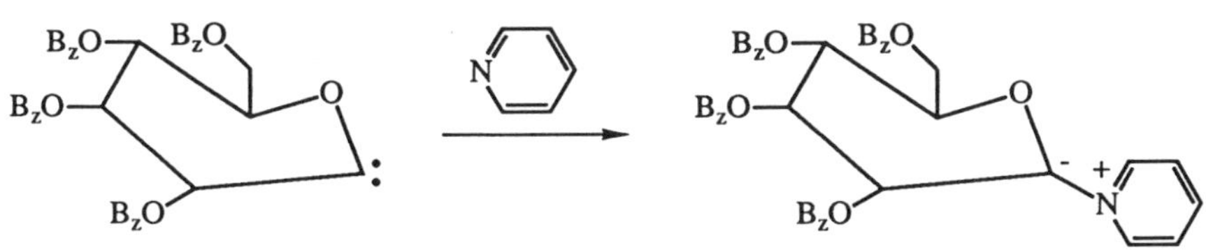

Scheme 22.

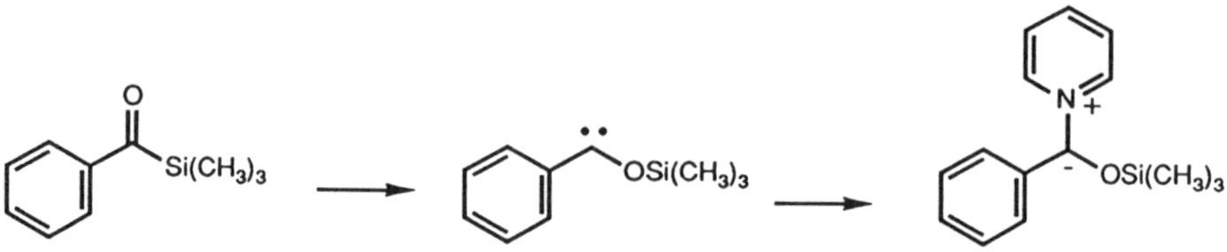

Scheme 23.

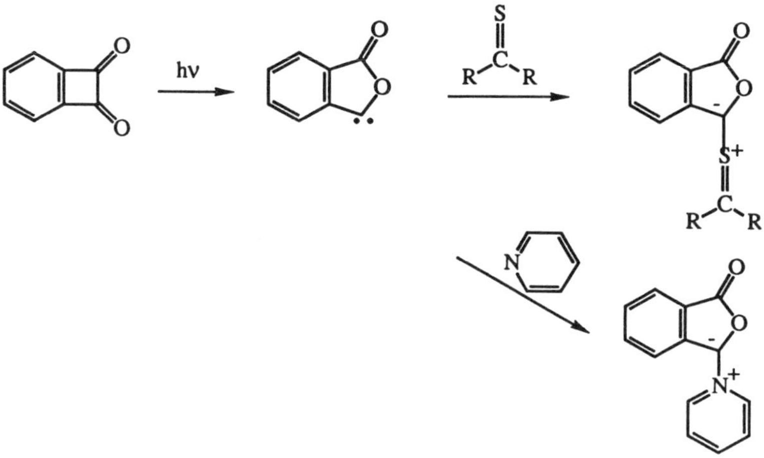

Scheme 24.

studied a siloxytrimethylcarbene by LFP techniques (Scheme 23), and Boate et al.[42] have trapped an oxacarbene with pyridine and with adamantane thione to produce ylides that are useful dynamical probes (Scheme 24).

D. Dialkylcarbenes

Dialkylcarbenes are not anticipated to have a useable chromophore in the UV–vis region and thus their dynamics may only be studied by LFP via a probe method. Although ethylidene (methylcarbene) is predicted to have a triplet ground state,[43] dialkylcarbenes constrained in a small ring are expected to have singlet ground states.[14,44] Two such carbenes, adamantanylidene (**22**) and homocubanylidene (**23**), will be considered herein along with simple alkyl and alkylarylcarbenes (in the next section) which have triplet ground states. Both of these carbenes are incorporated within rigid ring systems that enforce a relatively small angle at the carbene center; this seems to produce a singlet ground state in **22** and **23** that is based on the lack

22

23

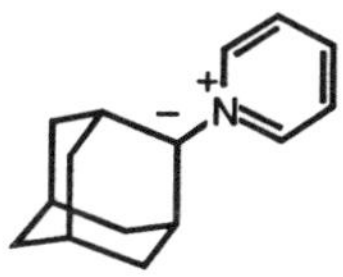

24

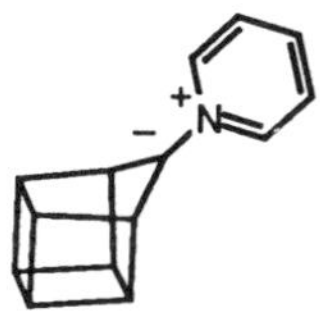

25

of reaction of these carbenes with oxygen. Carbenes **22** and **23** are quite long-lived, in contrast to other alkylcarbenes. Because their intramolecular rearrangements lead to highly-strained products, carbene **22** is long-lived and has been intercepted with a variety of chemical traps.[45]

Carbenes **22** and **23** can be trapped with pyridine to produce the corresponding ylides **24**[46] and **25**[47], whose transient spectra are shown in Figures 8 and 9. The spectra are attributed to the ylides because they are not formed in the absence of pyridine and their rates of formation are first order in pyridine concentration.

The ylide yields are also reduced in the presence of typical carbene quenchers such as methanol, *trans*-3-hexene, or triethylsilane. Adamantanylidene can also be

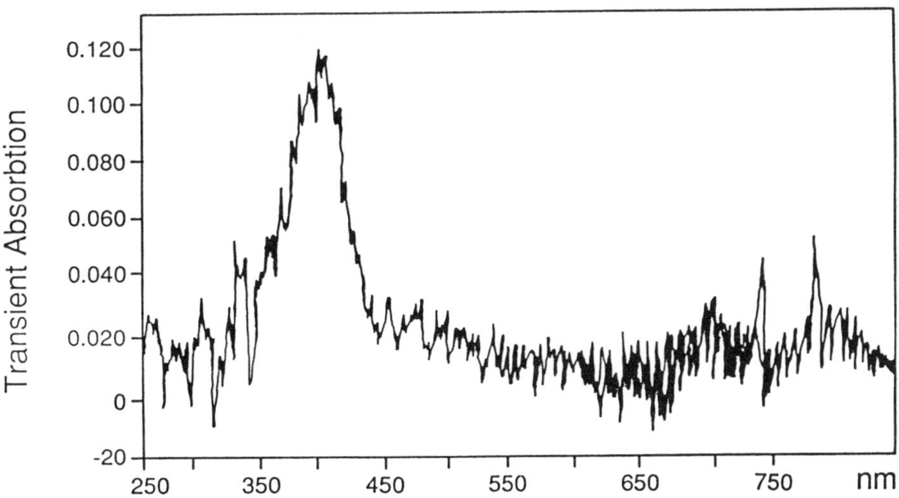

Figure 8. The transient absorption spectrum of the ylide derived from adamantanylidene and pyridine.

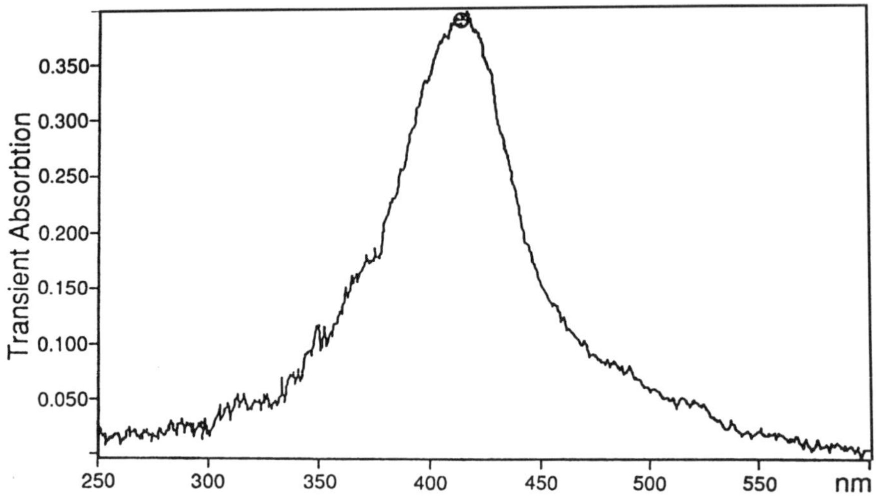

Figure 9. The transient absorption spectrum of the ylide derived from homocubanylidene and pyridine.

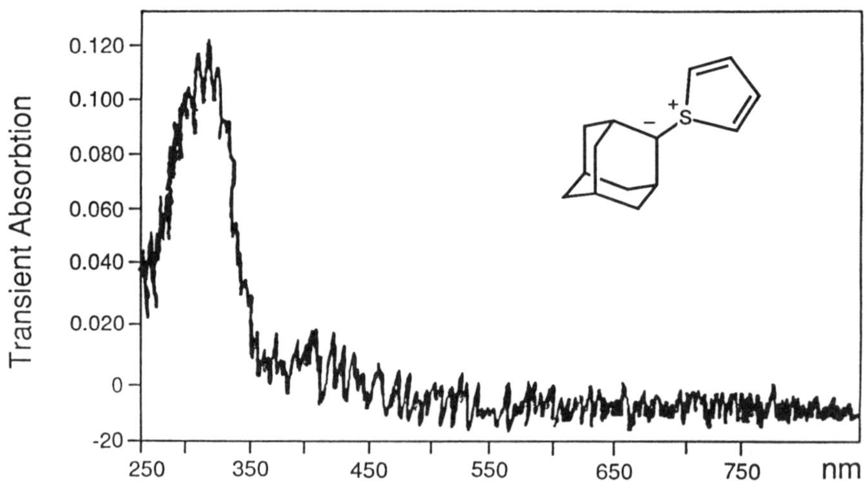

Figure 10. The transient absorption spectrum of the ylide derived from adamantanylidene and thiophene.

intercepted with thiophene to produce an ylide (see below) whose spectrum is shown in Figure 10.[46]

A Stern–Volmer analysis of the quenching of the ylides can be used to obtain the absolute rate constants for reaction of the carbenes with various quenchers (Table 3).[46] Adamantanylidene is sufficiently sluggish in its reactions that its absolute rate constants can also be measured by the direct method in which the increase in the rate of formation of the thiophene ylide (k'_{obsd}) is measured as a function of carbene quencher concentration (Eq. 3). As seen in Table 3, the rate constants for quenching adamantanylidene are systematically larger when measured by the Stern–Volmer method rather than by the direct method. In fact, certain quenchers (e.g., methanol) reduce the yield of ylide but cause no apparent increase in the observed rate of formation of the ylide. This means that there must be an intermediate present between the ground state of the diazirine and the ground state of the carbene which is capable of reacting with carbene quenchers (see Scheme 25).[48] Neither the growth nor the decay of these ylides, with or without substrates present, are affected by the presence of oxygen. This implies that **22** and **23** have singlet ground states and inaccessibly high energy triplet states, as triplet carbenes typically react with oxygen at a diffusion-controlled rate.[3]

Table 3. Absolute Rate Constants for the Reaction of Adamantanylidene with Various Quenchers in Benzene Solvent, Determined Using the Ylide Formed from Adamantanylidene and Thiophene as Probe[46]

Quencher	k $(M^{-1}\,s^{-1})^{a}$	k $(M^{-1}\,s^{-1})^{b}$
Thiophene		
TMS	3.01×10^{6c}	
benzene	2.38×10^{6d}	
Diazirine	2.87×10^{7}	
Pyridine	1.54×10^{6}	
n-Butylvinylether	3.80×10^{5}	4.04×10^{5}
Diethylfumarate	1.54×10^{6}	9.86×10^{6}
Triethylsilane	3.31×10^{5}	1.2×10^{6}
Diphenylsilane	2.09×10^{6}	1.2×10^{7}
Phenylsilane	4.19×10^{6}	2.2×10^{7}
Acetic acid	5.55×10^{9}	5.62×10^{9}
Acetic acid-*d*	4.81×10^{9}	4.74×10^{9}

Notes: [a]Direct method.
[b]Stern–Volmer method.
[c]Tetramethylsilane solvent.
[d]Benzene solvent.

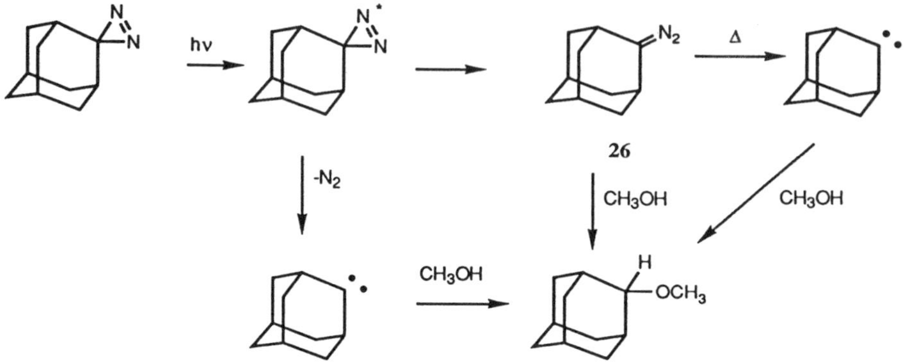

Scheme 25.

The Eaton group at Chicago[49] and the Jones group at Princeton[47,50] have demonstrated that homocubanylidene **23** can interconvert with its highly strained alkene isomer **27** (Scheme 26). Eaton has demonstrated that it is the α-carbon bond that migrates by means of a clever labeling experiment.[49] Jones and Chen have demonstrated that carbene alkene equilibrium is "complete" in benzene containing a $\leq 1M$ methanol trap by entry into the **23, 27** system from each direction with different precursors.[47] This seemed at first to be inconsistent with previously cited labeling studies; however, modeling of the reaction kinetics reveals that if **23** and **27** interconvert only two or three times prior to capture with methanol, equilibrium, as measured by the lack of a dilution effect, appears complete.[51] Only a few **23, 27**

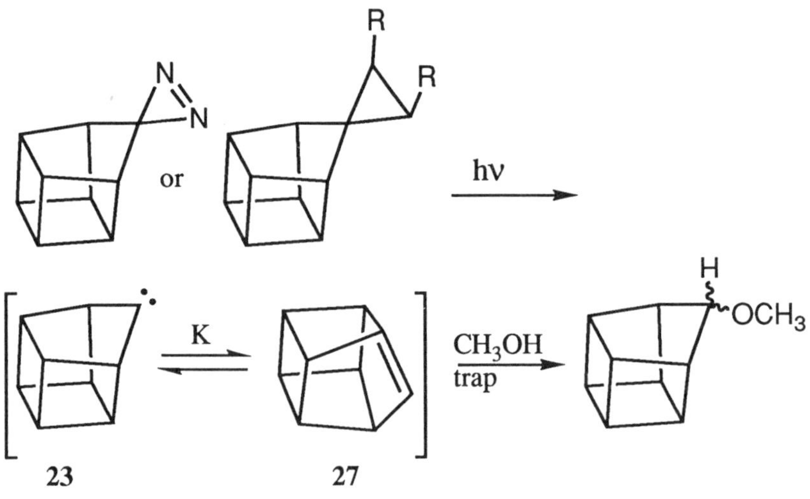

Scheme 26.

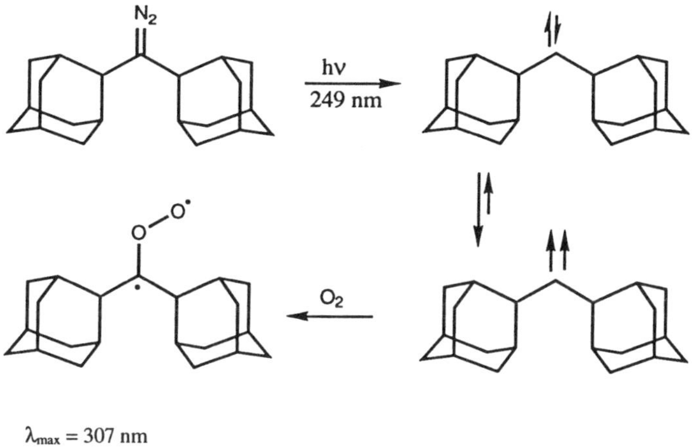

Scheme 27.

cycles of interconversion, however, is insufficient to allow a label to completely "wander" about the homocubyl skeleton.

It is possible to combine the product studies of Jones and Chen with the laser flash photolysis studies of White (utilizing pyridine ylide **25**) to deduce that equilibrium constant $K \approx 1$ at 25 °C in methylene chloride.[47] This result is consistent with Borden's *ab initio* calculations, which reveal that alkene **27** is highly pyramidalized.[52]

Morgan et al.[53] have used the ylide probe method to study diadamantylcarbene by LFP techniques. Laser flash photolysis of diadamantyldiazomethane (249 nm) in the presence of oxygen produces a carbonyl oxide (Scheme 27) presumably formed by interception of the triplet ground state of diadamantylcarbene.[54]

E. Simple Alkyl- and Alkylarylcarbenes

Dimethylcarbene (DMC) and methylcarbene (MC) may have triplet ground states[43]; however, they will be discussed at this time along with their more complex relatives **22** and **23** which are apparently singlet ground state species.

Dimethylcarbene and methylcarbene have not been detected by spectroscopy. This presumably is a consequence of their anticipated short lifetimes, which result from rapid rearrangements by H-migration. These rearrangements are either slow or reversible in **22** and **23** because of strain considerations. It is possible, however, to capture MC, albeit in low yield, with silanes and alkenes.[55] Thus, there would seem to be a chance to capture these carbenes with pyridine to produce UV–vis-active ylides. Recently, Modarelli and Platz[48,56] have used LFP techniques to study these systems. Laser flash photolysis of dimethyldiazirine in a pentane–

pyridine mixture produces the desired ylides. These workers have deduced the solution phase carbene lifetimes shown in Example 5 after assuming "appropriate" values of k_{pyr}.[57]

The kinetic isotope effect (KIE) for H(D) migration in DMC is surprisingly large for a transition state that is necessarily triangular. Nevertheless, a large KIE does explain the failure to trap methylcarbene with pyridine.[57] These trapping studies demonstrated that there must be barriers, possibly entropic in nature, to the rearrangement process, despite the extreme exothermicity. This finding is also in agreement with recent theoretical studies.[58]

The lifetime of DMC is surprisingly long for a species which can be trapped only in part. Using methanol trapping, Modarelli and Platz[48] demonstrated that the excited state of dimethyldiazirine decomposes to form both dimethylcarbene and propylene possibly in a concerted process, with almost equal efficiency, at 25 °C

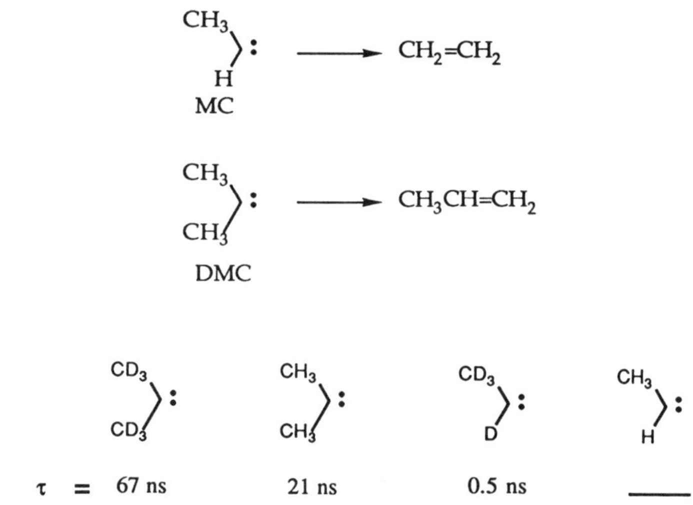

Example 5.

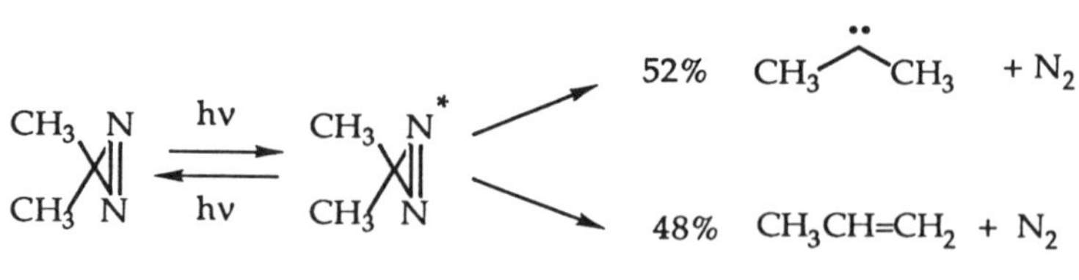

Scheme 28.

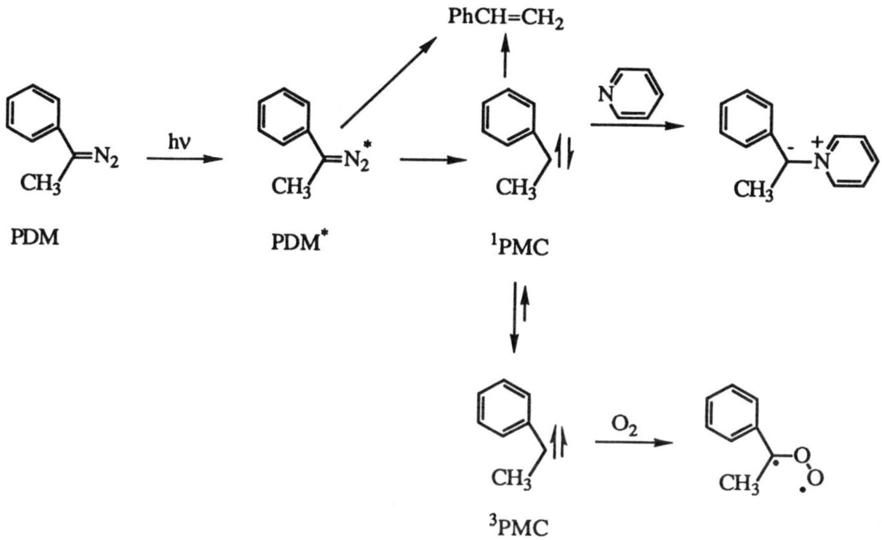

Scheme 29.

(Scheme 28). Thus, the failure to trap DMC in preparative chemistry in good yields arises in large part due to its inefficient formation from a standard precursor.

Sugiyama et al. have studied the triplet ground state species methylphenylcarbene (MPC) by laser flash photolysis of methylphenyldiazomethane (MPD) in the presence of pyridine (Scheme 29).[59] Using the pyridine ylide probe method, these workers were able to deduce the rate constants (at 20 °C) shown in Scheme 30.[59]

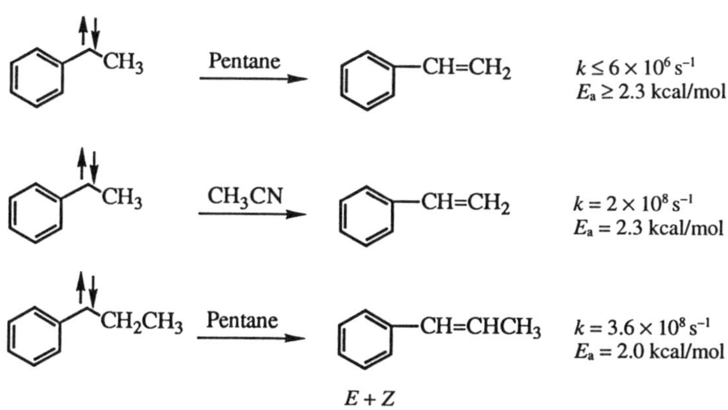

Scheme 30.

Example 6.

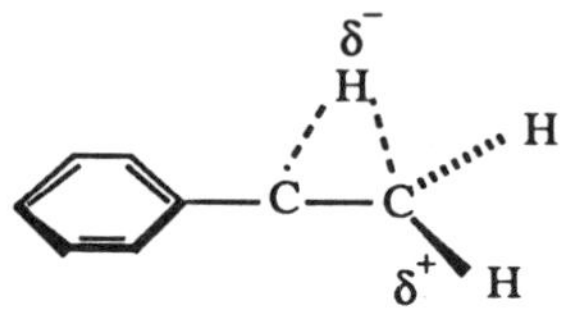

Example 7.

However, polar substituents (X = CF_3, OCH_3, H) in the *para* position on the phenyl ring do not increase the 1,2-hydrogen shift rate constant (Example 6). These observations led Sugiyama et al. to conclude that the transition state to hydrogen migration has considerable charge development (Example 7).[59] Similar conclusions were reported by Goodman[32] and Liu and Bonneau,[30] and were anticipated by theory.[58]

Sugiyama et al. concluded that singlet and triplet methylphenylcarbene are in rapid equilibrium, relative to the rate of their reactions with scavengers. If only the singlet carbene reacts with pyridine, the apparent bimolecular rate constant of reaction of the carbene with pyridine (k_{pyr}^{app}) can be equated to the equilibrium constant K (= [^{1}PMC]/[^{3}PMC]) as

$$k_{pyr}^{app} = k_{pyr}K \tag{5}$$

Typical values of k_{pyr} for singlet carbene reactions with pyridine are $1 \times 10^9\ M^{-1}\ s^{-1}$ and are insensitive to solvent polarity.[9] If k_{pyr} for ^{1}PMC is assumed to be $1 \times 10^9\ M^{-1}\ s^{-1}$, the equilibrium constants (K) and free energy gaps (ΔG) shown in Example 8 can be deduced in heptane at 298 K.

Çelebi et al.[60] have determined $\phi°_{sat}$ for alkylarylcarbene reactions with pyridine (Example 9). The results were very similar to those discussed earlier with alkylchlorocarbenes.[33]

X	*H*	*Br*	*OCH*$_3$	*CF*$_3$
K	0.020	0.0066	0.10	0.0026
ΔG_{298}(kcal/mol)	2.0	3.0	1.4	3.5

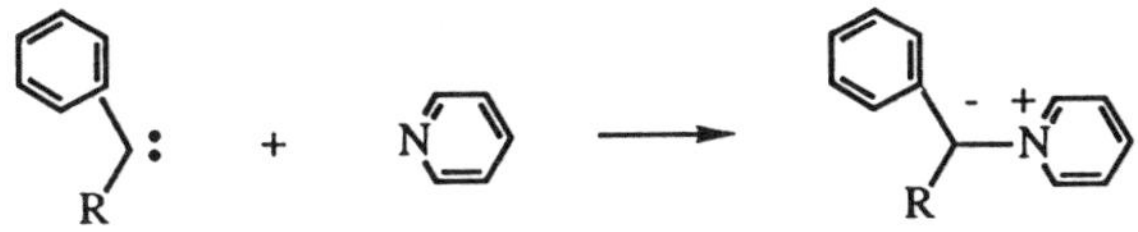

R	$\phi°_{sat}$
tert-butyl-	0.96
	1.0
CD_3	0.86
CH_3	0.72
CH_3CH_2	0.15
	0.06
$(CH_3)_2CH$	0.07

Example 9.

The data demonstrate that hydrogen migration complicates the photochemistry of alkylaryldiazomethanes in the same manner demonstrated for alkylchlorodiazirines and dialkyldiazirines. 1,2-Hydrogen migration dominates the excited-state chemistry of methylphenyl, ethylphenyl, and isopropylphenyldiazomethane (Scheme 31). Leyva[61] has shown that 1,2-carbon migration is important in the

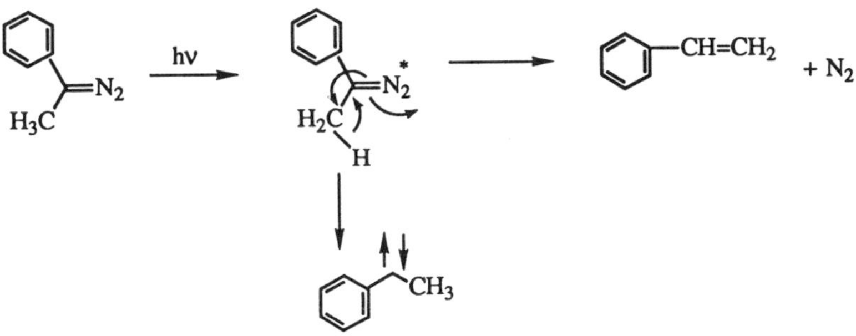

^{1}PMC

Scheme 31.

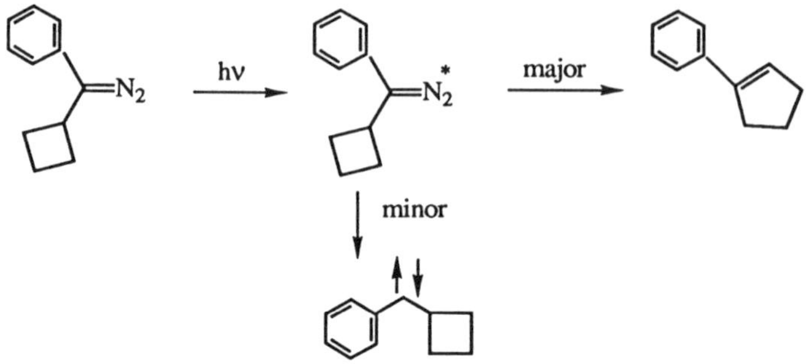

Scheme 32.

excited-state chemistry of the cyclobutyl analog (Scheme 32). Cyclopropylphenyldiazomethane and *tert*-butylphenyldiazomethane apparently do not undergo excited state rearrangements to a significant extent and therefore efficiently produce the alkylarylcarbenes. The results mirror those discussed earlier with alkylchlorodiazirines.[33]

The 351 nm XeF excimer laser line, pyridine as trap, and use of diazirine precursors is a nearly ideal combination for LFP. First, diazirines have absorption maxima near 351 nm, absorption minima between 300–320 nm, and essentially no absorption above 400 nm. This means that an optically dense sample can be prepared at the laser line and that there are monitoring windows at wavelengths both longer and shorter than 351 nm through which ylides can be observed. Second, pyridine does not absorb significantly at 351 nm and as a consequence does not screen the diazirine from the laser pulse at this wavelength. This can be a problem at 308 and 337 nm (XeCl excimer and N_2 lines), where pyridine has some absorbance, and is an insurmountable problem at 249 nm (KrF excimer), where pyridine absorbs strongly. As we have shown, other heterocycles, such as thiophene, produce ylides with different absorption maxima. These alternative carbene "developers" may be preferred in cases where pyridine's basicity is a complicating factor, and they allow some choice of detection wavelength. Thus, the ylide probe method is truly a flexible tool.

III. CARBENES WITH TRIPLET GROUND STATES BUT WITH SINGLET STATES THAT ARE EASILY ACCESSED

As mentioned previously, most hydrocarbon carbenes have triplet ground states. It is widely believed that in such systems photolysis of diazo compounds or diazirines produces the singlet state of the carbene. The bimolecular reactions of the singlet (e.g., insertion into C–H or O–H bonds, stereospecific addition to double bonds, and ylide formation) are very rapid ($k \approx 10^8$–10^{10} M^{-1} s^{-1}), but so is intersystem

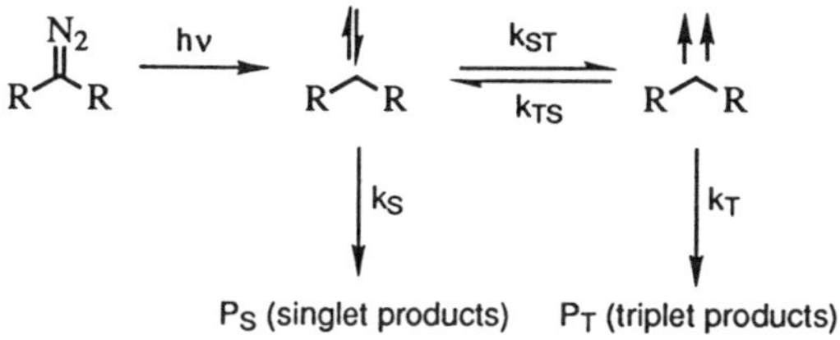

Scheme 33.

crossing ($k \approx 10^9$–10^{10} s^{-1}). If the singlet–triplet energy gap is small ($\Delta H_{ST} \approx 0$–2 kcal/mol), then the triplet can repopulate the low-lying singlet state at ambient temperature. Under these conditions, spin equilibration in both directions is faster than the chemical reaction of either spin state (Scheme 33).

One can realize predominantly singlet- or triplet-derived chemistry from a given spin-equilibrated carbene depending on the choice of quenching agents and the singlet–triplet equilibrium constant $K = k_{ST}/k_{TS}$.

A. 1- and 2-Naphthylcarbene

Prior to the 1976 publication of Closs and Rabinow's landmark paper,[1] nothing was known about the absolute rates of carbene reactions in solution. Closs and Rabinow used conventional flash photolysis to study diphenylcarbene (DPC), a species with a triplet ground state and a long lifetime (> 50 μs) in benzene. The latter point was crucial to their success because many carbenes have lifetimes shorter than 1 μs and cannot be studied with a conventional flash photolysis apparatus.

By the end of the seventies several groups had begun to glimpse the potential of LFP to elucidate carbene dynamics. Diphenylcarbene was restudied by LFP[8] and studies of chlorophenylcarbene (see Section II) were initiated.[13b,16] It was then natural to push on to other well-behaved and readily available diazo compounds such as 1- and 2-naphthyldiazomethane and 9-diazofluorene. Aromatic diazo compounds were chosen for study due to the expectation that the resultant carbenes would have suitable chromophores in the UV–vis spectral region. In addition, a well-established body of literature on classical substituent and product studies was available for many of these systems. What we did not, but probably should have anticipated was that the transient spectroscopy of these systems would be dominated by the transient spectra of secondary species such as ylides and radicals, derived from reaction of the naphthylcarbenes and fluorenylidene with solvent. The overabundance of transients produced by LFP led to some initial confusion as to the identities of the various absorbing species, but, far from being a nuisance, these extra transients were of chemical interest in their own right and ultimately proved to be more convenient for following carbene dynamics than the carbenes themselves.

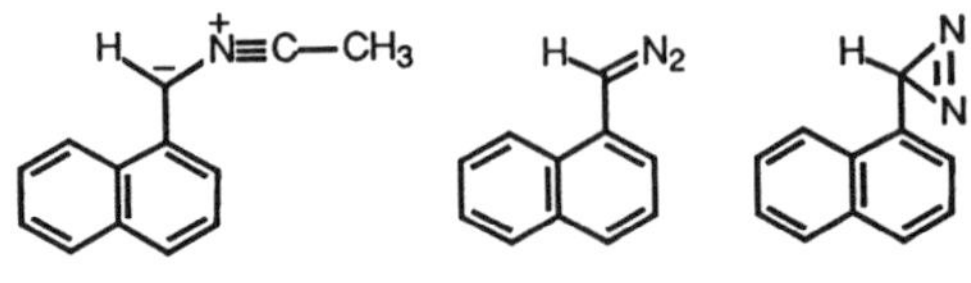

28 1-NDM 29

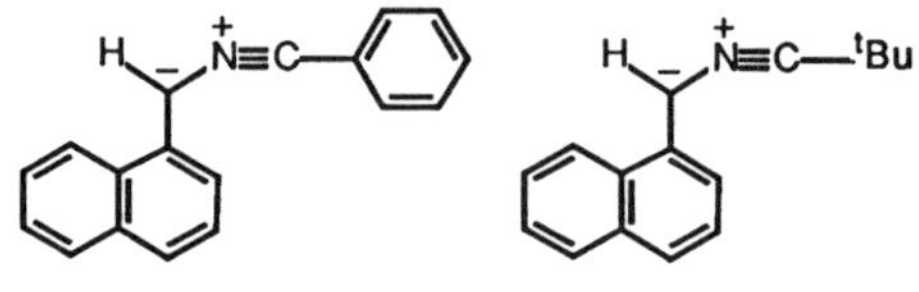

30 31

B. 1-Napthylcarbene–Nitrile Ylides

Of this class of ylides, nitrile ylide **28** has received the most attention.[62] It can be generated as a transient species by LFP of 1-naphthyldiazomethane (1-NDM), or of diazirine **29** in acetonitrile. Nitrile ylides **30** and **31** can be produced by LFP of

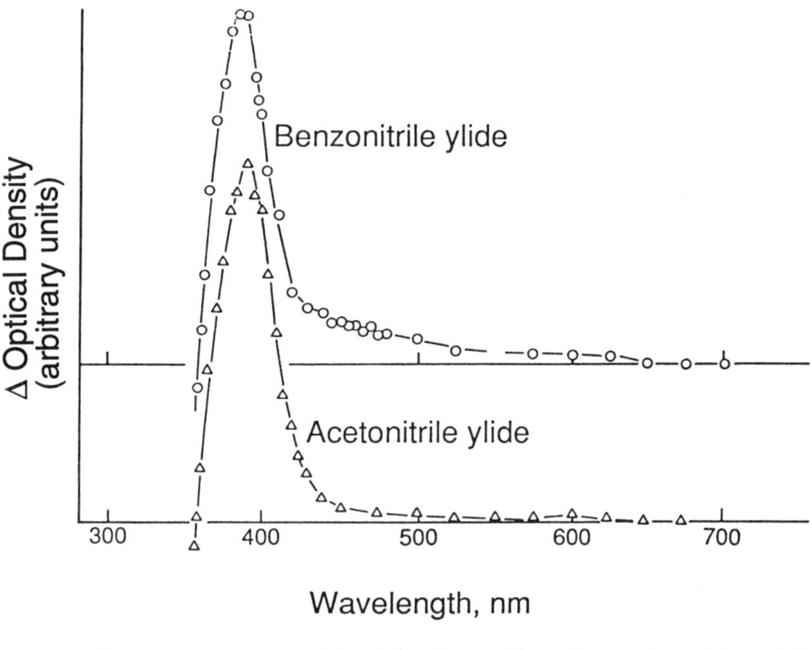

Figure 11. The transient spectra of the ylides derived from the reaction of 1-naphthylcarbene with acetonitrile and benzonitrile. (Reprinted with permission from *Chem. Phys. Lett.* **1983**, *97*, 446.)

1-NDM in benzonitrile and pivalonitrile, respectively. The transient spectra of the three nitrile ylides are very similar. The spectra of **28** and **30** are given in Figure 11. The assignment of the ~390-nm absorbing transient in acetonitrile was secured by the kinetics of its formation, reactivity patterns, chemical analysis, and its independent generation from a non-carbene precursor.

The observed rate of nitrile ylide formation is predicted and found to be dependent on the concentration of acetonitrile. A plot of the observed pseudo first-order rate constant for ylide formation versus acetonitrile concentration is linear (Figure 12). The slope of this plot is $k_Y = (4.6 \pm 0.2) \times 10^5\ M^{-1}\ s^{-1}$ at 300 K, which is the absolute rate constant for the reaction of triplet 1-naphthylcarbene with acetonitrile in benzene. An interpretation of this rate constant will be discussed later.

The chemistry of nitrile ylides has been developed independently by Huisgen, Padwa, and Schmid.[63] These intermediates can be generated by a variety of thermal and photochemical means and are known from competitive trapping studies to react efficiently with electron deficient olefins. The 390-nm absorbing transient produced by LFP of 1-NDM in acetonitrile has a lifetime in excess of several hundred microseconds. The decay of the transient, however, is accelerated by the presence of electron-deficient alkenes and under these circumstances can be described by a single exponential function. Plots of the observed pseudo first-order rate constants

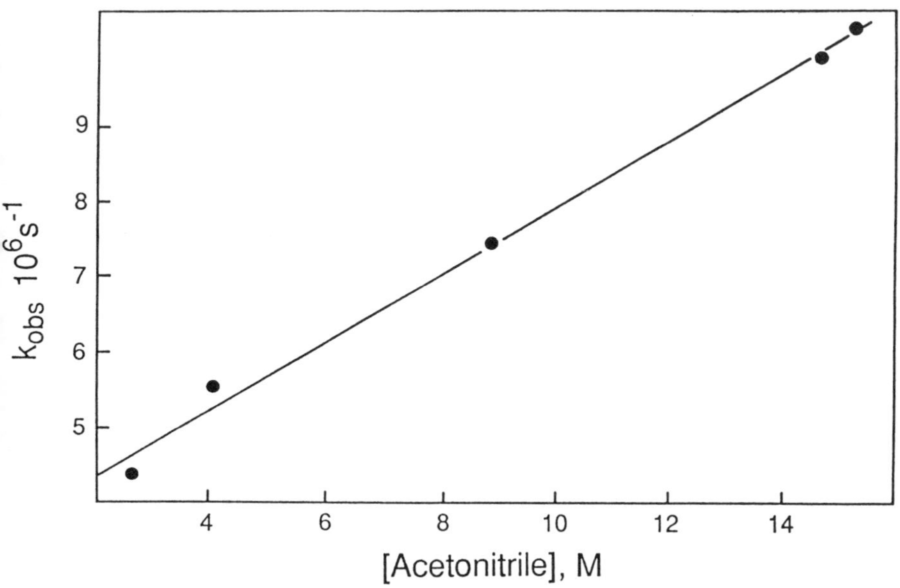

Figure 12. A plot of the observed rate of formation (k_{obsd}) of nitrile ylide **28** versus acetonitrile concentration. (Reprinted with permission from *Chem. Phys. Lett.* **1983**, *97*, 446.)

Table 4. The Absolute Rate Constants for the Reaction of 1-Naphthylcarbene-Acetonitrile Ylide **28** with Various Quenchers[62] in Acetonitrile

Quencher	*Rate Constant* (M^{-1} s^{-1})
Diethylfumarate	$(5.9 \pm 0.2) \times 10^8$
Diethylfumarate[a]	$(5.0 \pm 0.7) \times 10^8$
2,3-Dimethyl-2-butene	$(1.1 \pm 0.2) \times 10^5$
2,5-Dimethyl-2,4-hexadiene	$(3.5 \pm 0.5) \times 10^4$
cis-1,2-Dichloroethylene	$(2.3 \pm 0.2) \times 10^4$
trans-1,2-Dichloroethylene	$(3.8 \pm 0.4) \times 10^4$
Methanol	$(2.4 \pm 3.2) \times 10^3$
tert-Butanol	$(5.5 \pm 3.2) \times 10^3$
Acrylonitrile	$(5.4 \pm 0.2) \times 10^6$
Acrylonitrile[a]	$(4.8 \pm 0.3) \times 10^6$
2,5-Dimethyl-2,4-hexadiene	$(3.5 \pm 0.5) \times 10^4$

Note: [a]Azirine precursor

for decay of the 390-nm absorption band versus alkene concentration are linear with slopes equal to the absolute second-order rate constants for ylide quenching. As expected for a nitrile ylide, the 390-nm absorber react very rapidly with diethylfumarate, moderately fast with acrylonitrile, and quite sluggishly with tetramethylethylene (Table 4).

Perhaps the strongest piece of evidence that the 390-nm absorber is indeed nitrile ylide **28** is its generation from an independent precursor. Upon LFP of azirine **32** in acetonitrile, a transient spectrum is produced that is identical to the spectrum generated from 1-NDM in acetonitrile. Furthermore, the transients derived from azirine **32** and 1-NDM in acetonitrile each react with diethylfumarate with the same absolute rate constant, within experimental error (Table 4).

The final piece of supporting evidence that the 390-nm absorbing species is indeed nitrile ylide **28** comes from product studies. Upon photolysis of diazirine **29** in acetonitrile containing dilute acrylonitrile, adducts are formed that are derived

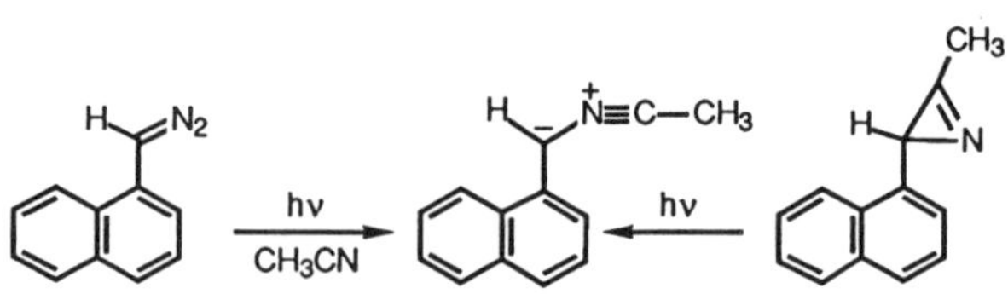

1-NDM **28** **32**

Scheme 34.

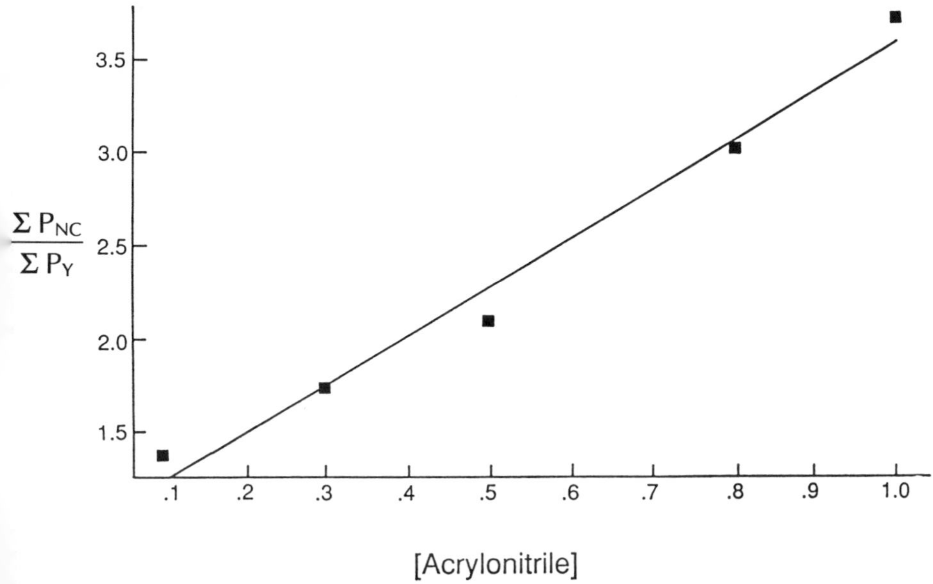

Figure 13. The mole ratio of 1-naphthylcarbene to nitrile ylide–derived products $\Sigma P_{NC}/\Sigma P_Y$ formed as a function of acrylonitrile concentration. (Reprinted with permission from *Tetrahedron Lett.* **1983**, *24,* 3955.)

from the trapping of both 1-naphthylcarbene (1-NC) and ylide **28**. Furthermore, as expected from the mechanism shown in Scheme 35, the ratio of carbene- to ylide-derived adducts increases with increasing concentration of acrylonitrile (Figure 13).

Upon LFP of 1-NDM in acetonitrile the 390-nm absorbing transient is formed in a pseudo first-order process with a time constant of 100 ± 10 ns. This value represents the lifetime of 1-naphthylcarbene in acetonitrile and is deserving of some comment. According to conventional wisdom, photolysis of a diazo compound leads to an excited singlet state that fragments to produce a singlet carbene via spin conservation. Although the rate of intersystem crossing from ^{1}1-NC to ^{3}1-NC has not been measured, the rate constant is probably similar to those of DPC and

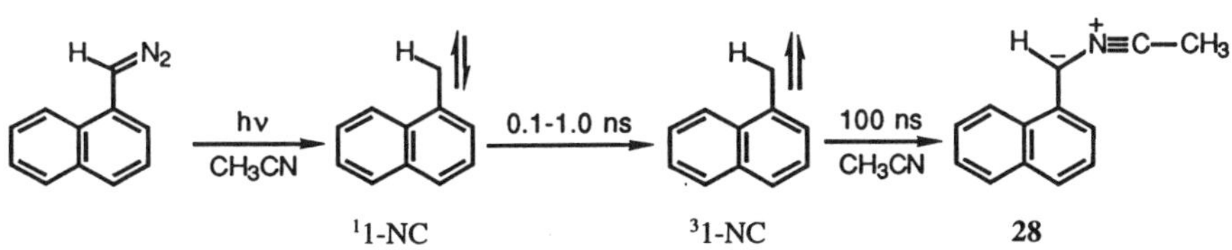

Scheme 35.

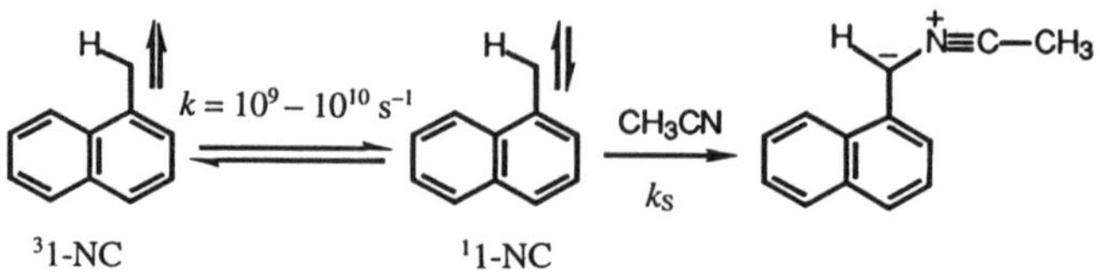

Scheme 36.

fluorenylidene (10^9–10^{10} s^{-1}). Thus, singlet to triplet ISC is almost certainly complete long before the singlet carbene reacts with solvent to produce ylide **28**.

How does ^{3}1-NC react to form ylide **28** in apparent violation of spin selection rules? The simplest answer is that the energy gap between ^{1}1-NC and ^{3}1-NC must be small enough to allow repopulation of the singlet state. Thus, ^{3}1-NC serves as a reservoir for the more reactive but higher energy singlet state (Scheme 36). According to this simple mechanism the rate constant for ylide formation (k_Y) is

$$k_Y = k_S K$$

where k_S is the rate constant of reaction of ^{1}1-NC with acetonitrile and K is the equilibrium constant given by the ratio of [^{1}1-NC]/[^{3}1-NC] at equilibrium. As the upper limit of k_S is diffusion control (~ 1×10^{10} M^{-1} s^{-1}), then

$$k_Y = 4.6 \times 10^5\ M^{-1}\ \text{s}^{-1} = k_S K \leq k_{\text{diff}} K$$

$$4.6 \times 10^{-5} \leq \text{K} \leq 1 \tag{6}$$

and the lower limit of K can be estimated. Therefore, the free energy separation between ^{1}1-NC and ^{3}1-NC must be less than 6 kcal/mol at room temperature. This analysis (first presented by Closs and Rabinow[1] in the context of DPC) has an appealing simplicity but it can be criticized for ignoring the possibility that ^{3}1-NC can react directly with acetonitrile to form ylide **28** by a surface crossing mechanism (cf. Section IV).

In principle the above analysis can be applied to any carbene quenching rate constant k_Q, which can be dissected into singlet and triplet components:

$$k_Q = k_T + k_S K \tag{7}$$

Laser flash photolysis experiments can not resolve k_Q into its component parts. This is usually accomplished by chemical analysis and intuition. The experimental fact is that singlet to triplet intersystem crossing in ^{1}1-NC is faster than interception by acetonitrile to give ylide ($k_{ST} > k_S[CH_3CN]$), and that ^{3}1-NC reacts with singlet carbene quenchers (e.g., acetonitrile, methanol). Thus, singlet and triplet 1-naphthylcarbene behave as a single kinetic entity in acetonitrile.

C. Nitrile Ylides as Probes of the Dynamics of 1-Naphthylcarbene

It is difficult but not impossible to observe ^{3}1-NC by LFP directly. Griffin and Horn[64] were able to detect the transient absorbance of ^{3}1-NC at 355 nm following flash photolysis in an alkane solution. This is a difficult region to monitor because it is near both the absorption maximum of 1-NDM and the laser excitation wavelength (337 nm). The nitrile ylide that absorbs at 390 nm is much more convenient for monitoring the dynamics of the carbene. As mentioned previously, it is possible to obtain the absolute rate constant of reaction of spin equilibrated 1-naphthylcarbene with a quencher (e.g., methanol; Scheme 37) by monitoring the decrease of the transient absorption of the probe ylide as a function of quencher concentration.

Analysis of the yield of nitrile ylide as a function of quencher concentration gives the Stern–Volmer $k_Q\tau$ values (where $\tau = 1/k_{obsd}$, see Eq. 2–4) that are listed in Table 5. Values of k_Q can be obtained by recalling that $\tau \approx 100$ ns in neat acetonitrile. Table 5 reveals that the absolute rate constants for the reaction of a spin-equilibrated mixture of the 1-naphthylcarbenes with alcohols and alkenes are on the order of

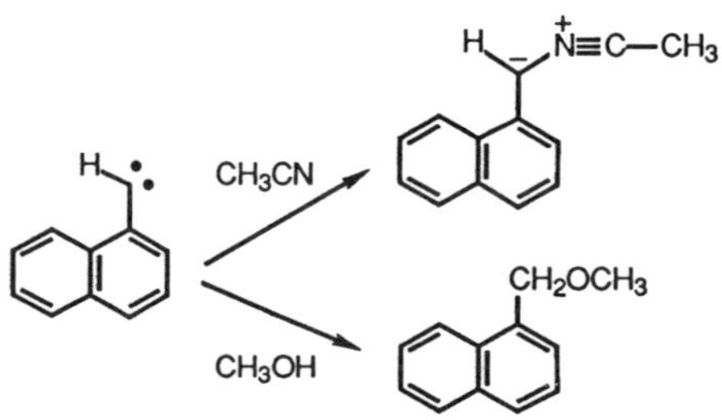

Scheme 37.

Table 5. Stern–Volmer Rate Constants for the Reaction of ^{1,3}1-NC with Quenchers using Acetonitrile Ylide **28** as Probe[62]

Quencher	$k_Q\tau$ (M^{-1})
Methanol	3.3
Methanol-*O*-d	2.2
tert-Butyl alcohol	0.8
Water	2.0
2,3-Dimethyl-2-butene	5.0
2,5-Dimethyl-2,4-hexadiene	2.8
cis-1,2-Dichloroethylene	2.5
trans-1,2-Dichloroethylene	5.9

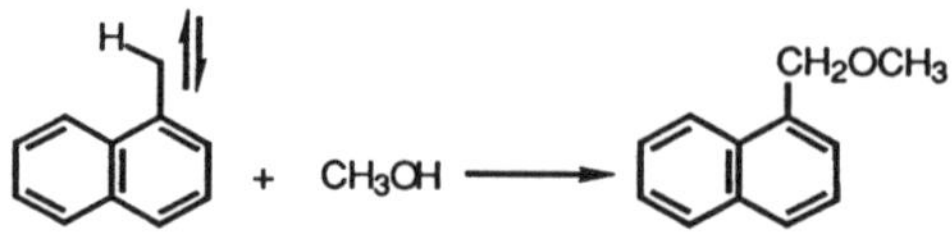

Scheme 38.

10^7 M^{-1} s^{-1}. Some of these quenchers (alcohols, water) are thought to be selective traps for the singlet state of the carbene (Scheme 38) and in these cases $k_Q = k_S K$.

In principle, however, both ^{1}1-NC and ^{3}1-NC can react with alkenes and dienes. In these cases $k_Q = k_T + k_S K$, where k_T is the absolute rate of reaction of ^{3}1-NC with the quencher. Laser flash photolysis can not reveal the relative magnitudes of k_T and $k_S K$, which must be obtained by a traditional product analysis.

D. The 1-Naphthylcarbene–Acetone Ylide

Upon LFP of 1-NDM in acetone a new transient is observed (λ_{max} = 640 nm), which is attributed to carbonyl ylide **33**.[62] The evidence for this assignment is

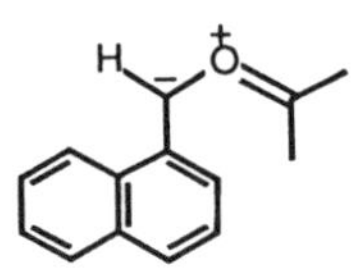

33

somewhat circumstantial. The assignment of this ylide is based on an analogy to the much more thoroughly investigated nitrile ylide systems, and the fact that related ylides are known to absorb in this region of the spectrum.[65] The lifetime of **33** is 2.3 μs at 300 K. The temperature dependence of its decay yields log (A/s^{-1}) = 6.9 and E_a = 1.5 kcal/mol.

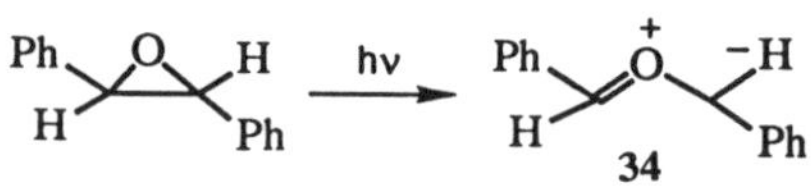

There are two observations worthy of special comment concerning the time evolution of the spectra of ylide **33**. The ylide is clearly formed in a biphasic manner following the laser pulse (Figure 14). The ylide is formed in a very fast process which follows the laser pulse, and in a slower process with a time constant similar to that observed in the formation of nitrile ylide **28** in acetonitrile. We speculate that photolysis of 1-NDM leads to fragmentation to ^{1}1-NC, which can either react immediately (k_S) with solvent (the fast component to formation of **33**) or intersys-

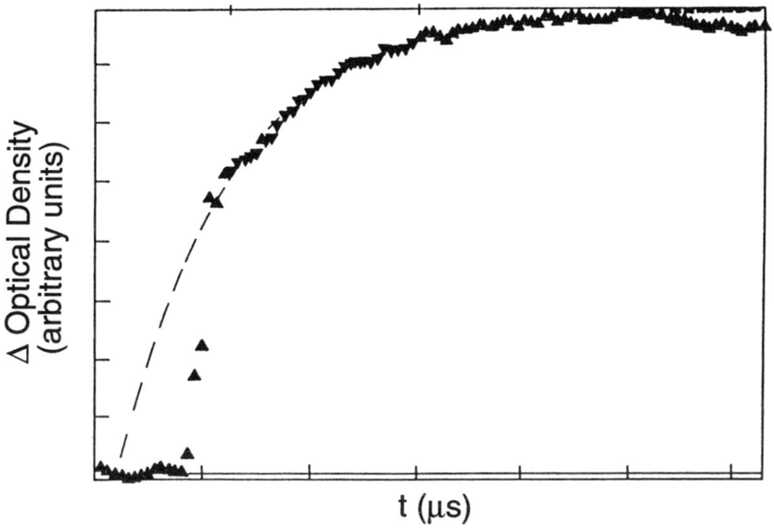

Figure 14. Biphasic formation of ylide **33** produced upon LFP of 1-NDM in acetone.

tem cross (ISC) to ^{3}1-NC (Scheme 39). The slow component to formation of **33** is the result of reconversion of ^{3}1-NC to ^{1}1-NC followed by capture ($k_S K$). The biphasic signal is due to the fact that in acetone, k_S [acetone] and k_{ST} are comparable, whereas in acetonitrile k_{ST} must be significantly larger than k_S [acetonitrile]. The second interesting feature is that the spectrum of **33** changes subtly with time. Das[65] has seen similar effects with ylide **34** and has attributed this to relaxation of the ylide geometry from its initially produced value. Several conformations are possible for **33** and there is no reason to expect that it is formed initially with the equilibrium conformational distribution.

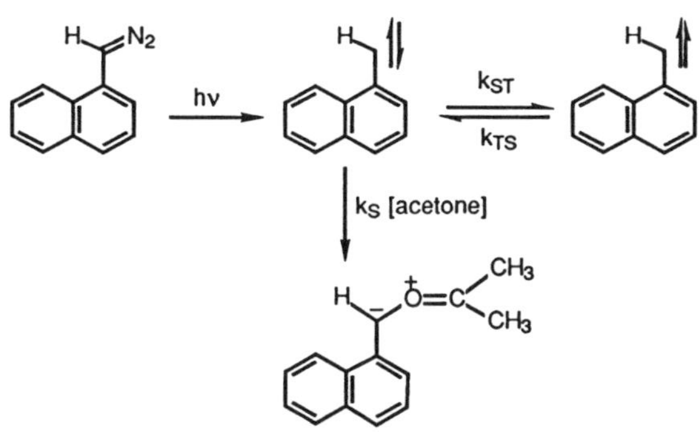

33

Scheme 39.

E. Miscellaneous Ylides Derived from 1-Naphthylcarbene

Laser flash photolysis[62] of 1-NDM in triethylamine produces two new transient absorption maxima at 363 and 485 nm (Figure 15). The short wavelength absorption is clearly that of radical **35** whereas the long wavelength band was tentatively assigned to ylide **36**. The putative ylide **36** is much shorter lived ($\tau = 380$ ns, 300 K) than radical **35**. The decay of **36** followed first-order kinetics between 225–294 K with log (A/s^{-1}) = 9.28 and $E_a = 3.9$ kcal/mol.

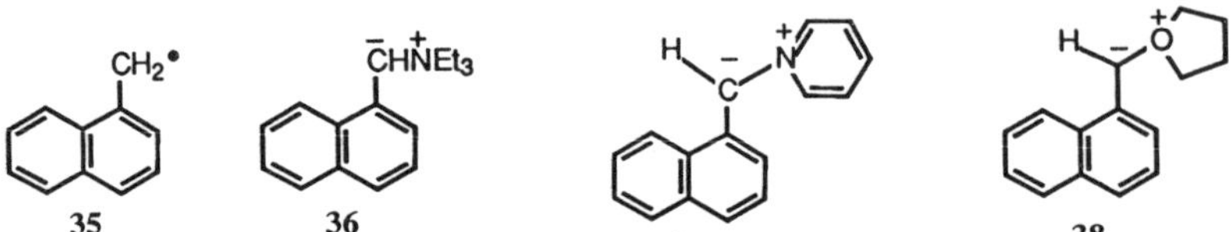

35 **36** **37** **38**

Laser flash photolysis[62] of 1-NDM in pyridine produced a new transient with λ_{max} = 535 nm and a lifetime in excess of 100 μs. This species reacted with diethylfumarate with a rate constant of $(5.3 \pm 1.0) \times 10^5$ M^{-1} s^{-1}. The 535-nm absorbing species was tentatively attributed to pyridinium ylide **37**. Laser flash photolysis of 1-NDM in tetrahydrofuran did not produce any new transient signals that could be ascribed to an oxonium ylide such as **38**.

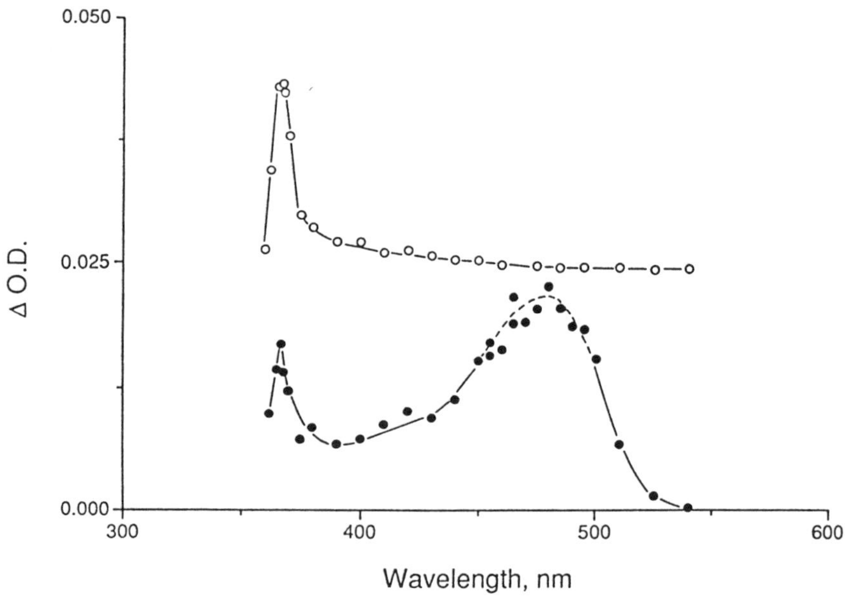

Figure 15. The transient absorption spectra produced upon LFP of 1-NDM in triethylamine. Spectra were recorded 100 ns (●) and 100 ms (○) after the laser pulse. The latter spectrum is offset by Δ O.D. = +0.025. (Reprinted with permission from *J. Am. Chem. Soc.* **1986**, *108*, 3928.)

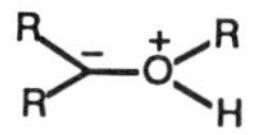

39

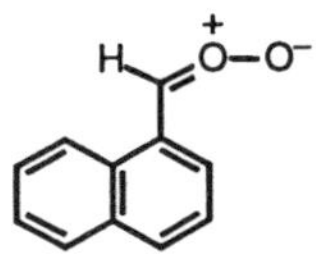

40

It is not clear whether the lifetime of **38** is too short ($\tau < 10$ ns) to permit its detection, whether it has a weak or inaccessible chromophore, or whether it is simply not formed. This is perhaps a significant omission as many workers (cf. Section IV) have postulated reversibly formed ylides (e.g., **39**) in the reaction of carbenes with alcohols and ethers.

Laser flash photolysis 1-NDM in aerated cyclohexane produces a new broadly absorbing transient (Figure 16). The rate of formation of the transient was first order in oxygen concentration [$k_{O_2} = (3.5 \pm 0.7) \times 10^9\ M^{-1}s^{-1}$]. These facts support the assignment of the 435-nm absorbing species as ylide **40**, a carbonyl oxide. The first half-life of the ylide was only a few microseconds. The ylide decayed by predominantly second-order kinetics with $2k/\varepsilon \approx 3.2 \times 10^6\ \varepsilon M^{-1}\ s^{-1}$, where ε is the molar extinction coefficient of **40**. The eventual product formed by the decay of **40** is 1-naphthaldehyde.

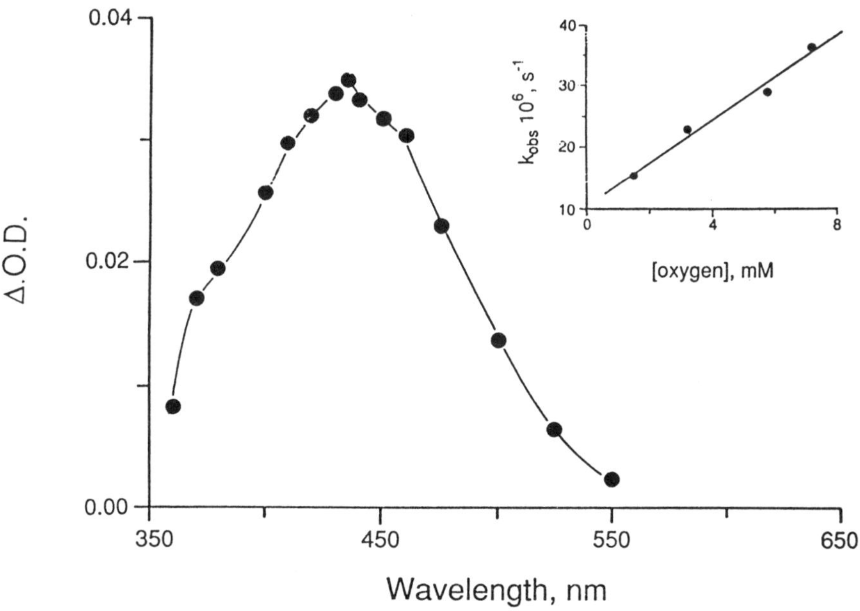

Figure 16. The transient absorption spectrum of carbonyl oxide **40** produced upon LFP of 1-NDM in aerated cyclohexane. (Reprinted with permission from *J. Am. Chem. Soc.* **1986**, *108*, 3928.)

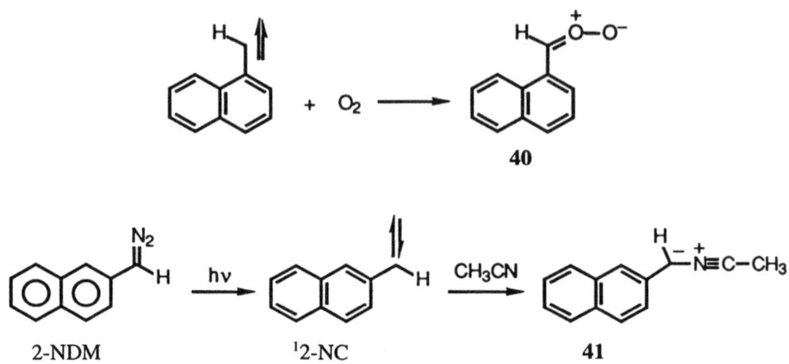

Scheme 40.

Carbonyl ylide **40** is formed directly from ^{3}1-NC in a spin-allowed reaction ($k_{O_2} = k_T$) (Scheme 40). According to theory, a collision between two triplet molecules will lead to a singlet state product with a probability of 1:9.[66]

F. 2-Naphthylcarbene Ylides

Relatively little work has been performed on ylides derived from 2-naphthylcarbene (2-NC). Laser flash photolysis of 2-NDM in acetonitrile produces a new transient λ_{max} = 372 nm that is attributed to nitrile ylide **41** (Scheme 40).[67] The absolute rate constant of reaction of spin-equilibrated 2-NC with acetonitrile is $(5.28 \pm 0.1) \times 10^{5}\ M^{-1}\ s^{-1}$. Horn and Chateauneuf[67] have found that it is possible to monitor the kinetics of ^{3}2-NC directly at 362 nm, so that there was no necessity to use ylides such as **41** as dynamical probes in this instance.

G. Fluorenylidene

Laser flash photolysis studies of fluorenylidene (Fl) are dominated by the reactions of this carbene with Lewis base solvents to form ylides.[68] The chemistry of fluorenylidene greatly resembles that of 1-NC. In fact, the two systems were studied concurrently by various groups and this work led to a common mechanistic viewpoint.

After some initial confusion, the dynamics of Fl in acetonitrile became well understood. Laser flash photolysis of 9-diazofluorene (DAF) in acetonitrile leads to fragmentation to form ^{1}Fl and molecular nitrogen. The singlet carbene has not yet been detected by spectroscopic techniques. Following a picosecond laser pulse, the absorption spectrum of ^{3}Fl grows in with a half-life of 0.28 ns ($k_{ST} = 2.46 \times 10^{9}$ s^{-1}). The transient spectrum of ^{3}Fl is in good agreement with a persistent spectrum of the carbene obtained in 2-methyltetrahydrofuran at 10 K. The lifetime of ^{3}Fl in acetonitrile is 17 ns and it reacts with solvent to form nitrile ylides (λ_{max} = 400 nm) and the 9-fluorenyl radical (λ_{max} = 500 nm) (Scheme 41). The lifetime of ^{3}Fl in acetonitrile-d_3 is increased to 28 ns by the kinetic isotope effect at room temperature.

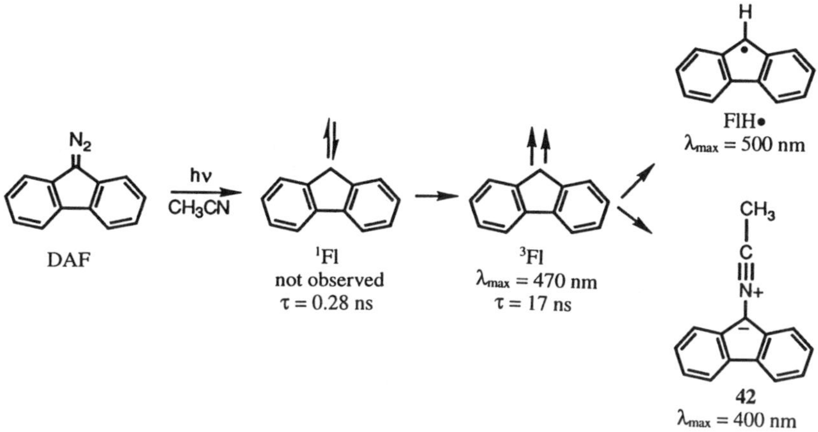

Scheme 41.

Association of the 400-nm absorbing species with nitrile ylide **42** is based upon evidence similar to that developed in the 1-naphthylcarbene–acetonitrile system. First, the lifetime of putative ylide **42** is longer than hundreds of microseconds in neat acetonitrile. The presence of electron-deficient alkenes substantially shortens the lifetime of **42** and yields the absolute rate constants of ylide quenching listed in Table 6. As expected, the more electron deficient the alkene the more effectively it traps ylide **42**.

Among the products isolated from preparative photolysis of 9-DAF in acetonitrile is the known azirine **43**, a compound that is easily visualized as arising from cyclization of ylide **42**.

Perhaps the most compelling evidence requiring association of the 400-nm absorption band with nitrile ylide **42** is the fact that LFP of azirine **43** produces the same transient spectrum observed upon LFP of 9-DAF in acetonitrile. Furthermore, within experimental error the 400-nm transient generated from either precursor reacted with dipolarophiles with identical rate constants (Table 6).

It is possible to observe other nitrile ylides as well. Laser flash photolysis of 9-DAF in benzonitrile and pivalonitrile produces **44** and **45**, respectively. The

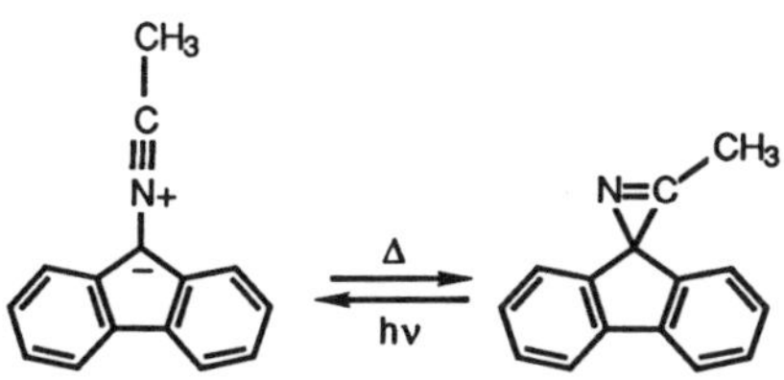

42 **43**

Table 6. Absolute Rate Constants for the Reaction of Ylide **42** (Derived from the Reaction of Fluorenylidene with Solvent) with Various Quenchers

Quencher[a]	*k* (M^{-1} s^{-1})	*log A* (M^{-1} s^{-1})	E_a (*kcal/mol*)
Maleic anhydride	9.8×10^8	—	—
Dimethyl fumarate	1.7×10^8	6.94	–1.65
Diethyl fumarate	1.6×10^8	6.66	–2.04
Fumaronitrile	5.7×10^7	7.47	–0.37
Fumaronitrile[b]	1.5×10^7	—	—
Diethyl maleate	1.9×10^6	6.28	0.0
2,3-Dimethyl-2-butene	1.3×10^5	—	—

Olefin	*k* (M^{-1} s^{-1}) *DAF*[d]	*k* (M^{-1} s^{-1}) *azirine 3*[e]
Maleic anhydride	9.8×10^8	*f*
3,4-Dimethyl maleic anhydride	6.2×10^5	5.3×10^5
Diethyl fumarate	9.1×10^7	1.0×10^8
Diethylketomalonate	6.3×10^8	4.5×10^8
Fumaronitrile	4.7×10^7	*f*
Methyl acrylate	2.0×10^6	*f*
Acrylonitrile	7.1×10^5	*f*

Notes: [a]Data of Griller et al.[68b]
[b]Pivalonitrile–fluorenylidene ylide.
[c]Data of Grasse et al.[68a]
[d]Ylide generated from photolysis of DAF and subsequent reaction of fluorenylidene with CH_3CN. Estimated error is ±10% of the reported value.
[e]Ylide generated directly from photolysis of azirine **43** in CH_3CN. Estimated error is ±10% of the reported value.
[f]Not determined.

absorption spectra of ylides **42**, **44** and **45** are all rather similar; however, the three ylides react with a common dipolarophile at significantly different rates. The fact that transient absorption maxima at 400 nm were seen only in nitrile solvents and not in other solvents such as alkanes, halocarbons, and aromatics was at the time an important clue that these transients were due to ylides rather than to singlet or triplet fluorenylidene.

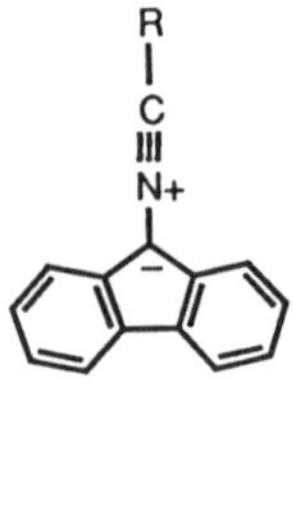

42: R = CH_3
44: R = C_6H_5
45: R = *tert*-butyl

The formation of nitrile ylide **42** is described by a simple exponential function, in contrast to the biphasic fast and slow growth seen with the 1-naphthylcarbene–acetone ylide. Thus, ^{1}Fl to ^{3}Fl intersystem crossing is faster than the reaction of ^{1}Fl with acetonitrile:

$$k_{ST} = 2.46 \times 10^9 \text{ s}^{-1} > k_S[\text{CH}_3\text{CN}] \tag{8}$$

If ^{3}Fl must revert to ^{1}Fl (as per ^{1,3}1-NC) before it can react with solvent to produce an ylide, then $k_{obsd} = k_S K$, and as τ (CD_3CN) = 28 ns, then 1/28 ns = $k_{obsd}[CD_3CN]$. The concentration of neat acetonitrile-d_3 is 19.1 *M*, so $k_S K = 1.87 \times 10^6\ M^{-1}\ s^{-1}$. Since the upper limit of k_S is diffusion controlled ($k_S \leq 5 \times 10^9\ M^{-1}\ s^{-1}$), the equilibrium constant for ^{1}Fl is bracketed at 300 K: $3.74 \times 10^{-4} < K < 1$. Thus, the equilibrium population of ^{1}Fl may be as much as 10 times larger than that of a singlet 1-naphthylcarbene. This line of reasoning allows us to estimate the free energy difference between singlet and triplet fluorenylidene as less than 4.7 kcal/mol at 300 K. Triplet fluorenylidene reacts more rapidly with methanol ($k_{obsd} = 8.6 \times 10^8\ M^{-1}\ s^{-1}$; see Table 7) than with acetonitrile. If one assumes that the mechanism of this reaction involves repopulation of ^{1}Fl, one can calculate that ΔH_{ST} in fluorenylidene may be only 1.1 kcal/mol.[68a]

The much shorter lifetime of ^{3}Fl than ^{3}1-NC in acetonitrile is seen to result from the smaller singlet–triplet free energy gap in the diarylcarbene. ^{3}Fl reacts more rapidly than ^{3}1-NC because the more reactive singlet state of this species is more accessible.

The fluorenylidene–acetonitrile ylide has been used extensively as a probe of the spin equilibrated ^{1,3}Fl mixture. Schuster and co-workers[69] have convincingly demonstrated by classical chemical means that spin equilibration in both directions is faster than the chemical reactions of either spin state. In the case of 1-naphthylcarbene, the ylide probe method was necessitated due to the unfavorable absorption maximum of ^{3}1-NC. This is not the case with ^{3}Fl (λ_{max} = 470 nm); with this carbene the ylide probe is necessitated by the short lifetime of the carbene. As ^{3}Fl decays with a half-life of only 17 ns in acetonitrile, and as a typical nanosecond-pulsed laser has a width of 8–10 ns, the addition of a quencher makes the decay of ^{3}Fl too fast to measure. It is possible, however, to measure the kinetics of disappearance of ^{3}Fl directly in hexafluorobenzene where the lifetime of ^{3}Fl is longer (90 ns).

The presence of a carbene quencher affects the ylide signal in two ways: the magnitude of the signal is reduced and the rate of formation of the signal is increased. One cannot measure the acceleration of the rate of formation of the ylide signal in acetonitrile for the aforementioned reasons. Thus, the absolute rate constants reported in Table 7 were determined by the Stern–Volmer method by measuring the changes in the yield of ylide signal as a function of carbene quencher. The data demonstrate the ^{1,3}Fl is substantially more reactive than ^{1,3}1-NC.

Schuster and co-workers have also used Stern–Volmer quenching of acetonitrile ylide **42** to obtain the absolute rate constant of reaction of spin equilibrated ^{1,3}Fl with alcohol, amine, and thiol nucleophiles.[69] The results are given in Table 8,

Table 7. Stern–Volmer Rate Constants for the Reaction of Fluorenylidene with Various Quenchers in Acetonitrile-d_3[68b]

Quencher	$k_Q\tau$			
CH_3OH	32.6 ± 3.8			
CH_3OD	23.5 ± 1.3			
CH_3CO_2H	21.6 ± 3.8			
	k_Q (M^{-1} s^{-1}, 300K)	E_a	*log A*	*Range (K)*
2,3-Dimethylbut-2-ene	$(5.1 \pm 0.6) \times 10^8$	1.93	10.23	265–303
Diethy fumarate	$(1.5 \pm 0.3) \times 10^9$	0.29	9.37	233–334
	k (M^{-1} s^{-1})			
Methanol	8.6×10^8			
Ethanol	7.3×10^8			
2-Propanol	5.2×10^8			
Diphenyl fumarate	15×10^8			
1,3-Pentadiene	10×10^8			
Styrene	9.4×10^8			
trans-trans-2,4-Hexadiene	7.8×10^8			
2,4-Dimethyl-2,4-hexadiene	6.7×10^8			
cis-2-Pentene	1.3×10^8			
trans-3-Pentene	1.4×10^8			
3,4-Dimethylmaleic anhydride	4.7×10^8			
2-Methyl-2-butene	3.5×10^8			
Cyclohexene	3.2×10^8			
cis-β-Methylstyrene	1.9×10^8			
trans-Dichloroethylene	0.28×10^8			
cis-Dichloroethylene	0.17×10^8			

Table 8. Absolute Rate Constants of the Reaction of Fluorenylidene with Nucleophiles in Acetonitrile[69]

Nucleophile	*Concentration*[a] *(M)*	*Trap yield %*[b]	k_{nuc}^{obs} (M^{-1} s^{-1})	k_{nuc}^{H}/k_{nuc}^{D} [c]	P_H/P_D [d]
CH_3OH	0.05 (0.5)	90	8.6×10^8	1.8 ± 0.04	2.3 ± 0.1 (1.4)
CH_3OD	0.1 (1.0)		4.7×10^8		
$(CH_3)_3COH$	0.4	87	1.6×10^8	3.8 ± 0.5	3.5 ± 0.3
$(CH_3)_3COD$	1.6		0.42×10^8		
$(CH_3)_3CNH_2$	3.5	60	6.3×10^7	0.94 ± 0.2	2.5 ± 0.3
$(CH_3)_3CND_2$	3.8		7.0×10^7		
$(CH_3)_3CSH$	0.56	78	1.1×10^9	0.92 ± 0.13	2.3 ± 0.2
$(CH_3)_3CSD$	0.55		1.2×10^9		
$CH_3(CH_2)_3SH$		55	1.4×10^9		
CH_3OH	0.1				
CH_3OD	0.4				
$(CH_3)_3COH$	0.4			2.4 ± 0.2	
$(CH_3)_3COD$	1.6				

Notes: [a]Concentration of reagents used in the determination of the product isotope effects.
[b]Traps are identified as the ether, sulfide, and secondary amine formed from the alcohol, thiol, and primary amine, respectively.
[c]Ratio of LFP determined rate constants for H- and D-containing substrates.
[d]Rate constant rates determined from competition product studies.

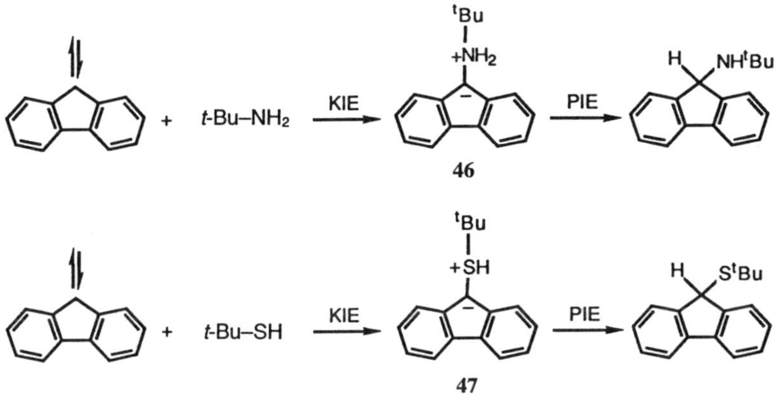

Scheme 42.

which includes both kinetic isotope effects and product isotope effects. The product isotope effects were determined by photolysis of 9-DAF in a mixture of the nucleophile and its deuterated modification, and analysis of the resulting mixture of adducts by ^{1}H NMR or mass spectroscopy. In the case of *tert*-butylamine and *tert*-butylthiol it was found that the kinetic isotope effects are significantly smaller than the product istope effects (PIE). In fact, there is essentially no kinetic isotope effect observed with either of these nucleophiles. The data show that the rate-determining and product-determining steps in the reactions of $^{1,3}Fl$ with these nucleophiles are not the same, and thus there must be intermediates present in these reactions, most likely ylides **46** and **47** (Scheme 42).

The data are consistent with *irreversible* formation of ylides **46** and **47**; however, a different conclusion was reached in the reaction of ^{1}Fl with alcohols. Schuster and co-workers found that the addition of methanol depresses the product isotope effect in the reaction of ^{1}Fl with a *tert*-butanol-*tert*-butanol OD mixture.[69] The PIE with *tert*-butanol was 3.5 ± 0.3 in the absence of methanol but only 2.4 ± 0.2 in the presence of methanol, which was interpreted as evidence of a *reversibly* formed alcohol ylide intermediate **48** (Scheme 43).

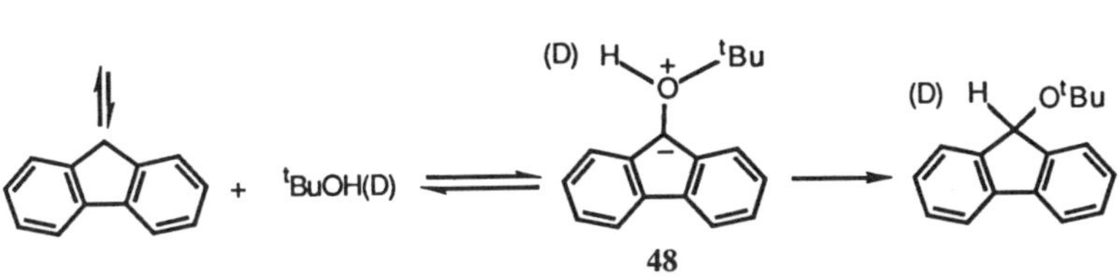

Scheme 43.

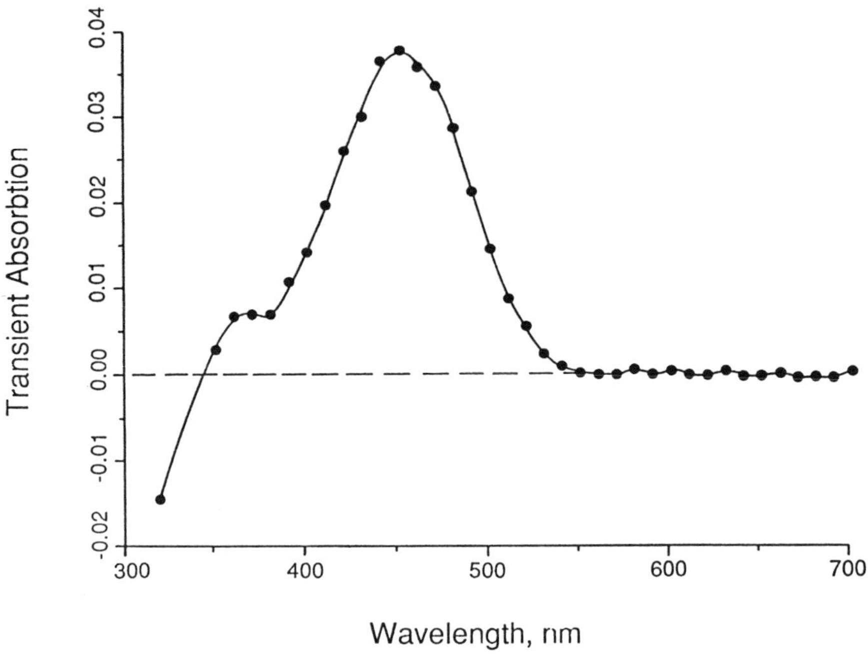

Figure 17. The transient absorption spectrum of fluorenone oxide **49** produced by LFP of 9-DAF in aerated fluorocarbon. (Reprinted with permission from *J. Am. Chem. Soc.* **1985**, *107*, 4617.)

H. Fluorenone Oxide

Laser flash photolysis of 9-DAF in oxygen-saturated 1,1,2-trichloro-trifluoroethane produces a new transient absorption at 450 nm (Figure 17).[70] The transient is attributed to carbonyl oxide **49** (Scheme 44) because its rate of formation is first order in the concentration of oxygen (Figure 18). The slope of this plot is $k_{O_2} = [(1.4 \pm 0.2) \times 10^9\ M^{-1}\ s^{-1}]$, and the reciprocal of the intercept of the plot yields the lifetime of ^{3}Fl in this solvent in the absence of oxygen, which in

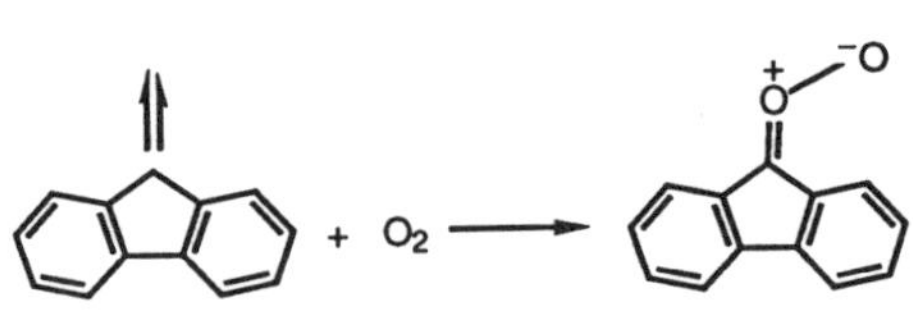

Scheme 44.

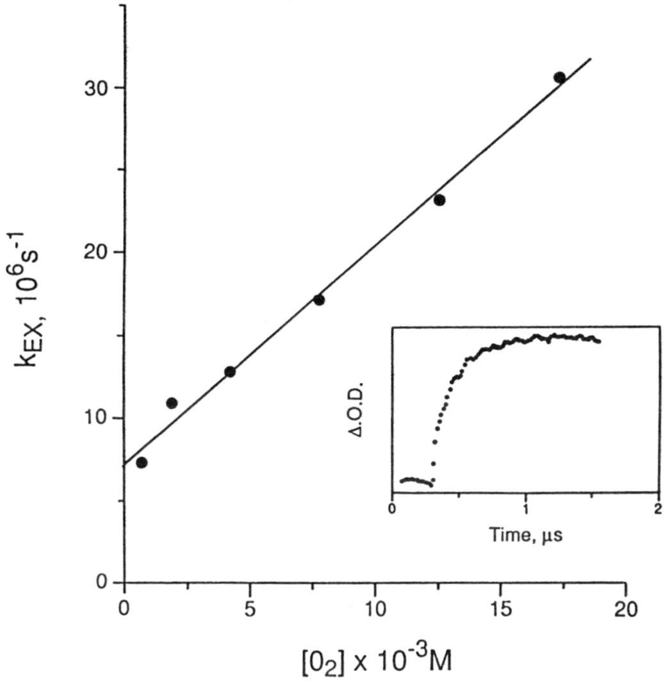

Figure 18. A plot of the observed pseudo first-order rate constant for the reaction of ^{3}Fl with oxygen. (Reprinted with permission from *J. Am. Chem. Soc.* **1985**, *107*, 4617.)

this case is 290 ns. The lifetime of ylide **49** in the absence of quenchers exceeds 100 μs.

Carbonyl oxide **49** can also be formed by the reaction of singlet oxygen with 9-DAF. In this experiment a solution of methylene blue, 9-DAF, and oxygen was excited with a 587-nm pulse of a flash-lamp-pumped dye laser. All of this light was absorbed by the methylene blue, which is known to be a sensitizer of singlet oxygen. These pulses produced a spectrum of fluorenone oxide identical to that produced by direct LFP of 9-DAF (587-nm LFP of 9-DAF alone does not produce ylide **49**). The formation of **49** produced by sensitized LFP was first order in the concentration of diazo compound. The quenching plot revealed that the rate of reaction of singlet oxygen with 9-DAF is $(6.6 \pm 0.4) \times 10^7$ M^{-1} s^{-1}. In this application, ylide **49** is actually a probe of the dynamics of singlet oxygen.

It is also possible to use fluorenone oxide **49** as a probe of the dynamics of fluorenylidene. In these experiments the yield of ylide **49** produced at a constant concentration of oxygen is measured upon LFP of 9-DAF as a function of carbene quencher concentration. Stern–Volmer analysis of the data yields the absolute rate constants of spin-equilibrated ^{1,3}Fl given in Table 9.

Table 9. Absolute Rate Constants of the Reaction of ^{1,3}Fl with Various Quenchers (Obtained using Fluorenone Oxide **49** as Probe in Freon-113 at 300 K)[70]

Substrate	k (M^{-1} s^{-1})
Diethyl ether	$(1.3 \pm 0.2) \times 10^8$
1,4-Cyclohexadiene	$(1.1 \pm 0.2) \times 10^9$
β-Pinene	$(1.2 \pm 0.4) \times 10^9$
Isooctane	$(9 \pm 1) \times 10^6$

Table 10. The Absolute Rate Constants of the Reaction of Fluorenone Oxide with Various Aldehyde Quenchers[70]

Substrate	k (M^{-1} s^{-1})
Acetaldehyde	$(1.21 \pm 0.04) \times 10^9$
Phenylacetaldehyde	$(5.0 \pm 1.0) \times 10^8$
n-Octanal	$(4.5 \pm 0.5) \times 10^7$
5-Norbornene-2-carboxaldehyde	$\sim 1.5 \times 10^7$

Table 11. The Lifetime of Fluorenone Oxide **49** as a Function of Solvent[70]

Solvent	τ (μs)
CH_2Cl_2	9
Freon-113	370
Cyclohexane	≥ 500
Acetone	26
Benzene	≥ 800
Toluene	270
Chlorobenzene	≥ 500
CCl_4	35
2-Propanol	≥ 55
Acetonitrile	35

Carbonyl oxides have long been postulated to be important intermediaries in ozonolysis reactions and are often referred to as "Criegee intermediates."[71] Scaiano and co-workers[70] have monitored the dynamics of the reaction of fluorenone oxide with aldehydes; the absolute rate constants are given in Table 10. The lifetimes of **49** in various solvents are given in Table 11.

I. Fluorenylidene–Carbonyl Ylides

Laser flash photolysis of 9-DAF in acetone produces the transient spectrum of the 9-fluorenyl radical and carbonyl ylide **50**.[72] In acetone-d_6 fluorenyl radical formation has been almost completely suppressed (Figure 19). The transient spectrum of **50** is consistent with the spectra of other carbonyl ylides produced in solution[65] and in low temperature matrices by photolysis of epoxides and oxadiazolines.[73]

A number of other ketones produce carbonyl ylides upon LFP of 9-DAF (Table 12). Ylides could not be detected upon LFP of 9-DAF in diisopropyl ketone, acetaldehyde, and ethyl acetate.

It was possible to obtain the rate constant of reaction of ^{1,3}Fl with acetone in acetonitrile-d_3 by measuring the rate of growth of the ylide at 640 nm as a function of acetone concentration. Also, it was possible to obtain this value by Stern–Volmer quenching of the acetonitrile-d_3 derived ylide, **42**, by acetone. In this approach a nitrile ylide is used as a probe of the dynamics of a carbonyl ylide–forming reaction. The absolute rate constant for reaction of ^{1,3}Fl with acetone is $1 \times 10^7\ M^{-1}\ s^{-1}$.

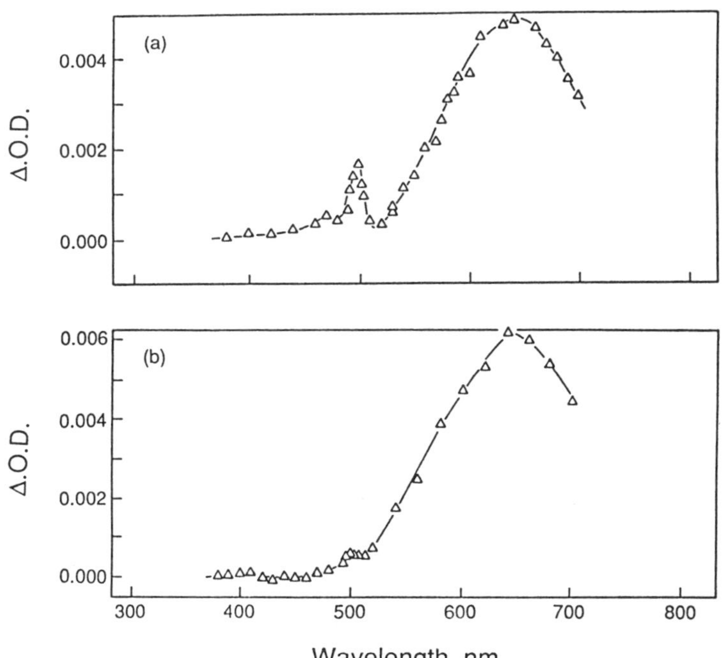

Figure 19. **A.** The transient spectrum of the fluorenylidene–acetone derived ylide and the 9-fluorenyl radical. **B.** The transient spectrum of the fluorenylidene–acetone-d_6 derived ylide. (Reprinted with permission from *J. Am. Chem. Soc.* **1982**, *104*, 663.)

Table 12. Ketones That React with Fluorenylidene to Form Ylides.[72]

Ketones	λ_{max} (nm)	$\tau_Y{}^a$ (μs, 27 °C)	*Intensity ratio (ylide/fluorenyl)*	$10^{-7}k_Q$ ($M^{-1}\,s^{-1}$)
Acetone	640	3.4	3.1	1.0 ± 0.5
				1.3 ± 0.1
Acetone-d_6	640	4.3	12.4	0.5 ± 0.2
				0.99 ± 0.01
Methyl ethyl ketone	650	1.2	0.5	
Methyl *tert*-butyl ketone	640		1.0	
Cyclobutanone	620	3.6	0.6	
Cyclopentanone	640	6.7	1.1	
Cyclohexanone	650	1.9	0.3	
Cycloheptanone	680	5.1	0.6	

Note: ${}^a\tau_Y$ = Ylide lifetime

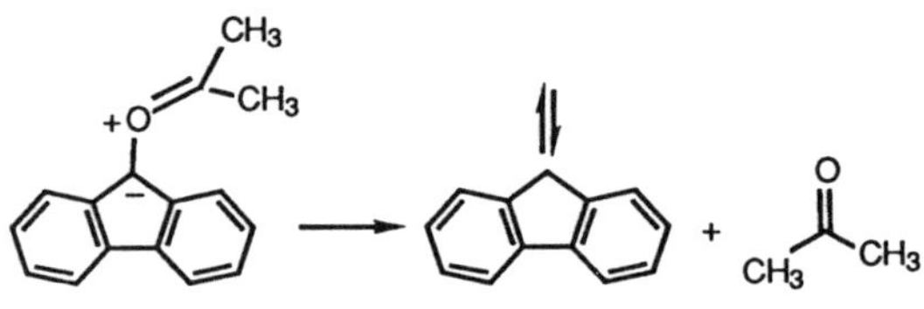

Scheme 45.

The lifetime of the carbonyl ylide **50** (Scheme 45) was measured in neat acetone and in acetonitrile containing acetone or acetone-d_6. The ylide lifetime in acetonitrile is not a function of the concentration of acetone. This indicates that the reverse reaction, and the reaction of the ylide with acetone, are probably unimportant.

The lifetime of **50** in acetone-d_6 was measured between 270–329 K to give Arrhenius parameters of $A = 10^{13.26 \pm 0.4}\ s^{-1}$ and $E_a = 10.96 \pm 0.12$ kcal/mol. The absolute rate constants for intermolecular capture of various carbonyl ylides derived from fluorenylidene are given in Table 13.

Scaiano et al.[74] have studied the photochemistry of carbonyl and acetonitrile ylides **50** and **42** by means of a "two-laser" experiment. In this experiment, ylide **50** is generated in the usual way by LFP of 9-DAF in acetone; however, 1.5 μs after the "synthesis" laser pulse the sample is exposed to a second laser pulse (λ = 591 nm, ~250 ns). The second laser pulse is absorbed only by the ylide and destroys 80% of the absorption of **50** (Figure 20). It was felt that cyclization to epoxide **51** was the mode of disappearance of ylide **50** (Scheme 46).

Table 13. The Absolute Rate Constants of the Reaction of Carbonyl Ylides Derived from Fluorenylidene with Various Quenchers (27 °C)[72]

Ketone precursor	*Quencher*	k_Y (M^{-1} s^{-1})
Acetone	Diethyl fumarate	$(4.47 \pm 0.30) \times 10^7$
Acetone	Fumaronitrile	$(1.45 \pm 0.02) \times 10^7$
Acetone	Oxygen	$(7.24 \pm 0.43) \times 10^7$
Methyl ethyl ketone	Diethyl fumarate	$(3.14 \pm 0.30) \times 10^7$
Cyclobutanone	Diethyl fumarate	$(3.30 \pm 0.30) \times 10^8$
Cyclopentanone	Diethyl fumarate	$(3.75 \pm 0.24) \times 10^7$
Cyclohexanone	Diethyl fumarate	$(2.06 \pm 0.02) \times 10^7$
Cycloheptanone	Diethyl fumarate	$(4.40 \pm 0.40) \times 10^6$

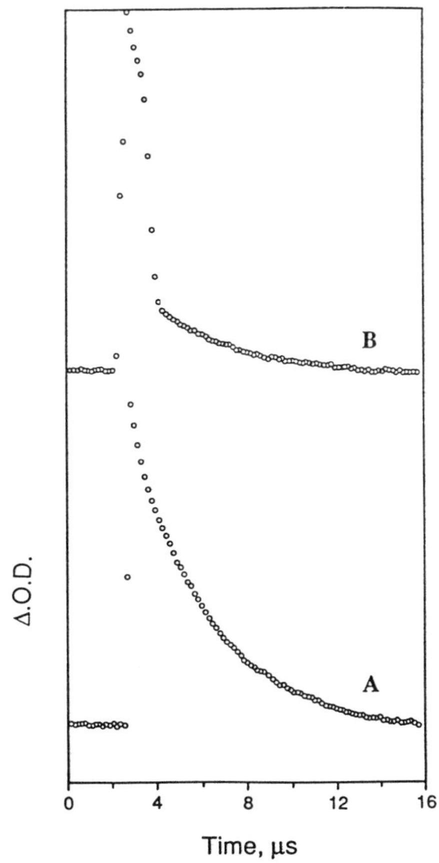

Figure 20. Photolytic bleaching of the fluorenylidene–acetone ylide **50** by means of a "two-laser" experiment (see text). Trace **A**: unperturbed decay of ylide **50**. Trace **B**: photolysis of ylide **50** (at ca. 4 μs) showing depletion of absorbance. (Reprinted with permission from *J. Am. Chem. Soc.* **1985**, *107*, 7204.)

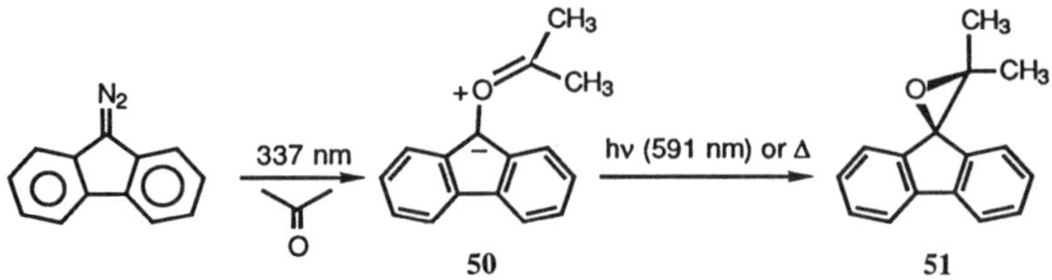

Scheme 46.

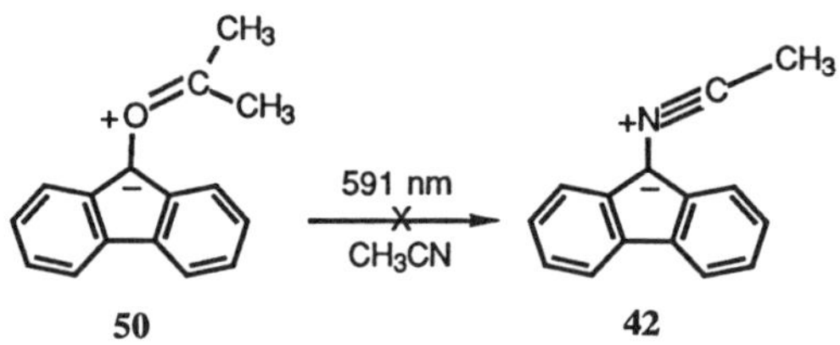

Scheme 47.

Scaiano et al.[74] concluded that photolysis of the ylide does not lead to reversion to fluorenylidene and acetone, as photolysis of ylide **50** in the presence of acetonitrile does not produce nitrile ylide **42** (Scheme 47).

These workers demonstrated the "surprising photostability" of nitrile ylide **42**. Laser flash photolysis of 9-DAF in acetonitrile produced ylide **42**, which was subjected to a second laser pulse (429 nm, 250 ns) a few μs after the initial "synthesis" laser pulse. In contrast to carbonyl ylide **50**, nitrile ylide **42** was not photochemically consumed.

J. Fluorenylidene–Sulfur Ylides

Laser flash photolysis of 9-DAF in acetonitrile containing sulfides and disulfides produces new transients with $\lambda_{max} = 350–450$ nm (Figure 21).[75] Rate constants for the formation of these transients were obtained by monitoring their pseudo first-order rate of formation as a function of sulfur compound concentration and are reported in Table 14.

McGimpsey and Scaiano[76] have found that ^{1,3}Fl reacts with thioketones to form ylides **52** and **53**.

The ylides are formed at a nearly diffusion controlled rate (k_Y), persist for many microseconds and absorb strongly in the visible. It is not clear whether these ylides are formed from ^{1}Fl ($k_Y = k_S K$) or by stepwise addition of ^{3}Fl to the thioketone followed by intersystem crossing to the ylide ($k_Y = k_T$) (Scheme 48). This will be discussed more fully in Section IV.

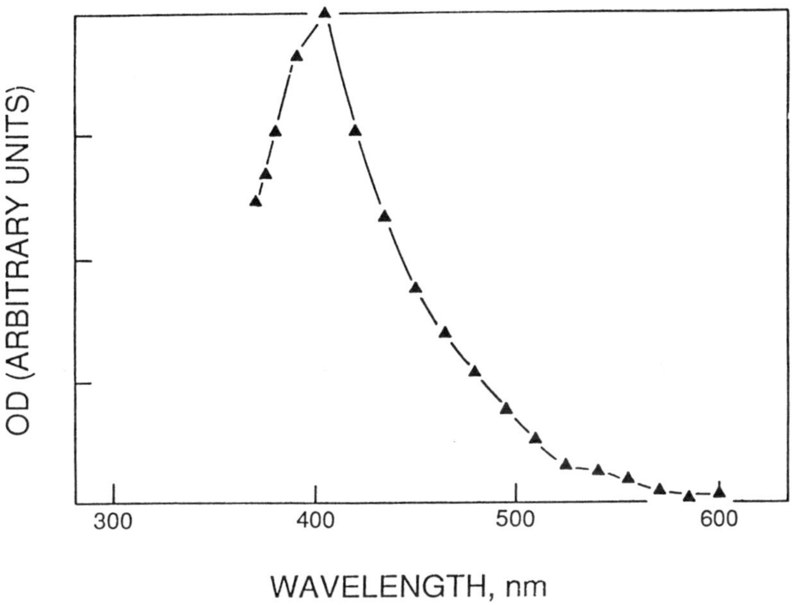

Figure 21. Transient absorption spectra of fluorenylidene–sulfur derived ylides. (Reprinted with permission from *J. Am. Chem. Soc.* **1986**, *108*, 3024.)

Table 14. Absolute Rate Constants for the Reaction of Fluorenylidene with Sufides and Disulfides at 298 K[75]

Substrate	λ_{max} *(nm)*	k_2 $(M^{-1}\ s^{-1})$
MeSSMe[a]	443	$(2.0 \pm 0.1) \times 10^8$
t-BuSS-*t*-Bu[a]	404	$(4.0 \pm 0.2) \times 10^8$
PhSPh[a]	384	$(3.1 \pm 0.3) \times 10^8$
MeSSMe[b]	450	~10^9
EtSSEt[b]	449	~8×10^8
i-PrSS-*i*Pr[b]	449	~4×10^8
MeSMe[b]	~357	~10^9

Notes: [a]Acetonitrile as solvent.
[b]Benzene as solvent.

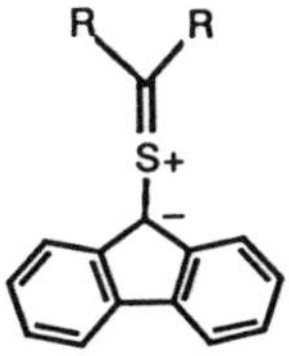

	$k_Y(M^{-1}s^{-1})$	$t_{decay}(\mu s)$	$\lambda_{max}(nm)$
52: R = R = *tert*-butyl	$(1.1 \pm 0.1) \times 10^9$	>100	700
53: R,R = 2-adamantyl	$(6.4 \pm 1.1) \times 10^9$	18.5	586

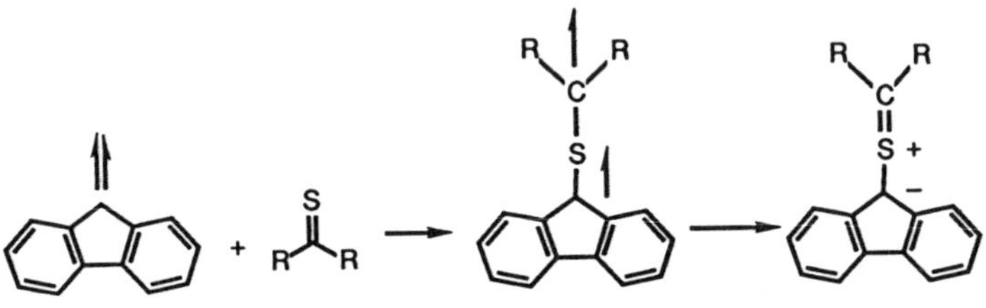

Scheme 48.

IV. CARBENES WITH WELL-SEPARATED TRIPLET GROUND STATES

A. Methylene

As we mentioned at the outset, methylene, the simplest divalent carbon species, has a triplet ground state.[11] The lowest lying singlet state is known to be about 9 kcal/mol above the ground state, in the gas phase. If this value holds in solution, a chemically insignificant amount of $^1CH_2$ would be present at equilibrium; however, due to the anticipated extreme reactivity of $^1CH_2$ and $^3CH_2$, one would expect chemical reactions of either state to easily outpace spin-state equilibration. In Scheme 49, k_S and k_T are the absolute rate constants for trapping singlet and triplet methylene, respectively, with substrate M. The reader will note that this is exactly the opposite of the situation with the 1- and 2-naphthylcarbenes, fluorenylidene, and, as we will see, diphenylcarbene. The change reflects the larger energy gap

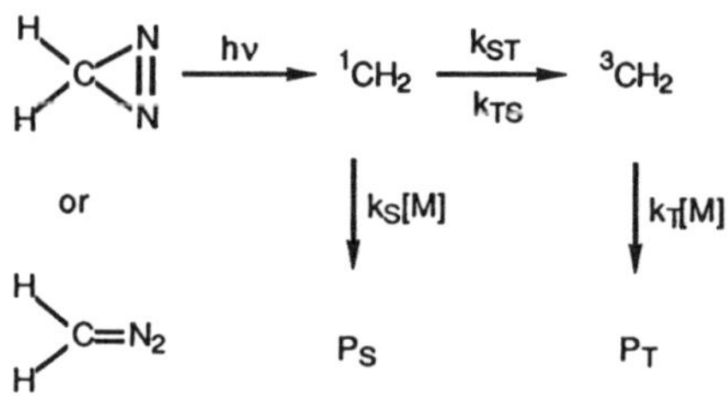

Scheme 49.

Table 15. Absolute Rate Constants for the Reaction of the Methylene–Acetonitrile Ylide **54** with Various Quenchers[77]

Quencher	k_Q (M^{-1} s^{-1})	k_{rel} *(time resolved)*[a]	k_{rel} *(product analysis)*[b]
Fumaronitrile	$(2.2 \pm 0.1) \times 10^9$	$(3.9 \pm 0.3) \times 10^2$	$(4.3 \pm 0.6) \times 10^2$
Dimethyl acetylenedicarboxylate	$(1.5 \pm 0.1) \times 10^8$	27 ± 2	42 ± 10
Acrylonitrile	$(5.6 \pm 0.2) \times 10^6$	1.0	1.0
Methyl acrylate	$(3.3 \pm 0.1) \times 10^6$	0.6 ± 0.1	0.7 ± 0.1
Methyl propiolate	$(4.4 \pm 0.3) \times 10^5$	0.08 ± 0.01	0.10 ± 0.01

Notes: [a]Relative rate constants from time-resolved laser flash analysis of quenching **54** in air-saturated acetonitrile at 25 °C.

[b]Relative from product ratios produced in steady state photolysis in air-saturated acetonitrile solutions at 0 °C.

between singlet and triplet methylene and the higher absolute reactivity of the parent carbene.

Under these conditions ylide formation will only be possible by interception of $^1CH_2$ prior to intersystem crossing to triplet methylene. This has been accomplished by Turro et al.,[77] who studied the LFP of diazirine or diazomethane in acetonitrile. Flash photolysis in this solvent produces a new transient absorption with λ_{max} = 280 nm. This transient was attributed to nitrile ylide **54** because it reacts very rapidly with typical dipolarophiles (Table 15), and because the relative rates of ylide quenching determined by LFP agree well with the chemically determined relative reaction rates. The chemically determined relative reactivities could be determined either by photolysis of diazirine in acetonitrile or thermally from decomposition of silane **55** with silver fluoride (Scheme 50).

The absolute rate of formation of ylide **54** in neat acetonitrile was too fast to resolve by nanosecond spectroscopy. This of course does not prohibit Stern–Volmer

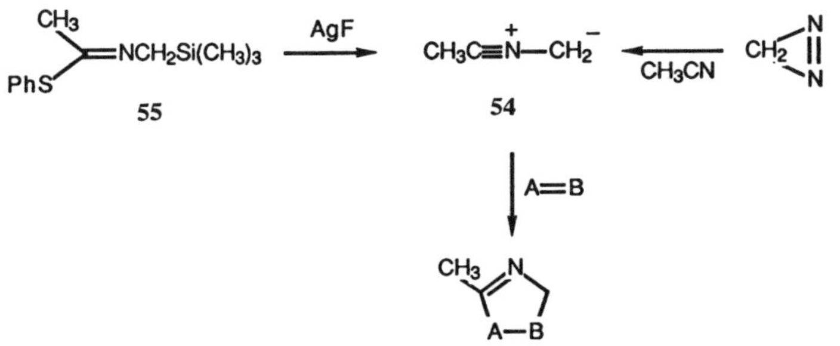

Scheme 50.

analysis of the ylide to obtain the relative reactivities of $^1CH_2$ towards a variety of substrates. The reduction in the yield of ylide signal as a function of quencher leads to the Stern–Volmer data of Table 16 where k_S is the absolute rate constant of reaction of $^1CH_2$ with quencher and τ ($=1/k_{obsd}$) is the lifetime of $^1CH_2$ in neat acetonitrile in the absence of quencher (see Eqs. 2–4). Table 16 reveals that, as expected, $^1CH_2$ reacts with little to no selectivity with a wide variety of functional groups. Thus, $^1CH_2$ reacts at a nearly diffusion-controlled rate with all of the substrates of Table 16.

The extraordinary reactivity of singlet methylene has frustrated attempts at finding a solvent that is relatively inert to this reagent; however, Turro et al.[77] have found that perfluorohexane serves this function admirably. It can dissolve both diazirine and chloroform (a diagnostic carbene quencher), yet the C–F bond is sufficiently strong to withstand reaction with singlet methylene. Chloroform is a diagnostic carbene quencher because it reacts by different pathways with $^1CH_2$ and $^3CH_2$ to give different products.

By using CIDNP techniques, Roth[78] has found that $^1CH_2$ abstracts chlorine atoms from chloroform,

$$^1CH_2 + CHCl_3 \rightarrow \dot{C}H_2Cl + \dot{C}HCl_2 \rightarrow \underset{\mathbf{56}}{CH_2ClCHCl_2} \tag{9}$$

Table 16. Stern–Volmer Constants for the Reaction of Methylene with Various Quenchers in Acetonitrile at 298 K[77]

Quencher	$k_5\tau$
Carbon tetrachloride	1.9 ± 0.1
Chloroform	1.62 ± 0.08
Chloroform-*d*	1.67 ± 0.08
Methylenechloride	1.18 ± 0.02
Methanol	1.89 ± 0.08
Methanol-d_1	1.49 ± 0.08
Methanol-d_4	1.53 ± 0.08
Water	1.1 ± 0.1
Deuterium oxide	0.88 ± 0.07
Isoprene	2.2 ± 0.02
cis-2-Pentene	1.54 ± 0.08
trans-2-Pentene	1.37 ± 0.07
Benzene	0.88 ± 0.05
Hexafluorobenzene	0.1 ± 0.1
Ether	1.21 ± 0.05
Tetrahydrofuran	1.12 ± 0.05

whereas triplet methylene prefers to abstract hydrogen atoms from this substrate.

$$^{3}CH_2 + CHCl_3 \rightarrow \dot{C}H_3 \;+\; \dot{C}Cl_3 \rightarrow \underset{\mathbf{57}}{CH_3CCl_3} \tag{10}$$

Turro et al.[77] have elegantly used this reaction along with the Stern–Volmer data to obtain the absolute rate constant of intersystem crossing (k_{ST}) of $^{1}CH_2$ in solution. Photolysis of diazirine in the presence of 5–6 *M* chloroform produces mainly **56**, the product of singlet carbene capture; but upon dilution with perfluorohexane, increasing amounts of triplet-derived product **57** are observed. This is consistent with irreversible singlet to triplet carbene intersystem crossing. In fact, a plot of the ratio of singlet- to triplet-derived products is linear, with a slope of k_S/k_{ST} according to the mechanism of Scheme 49. Making the assumption that $k_S = k_{diff} = 5 \times 10^9$ M^{-1} s^{-1} in perfluorohexane produces a value of $k_{ST} = 8 \times 10^8$ s^{-1}. This value is at least one order of magnitude lower than the value expected from extrapolation of gas-phase dilution studies to the solution phase.[79] This implies that there is strong solvation of $^{1}CH_2$ in perfluorohexane, perhaps due to complexation of the empty orbital of the carbene with a lone pair of electrons on fluorine. The intersystem crossing rate in methylene is also about an order of magnitude smaller than the values measured for fluorenylidene[68a] and diphenylcarbene.[80]

B. Diphenylcarbene and Derivatives

The chemistry and spectroscopy of methylene and the arylcarbenes differ in some important aspects. Triplet methylene has been detected by matrix EPR spectroscopy,[81] and both singlet and triplet methylene have been characterized by high resolution gas-phase spectroscopy.[11,82] Although triplet arylcarbenes have been detected by matrix and time resolved solution-phase spectroscopy, the low lying singlet state of a ground state triplet arylcarbene has never been observed by any physical method.

As discussed, singlet and triplet methylene give distinct chemistries. Photolysis of diazirine or diazomethane in neat quencher yields products that trace their origin to $^{1}CH_2$. On the other hand, triplet-sensitized photolysis of diazirine in neat quencher or direct photolysis of diazomethane in dilute quenchers produces species that are derived from $^{3}CH_2$; this demonstrates that there are two chemically distinct intermediates that can be formed in solution. This is not the case with diphenylcarbene,[83] phenylcarbene,[84] and perhaps fluorenylidene.[85] The distribution of products formed upon photolysis of diphenyldiazomethane in *cis*- and *trans*-2-butene is invariant with changes in the concentration of alkene. Similar results are obtained upon photolysis of 9-diazofluorene in a variety of solvents with the notable exception of hexafluorobenzene, where a dilution effect was observed. No olefin dilution effects are observed upon photolysis of phenyldiazomethane. Furthermore, direct- and triplet-sensitized photolyses of α-*n*-butylphenyldiazomethane give the same distribution of products.[86]

These observations can be reconciled with the idea that the singlet–triplet splitting in the arylcarbenes is much smaller than that in methylene, and that spin equilibration in both directions is faster than chemical reactions of either spin state:

$$k_{ST} > k_S[M]$$

$$k_{TS} > k_T[M] \qquad (11)$$

The ultimate product distribution observed therefore depends on the equilibrium constant *K* and the relative reactivity of each spin state.

The reader will note that the above is not the only possible interpretation. Another mechanism that is formally consistent with the data is one in which only a single reactive intermediate is present in solution. The intermediate must be capable of reacting via pathways normally associated with "pure" singlet or triplet states. This mechanism is generally not preferred (despite its greater simplicity) largely due to the analogy to methylene, where the chemical and spectroscopic data require the presence of two intermediates of different spin multiplicity.

The behaviors of phenylcarbene (PC), 1- and 2-NC, and Fl are frequently dominated by the higher energy but more reactive singlet state. In cyclohexane, for example, the chemistry of PC, 1- and 2-NC, and Fl is almost entirely due to C–H insertion[87] reactions of the singlet carbene rather than to hydrogen atom abstraction reactions of the ground triplet state. Upon generation in *cis* or *trans*-2-butene, the cyclopropanation reactions of PC[84] and Fl[85] are highly stereospecific, again pointing to predominantly singlet chemistry. The chemistry of 1- and 2-NC in *cis* or *trans*-2-butene is not known at this time. It is unsurprising then that 1- and 2-NC and Fl react rapidly through their easily accessible singlet states with acetone and acetonitrile to form ylides.

Laser flash photolysis of diphenyldiazomethane in acetone or acetonitrile does not produce an ylide. There is nothing inherently unfavorable about ylide **58**, and indeed an ylide of this type is formed upon generation of DPC in a frozen polycrystal containing benzophenone at 77 K. The ylide may be formed by photoexcitation of matrix isolated DPC (Scheme 51).[73,88]

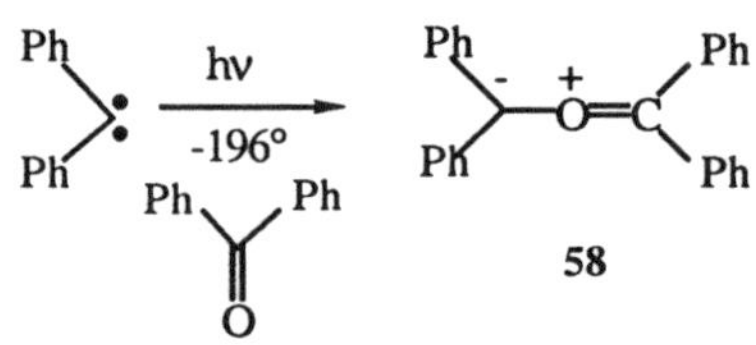

58

Scheme 51.

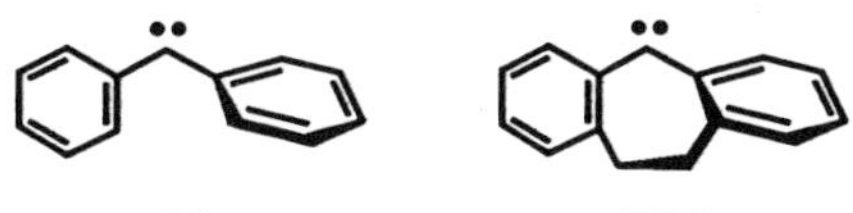

Example 10.

The failure of DPC to form ketone or nitrile ylides in solution means that the reactions of ^{3}DPC (probably H-atom abstraction) are faster than those of the low lying singlet state in these solvents ($k_T >> k_S K_{ST}$); this would be consistent with the chemistry of DPC in alkanes and alkenes. Generation of DPC in cyclohexane leads to nearly exclusive triplet chemistry,[87] and the cyclopropanation reactions of DPC with *cis*- and *trans*-2-butene are much less stereospecific than those of Fl, PC, or 1- and 2-NC.[83] Thus, it would appear to be quite common for DPC to react predominantly through its triplet state. One could argue that the reluctance of DPC to form a carbonyl or nitrile ylide is due to steric problems caused by the two phenyl moieties. The solution or gas-phase geometries of singlet and triplet DPC are unknown, but matrix EPR work indicates that ^{3}DPC is both bent and nonplanar[88] (Example 10).

Dibenzocycloheptadienylidene (DBC) also fails to form ylides with acetone or acetonitrile[89] and it is probably less sterically encumbered than DPC because the rings are tethered by an ethano bridge. Furthermore, acetonitrile is a small linear molecule which should not be subject to great steric difficulties. Thus, the weight of evidence would seem to suggest that the singlet states of DPC and DBC are higher in energy and less accessible than the singlet states of PC, 1- and 2-NC, and Fl.

DPC and DBC react with methanol by formal OH-bond insertion, a process which would seem to be associated with singlet rather than triplet carbene behavior. (For a detailed discussion see Chapter 1 of this volume.) One would predict that ^{3}DPC should react with methanol by hydrogen atom abstraction and radical recombination (Scheme 52). Furthermore, although it is not the only possible mechanism, one might imagine that the reaction of ^{1}DPC with methanol proceeds via an ylide such as **59** (Scheme 53).

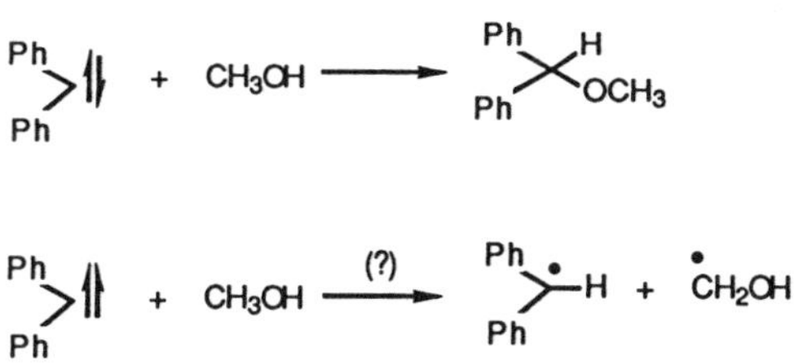

Scheme 52.

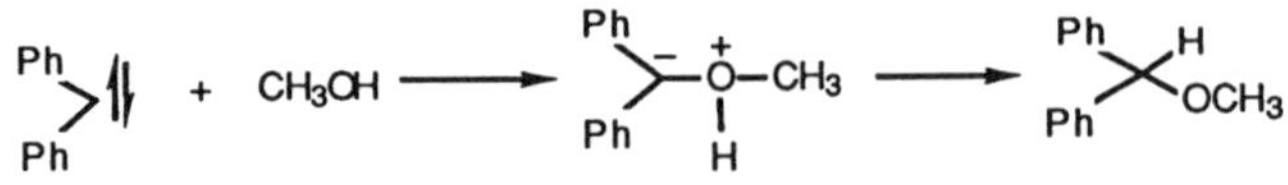

59

Scheme 53.

Closs and Rabinow were the first to realize that the kinetics of the reaction of ^{3}DPC with methanol could be used to obtain a measure of the singlet–triplet energy gap.[1] They found that the transient absorbance of ^{3}DPC is quenched by methanol according to the following rate law:

$$\frac{-d[^3\mathrm{DPC}]}{dt} = k[^3\mathrm{DPC}][\mathrm{CH_3OH}] \tag{12}$$

As ^{1,3}DPC equilibration is very rapid and assuming it is ^{1}DPC which reacts with methanol

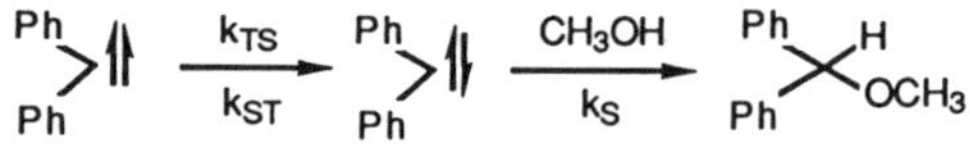 (13)

then $k = k_S K$. The value of k in benzene at 25 °C is $6.8 \times 10^6\ M^{-1}\ \mathrm{s}^{-1}$, and since $k_S \leq k_{diff} = 5 \times 10^9\ M^{-1}\ \mathrm{s}^{-1}$, a lower limit of K can be calculated. This yields a maximum free energy separation of 3–4 kcal/mol between singlet and triplet DPC at room temperature. Assuming further that the only entropic difference between the spin states is due to multiplicity, one obtains a maximum enthalpic splitting (ΔH_{ST}) of about 3 kcal/mol.

Turro and co-workers repeated and expanded upon the Closs and Rabinow experiment. Their analysis indicated that ΔH_{ST} (acetonitrile) is 2.5 ± 0.3 kmol and ΔH_{ST} (isooctane) $= 4.05 \pm 0.15$ kcal/mol.[90]

Griller et al. studied the temperature dependence of the reaction of ^{3}DPC with methanol.[91] They expected to find activation energies of 5–7 kcal/mol ($\Delta H_{ST} + \Delta H_{diff}$). The measured activation energies were actually much lower (Table

Table. 17. Temperature Dependence of the Quenching of Triplet Diphenylcarbene by Methanol[91]

T (K)	Solvent	E_a (kcal/mol)	$\log (A/M^{-1}\ \mathrm{s}^{-1})$
248–342	Isooctane	3.61 ± 0.36	10.51 ± 0.27
255–320	Chlorobenzene	2.48 ± 0.44	9.30 ± 0.34
242–336	Acetonitrile	1.66 ± 0.20	8.62 ± 0.16
282–328	Benzene	1.17 ± 0.36	8.17 ± 0.26

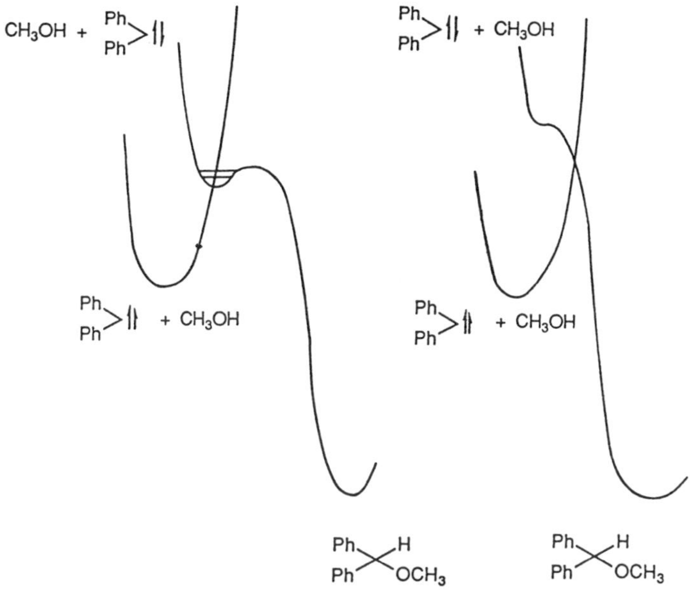

Figure 22. Classical (*left*) and surface crossing (*right*) potential energy surfaces for the reaction of diphenylcarbene with methanol.

17). This observation, along with certain other kinetic anomolies, led them to consider the possibility of a surface crossing mechanism. This is a mechanism in which the intersystem crossing does not occur at the minimum of isolated ^{3}DPC, but at a place on the potential surface where the triplet carbene is interacting with solvent (Figure 22).

An alternative explanation of the unexpectedly low activation energies of Table 17 involves the reversibly formed ylide intermediate **59** (Scheme 54). Under these circumstances, $k \neq k_S K$ but is equal to a more complex function involving the formation and decay terms of ylide **59**.

Other interpretations are possible. The classical mechanism can be reconciled with the data if the activation energy for the reaction of singlet diphenylcarbene with methanol (by whatever mechanism) is negative. This could be the result of an entropic barrier to this reaction.

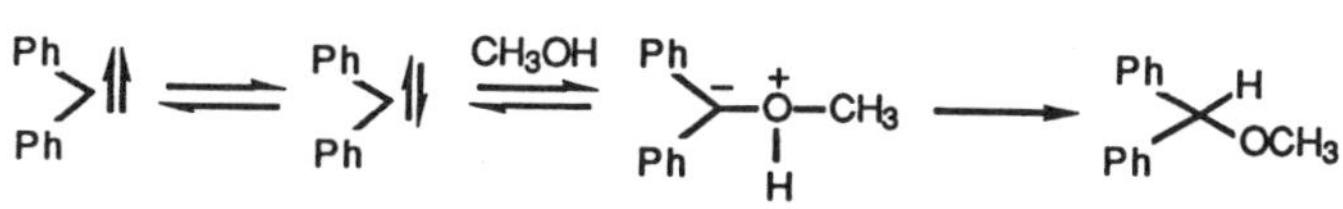

59

Scheme 54.

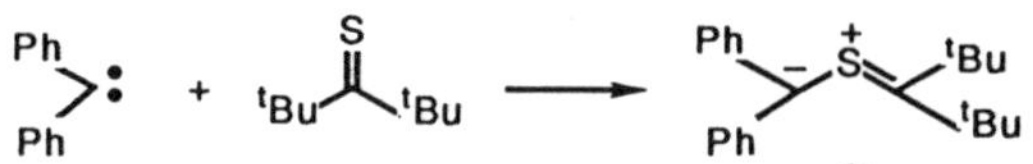

60
λ_{max} = 580 nm

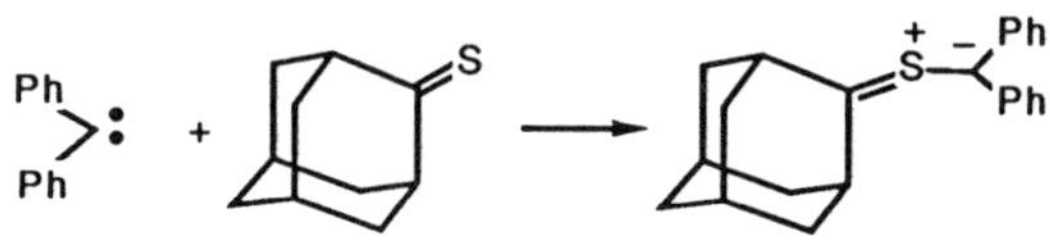

61
λ_{max} = 500 nm

Scheme 55.

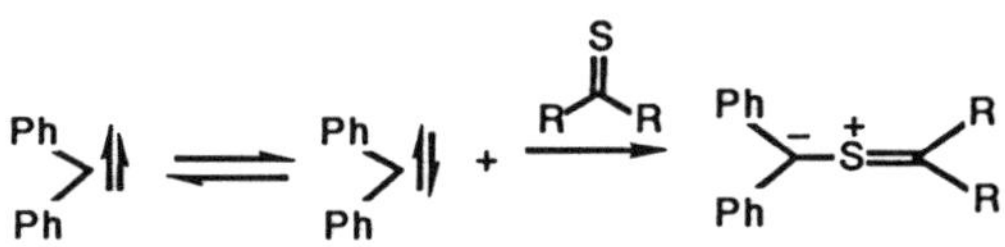

Scheme 56.

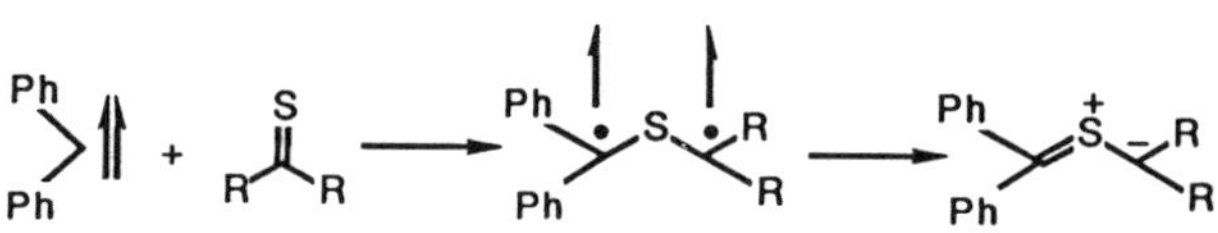

Scheme 57.

The possible intermediacy of **59** stimulated two attempts to prepare spectroscopically observable ylides derived from DPC. McGimpsey and Scaiano have intercepted ^{3}DPC with di-*tert*-butylthioketone and adamantanethione (Scheme 55).[76] The ylides are formed very rapidly. Di-*tert*-butyl thione ylide **60** is formed with $k = 9.1 \pm 0.7 \times 10^7\ M^{-1}\ s^{-1}$ and has a lifetime of 44 μs. The adamantane thione ylide **61** is formed with $k = 1.3 \times 10^9\ M^{-1}\ s^{-1}$ and has a lifetime of 26 μs. The latter ylide-forming reaction occurs at nearly a diffusion-controlled rate. This is clearly inconsistent with a ΔH_{ST} of 2.5–4 kcal/mol if the reaction involves ^{1}DPC (Scheme 56). One can argue, however, that the reaction involves triplet carbene addition to the thione followed by intersystem crossing to the ylide (Scheme 57); Yet, one is more reluctant to advance this argument with ylide **62**—which is formed upon LFP of DPDM in pyridine (Scheme 58)[9]—as the radical-like addition of ^{3}DPC to a pyridine ring (Scheme 59) has no benzene analog (Scheme 60).

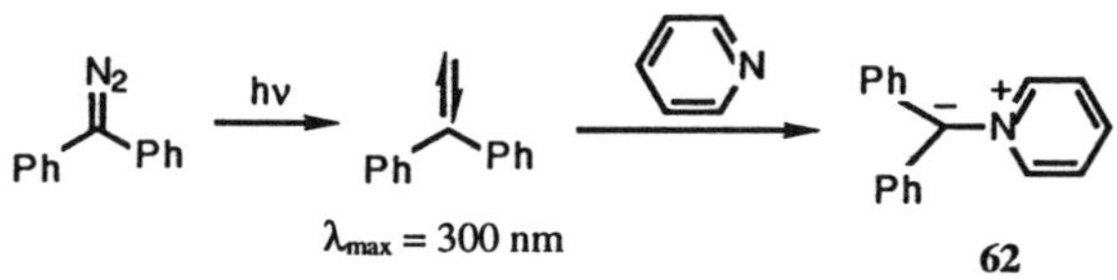

62

λ_{max} = 500 nm

Scheme 58.

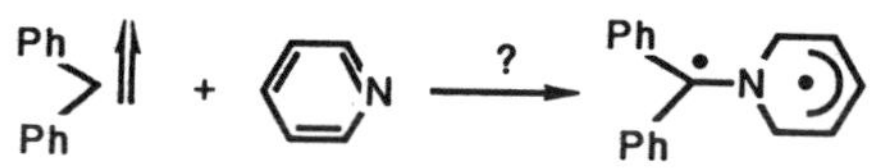

Scheme 59.

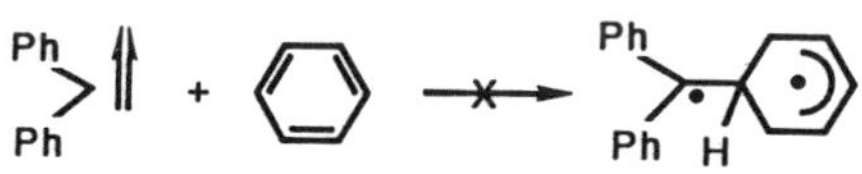

Scheme 60.

The long wavelength absorption band of ylide **62** probably involves promotion of the ylide to an excited state with singlet biradical character (Scheme 61). Thus, the transition energy (57.3 kcal/mol) is probably a reasonable estimate of the energy separation between pyridinum ylide **62** and the putative triplet biradical ^{3}**63.**

Pyridinium ylide **62** is very long-lived and is probably a prototype for an irreversibly formed ylide. Recently, Jones[9] has found that the activation energy for ylide formation is about 2 kcal/mol in chlorobenzene and heptane. This value is not unlike that measured for the reaction of ^{3}DPC with methanol. This points to a similarity of mechanism in the reaction of ^{3}DPC with methanol and pyridine and reduces the urgency to invoke reversibly formed ylide intermediates in the reaction of ^{3}DPC with methanol. The low activation energies observed in the reaction of ^{3}DPC with pyridine and methanol are clearly inconsistent with the reported measurements of ΔH_{ST} of this carbene if the simple kinetic models used are appropriate.[91] The mechanism of reaction of ^{3}DPC with typical singlet quenchers may be more complex than a simple ^{1,3}DPC preequilibrium,[1,90] and other nontraditional mechanisms, such as Griller–Nazran–Scaiano[91] surface crossing or nontraditional kinetic expressions,[92] must be considered. It seems appropriate that the first carbene studied by solution flash photolysis in the seminal work of Closs and

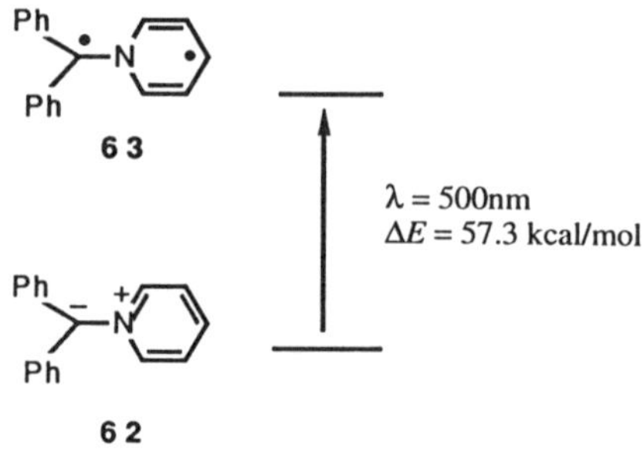

Scheme 61.

Rabinow[1] continues to challenge new techniques with its rich and complex chemistry.

Recent studies[93] demonstrate that singlet diphenylcarbene is basic and abstracts protons from water to form the benzhydryl radical-hydroxide ion pair. Thus, it seems unlikely that singlet diphenylcarbene and methanol react to form an ylide.

V. OUTLOOK

Early LFP studies of carbenes were confined to the intermolecular reactions of aryl- and diaryl-substituted species. Alkyl-substituted carbenes were not studied because they were not expected to have useful chromophores. It is now apparent that the capture of alkyl, diakyl, and alkylhalo carbenes with pyridine or other heterocycles to form an ylide is a general reaction. The ylides have excellent chromophores and have lifetimes that are much longer than those of the carbenes from which they are derived. These two properties make them ideal probes of the dynamics of carbenes whose chromophores are nonexistent, weak, or simply inconvenient. Thus, it can be confidently predicted that ylide probes will play a major role in obtaining the bimolecular rate constants of alkyl, diakyl, and alkylhalocarbenes. The ylide probe method can also reveal the absolute kinetics of the intramolecular reactions of these carbenes and allow separation of excited state precursor and ground state carbene chemistry. This is particularly exciting as these rearrangements are among the most fundamental of all carbene processes.

ACKNOWLEDGMENTS

Support of this work in Columbus, Ohio by the National Science Foundation (CHE-8814950) is gratefully acknowledged. The authors also which to acknowledge the friendship and expertise of their co-workers, Linda Hadel, Vince Maloney, J. C. Scaiano, David Griller, Michelle Jones, Walter White, Scott Morgan, Dave Modarelli and N. Soundararajan, who made much of this work possible, and to acknowledge Professors Jones and Moss for enlightening discussions.

REFERENCES

1. Closs, G. L.; Rabinow, B. E. *J. Am. Chem. Soc.* **1976**, *98*, 8190.
2. (a) Scaiano, J. C. *Acc. Chem. Res.* **1982**, *15*, 252; (b) Wang, Y.; Eisenthal, K. B. *J. Chem. Ed.* **1982**, *59*, 482; (c) Eisenthal, K. B. In *Ultrashort Light Pulses*; Shapiro, S. Ed.; Springer-Verlag, Berlin, 1977, Chapter 5.
3. (a) Griller, D.; Nazran, A. S.; Scaiano, J. C. *Tetrahedron* **1985**, *41*, 1525; (b) Schuster, G. B. *Adv. Phys. Org. Chem.* **1986**, *22*, 311; (c) Eisenthal, K. B.; Moss, R. A.; Turro, N. J. *Science* **1984**, *225*, 4669; (d) Platz, M. S.; Maloney, V. M. M. In *Kinetics and Spectroscopy of Carbenes and Biradicals*; Platz, M. S., Ed.; Plenum, New York, NY, 1990.
4. Rabek, J. F. *Experimental Methods in Photochemistry and Photophysics*, Part 2, Wiley, New York, 1990; pp 822–853.
5. (a) Hodgson, B. W.; Keene, J. P. *Rev. Sci Inst.* **1978**, *43*, 493; (b) Huiid, T.; Nielsen, S. O. *Rev. Sci. Inst.* **1972**, *43*, 1198; (c) Taylor, W. B.; LeBlanc, J. C.; Whillans, D. W.; Herbert, M. A.; Johns, H. E. *Rev. Sci. Inst.* **1972**, *43*, 1797.
6. Engstrom, R. W. *Photomultiplier Handbook*; RCA, Solid State Division, Lancaster, PA, 1980; 17604.
7. (a) Porter, G., Windsor, M. *Nature* **1957**, 180; (b) Norman, I.; Porter, G. *Proc. R. Soc. (London) Ser. A.* **1957**, *230*, 399.
8. Hadel, L. M.; Platz, M. S.; Scaiano, J. C. *J. Am. Chem. Soc.* **1984**, *106*, 283–287.
9. (a) Jones, M. B.; Platz, M. S. *J. Org. Chem.* **1991**, *56*, 1694; (b) Jones, M. B.; Platz, M. S. *Tetrahedron Lett.* **1990**, *31*, 953.
10. Zugravescu, I.; Petrovanu, M. *N-Ylid Chemistry*; McGraw-Hill, New York, 1976.
11. (a) Leopold, D. G.; Murray, K. K.; Miller, A. E. S.; Lineberger, W. C. *J. Chem. Phys.* **1984**, *83*, 4849; (b) Shavitt, I. *Tetrahedron* **1985**, *41*, 1531 (and references therein).
12. (a) Trozzolo, A. M.; Wasserman, E. *Carbenes*, Vol. 2; Moss, R. A.; Jones, M., Jr.; Eds., Wiley, New York, 1975, p 185; (b) Wasserman, E. *J. Chem. Phys.***1964**, *41*, 1763; (c) Wasserman, E.; Hutton, S. *Acc. Chem. Res.* **1977**, *10*, 27.
13. (a) Murray, K. K.; Leopold, D. G. J.; Miller, T. M.; Lineberger, W. C. *J. Chem. Phys.* **1988**, *89*, 5442; (b) Moss, R. A. and Turro, N. J. In *Kinetics and Spectroscopy of Carbenes and Biradicals*; Platz, M. S., Ed., Plenum, New York, 1990.
14. Hoffmann, R.; Zeiss, G. D.; VanDine, G. W. *J. Am. Chem. Soc.* **1968**, *90*, 1485.
15. (a) Kirmse, W. *Carbene Chemistry*; Academic Press, New York, 1970; (b) *Carbenes*, Vols. 1 and 2, Moss, R. A., Jones, M., Jr., Eds.; New York, 1973, 1975.
16. Gould, I. R.; Turro, N. J.; Butcher, J. Jr.; Doubleday, C. Jr.; Hacker, N. P.; Lehr, G. F.; Moss, R. A.; Cox, D. P.; Guo, W.; Munjal, R. C.; Perez, L. A.; Fedorynski, M. *Tetrahedron* **1985**, *41*, 1587. See also Ref. 13b.
17. Jackson, J. E.; Soundararajan, N.; Platz, M. S. *J. Am. Chem. Soc.* **1988**, *110*, 5595.
18. Soundararajan, N.; Jackson, J. E.; Platz, M. S. *Tetrahedron Lett.* **1988**, *29*, 3419.
19. Winnewisser, M.; Möller, K.; Gambi, A. In *Diazirines*, Vol. 1; Liu; M. T. H., Ed.; CRC Press, Boca Raton, FL, 1987, p. 19.
20. Soundararajan, N.; Platz, M. S.; Jackson, J. E.; Doyle, M. P.; Oon, S.-M.; Liu, M. T. H.; Anand, S. M. *J. Am. Chem. Soc.* **1988**, *110*, 7143.
21. (a) Griller, D.; Liu, M. T. H.; Montgomery, C. R.; Scaiano, J. C.; Wong, P. C. *J. Org. Chem.* **1983**, *48*, 1359; (b) Griller, D.; Liu, M. T. H.; Scaiano, J. C. *J. Am Chem. Soc.* **1982**, *104*, 5549.
22. (a) Jones, M. B.; Jackson, J. E.; Soundararajan, N.; Platz, M. S. *J. Am. Chem. Soc.* **1988**, *110*, 5597; (b) Jones, M. B.; Maloney, V. M.; Platz, M. S. *J. Am. Chem. Soc.* **1992**, *114*, 2163.
23. (a) Small, R. D., Jr.; Scaiano, J. C. *J. Phys. Chem.* **1977**, *81*, 828; (b) Small, R. D., Jr.; Scaiano, J. C. *Chem. Phys. Lett.* **1977**, *50*, 431; (c) *Chem. Phys. Lett.* **1978**, *59*, 246.
24. Lexa, D.; Savéant, J.-M.; Su, K.-B.; Wang, D.-L. *J. Am. Chem. Soc.* **1988**, *110*, 7617.

25. Jackson, J. E.; Soundararajan, N.; Platz, M. S.; Doyle, M. P.; Liu, M. T. H. *Tetrahedron Lett.* **1989**, *30*, 1335.
26. Houk, K. N.; Rondan, N. G.; Mareda, J. *Tetrahedron*, **1985**, *41*, 1555.
27. Personal communication from Professor R. A. Moss.
28. (a) Mansoor, A. M.; Stevens, I. D. R. *Tetrahedron Lett.* **1966**, *16*, 1733; (b) Kirmse, W.; Buschoff, M. *Angew Chem Int. Ed.* **1965**, *4*, 692; (c) Frey, H. M.; Scaplehorn, W. *J. Chem. Soc. (A)* **1966**, 968; (d) Chang, K. T.; Shechter, H. *J. Am. Chem. Soc.* **1979**, *101*, 5082; (e) One can also argue that these differences arise from a common intermediate subject to dynamical effects; see Newman-Evans, R. H.; Simon, R. J.; Carpenter, B. K. *J. Org. Chem.* **1990**, *55*, 695; (f) Frey, H. M.; Stevens, I. D. R. *J. Chem. Soc.* **1965**, 1700; (g) Figuera, J. M.; Pérez, J. M.; Tobar, A. *J. Chem. Soc. Perkin Trans 1* **1978**, *74*, 809; (h) Friedman, L.; and Shechter, H. *J. Am. Chem. Soc.* **1959**, *81*, 5512; (i) Frey, H. M. *Adv. Photochem.* **1966**, *4*, 225.
29. (a) Moss, R. A.; Ho, G. J. *J. Am. Chem. Soc.* **1990**, *112*, 5642; (b) Moss, R. A.; Munjal, R. C. *J. C. S. Chem. Comm.* **1978**, 775.
30. (a) Liu, M. T. H.; Bonneau, R. *J. Am. Chem. Soc.* **1989**, *14*, 6873; (b) Bonneau, R.; Liu, M. T. H.; Subramanian, R.; Linkletter, B. *Laser Chem.* **1990**, *10*, 267; (c) Liu, M. T. H.; Bonneau, R. *J. Am. Chem. Soc.* **1990**, *112*, 3915; (d) Liu, M. T. H.; Bonneau, R. *J. Phys. Chem.* **1990**, *93*, 7298; (e) Liu, M. T. H.; Bonneau, R.; Rayez, M. *J. Am. Chem. Soc.* **1989**, *111*, 5973.
31. Chateauneuf, J. E.; Johnson, R. P.; Kirchoff, M. M. *J. Am. Chem. Soc.* **1990**, *112*, 3217.
32. LaVilla, J. A.; Goodman, J. L. *J. Am. Chem. Soc.* **1989**, *111*, 6877.
33. White, W. R. III; Platz, M. S. *J. Org. Chem.* **1992**, *57*, 2841.
34. (a) Moss, R. A.; Ho, G.-J.; Shen, S.; Krogh-Jespersen, K. *J. Am. Chem. Soc.* **1990**, *112*, 1638; (b) Ho, G. J.; Krogh-Jespersen, K. *J. Am. Chem. Soc.* **1989**, *111*, 6875.
35. (a) Moss, R. A.; Fantina, M. E.; Munjol, R. C. *Tetrahedron Lett.* **1979**, *20*, 1277; (b) Moss, R. A.; Ho, G. J. *Tetrahedron Lett.* **1990**, *31*, 1225; (c) Moss, R. A.; Ho, G. J.; Liu, W. *J. Am. Chem. Soc.* **1992**, *114*, 959.
36. (a) Benson, S. W. *Thermochemical Kinetics*; Wiley, New York, 1968; (b) Benson, S. W.; Cruickshank, F. R.; Golden, D. M.; Haugen, G. R.; O'Neal, H. E.; Rodgers, A. S.; Shaw, R.; and Walsh, R. *Chem. Rev.* **1969**, *69*, 279.
37. (a) Liu, M. T. H.; Murray, S. K.; Zhu, J. *Chem. Comm.*; (b) Stevens, I. D. R.; Liu, M. T. H.; Soundararajan, N.; Paike, N. *Tetrahedron Lett.* **1989**, *30*, 48; (c) Liu, M. T. H.; Subramanian, R. J. Phys. Chem. **1986**, *90*, 75; (d) Tomioka, H.; Huyashi, N.; Izawa, Y. *J. Am. Chem. Soc.* **1984**, *106*, 454; (e) Liu, M. T. H.; Subramanian, R. *J. Phys. Chem.* **1986**, *90*, 75; (f) Stevens, I. D. R.; Liu, M. T. H.; Soundararajan, N.; Parke, N. *Tetrahedron Lett.* **1989**, *30*, 48; (g) Tomioka, H.; Hayashi, N.; Izawa, Y.; Liu, M. T. H. *J. Chem. Soc. Chem. Comm.* **1984**, 476.
38. Jackson, J. E.;Soundararajan, N.; White, W.; Liu, M. T. H.; Bonneau, R.; Platz, M. S. *J. Am. Chem. Soc.* **1989**, *111*, 6874.
39. Warner, P. M. *Tetrahedron Lett.* **1984**, *25*, 4211.
40. Vasella, A.; Briner, K.; Soundararajan, N.; and Platz, M. S. *J. Org. Chem.* **1991**, *56*, 4741.
41. Perrin, H. M.; White, W. R.; and Platz, M. S. *Tetrahedron Lett.* **1991**, *32*, 4443.
42. Boate, D. R.; Johnston, L. J.; Kwong, P. C.; Lee-Ruff, E.; Scaiano, J. C. *J. Am. Chem. Soc.* **1990**, *112*, 8858.
43. Gallo, M. M.; Schaefer, H. F. III; *J. Am. Chem. Soc.* (submitted.)
44. Backes, J.; Brinker, U. H. In *Carbene(oide), Carbine* Houben-Weyl (Methoden der Organischen Chemie); Regitz, M., Ed.; Thieme, Stuttgart, 1989, Vol. E19b, pp 391–510; (b) ibid, pp 511–541.
45. (a) Moss, R. A.; Chang, M. J. *Tetrahedron Lett.* **1981**, *38*, 349; (b) Adding, A. C.; Strating, J.; Wynberg, H.; Schlatmann, J. L. M. A. *Chem. Comm.* **1966**, 657; (c) Isaev, S. D.; Yuchenko, A. G.; Stepanov, F. N.; Kolyada, G. G.; Novikov, S.; Karpentko, N. F. *J. Org. Chem. USSR* (Engl. trans.) **1973**, *9*, 724; (d) Bayley, H.; Knowles, J. R. *Biochemistry* **1978**, *17*, 2420; **1980**, *19*, 3883.
46. Morgan, S. C.; Jackson, J. E.; Platz, M. S. *J. Am. Chem. Soc.* **1991**, *113*, 2783.

47. White, W. R.; Platz, M. S.; Chen, N.; Jones, M., Jr. *J. Am. Chem. Soc.* **1990**, *112*, 7794; (b) Chen, N., Jones, M., Jr.; White, W. R.; Platz, M. S *J. Am. Chem. Soc.* **1991** *113*, 4981.
48. Modarelli, D.; Morgan, S.; Platz, M. S. *J. Am. Chem. Soc.* **1992**, *27*, 7034.
49. (a) Eaton, P. E.; Hoffmann, K.-L. *J. Am. Chem. Soc.* **1987**, *109*, 5285; (b) Eaton, P. E.; White, A. *J. Org. Chem.* **1990**, *55*, 1321; (c) Eaton, R. E.; Appell, R. B. *J. Am. Chem Soc.* **1990**, *112*, 4055.
50. (a) Chen, N.; Jones, M., Jr. *J. Phys. Org. Chem.* **1988**, *1*, 305; (b) Chen, N.; Jones, M., Jr. *Tetrahedron Lett.* **1989**, *30*, 6969.
51. (a) White, W. R. Ph.D. Thesis, The Ohio State University, 1992; (b) Eaton, D. Unpublished research at The University of Chicago.
52. Borden, W. T. *J. Am. Chem. Soc.* **1992**, *27*, 2719.
53. Morgan, S.; Platz, M. S.; Jones, M, Jr.; Myers, D. R. *J. Org. Chem.* **1991**, *56*, 1351.
54. Myers, D. R.; Senthilnathan, V. P.; Platz, M. S.; Jones, M., Jr. *J. Am. Chem. Soc.* **1986**, *108*, 4232.
55. (a) Kramer, K. A. W.; Wright, A. N. *Tetrahedron Lett.* **1962**, 1095; (b) Frey, H. M. *J. Chem. Soc.* **1962**, 2293.
56. Modarelli, D. A.; Platz, M. S. *J. Am. Chem. Soc.* **1991**, *113*, 8985.
57. (a) Modarelli, D. A.; Platz, M. S. *J. Am. Chem. Soc.* **1992**, *114*, 7034; (b) Modarelli, D. A.; Platz, M. S. *J. Am. Chem. Soc.* **1990**, *115*, 470.
58. (a) Evanseck, J. D.; Houk, K. N. *J. Phys. Chem.* **1990**, *94*, 5518; (b) Evanseck, J. D.; Houk, K. N. *J. Am. Chem. Soc.* **1990**, *112*, 9148.
59. Sugiyama, M. H.; Çelebi, S.; Platz, M. S. *J Am. Chem. Soc.* **1992**, *114*, 966.
60. Çelebi, S.; Modarelli, D. A.; and Platz, M. S. *J. Am. Chem. Soc.* (submitted).
61. Leyva, C. Unpublished research at The Ohio State University.
62. (a) Barcus, R. L.; Hadel, L. M.; Johnston, L. J.; Platz, M. S.; Savino, T. G.; Scaiano, J. C. *J. Am. Chem. Soc.* **1986**, *108*, 3928; (b) Hadel, L. M.; Platz, M. S.; Scaiano, J. C. *Chem. Phys. Lett.* **1983**, *97*, 446; (c) Barcus, R. L.; Wright, B. B.; Platz, M. S.; Scaiano, J. C. *Tetrahedron Lett.* **1983**, *24*, 3955.
63. (a) Huisgen, R. *Angew Chem.* (Int. Ed. Engl.) **1963**, *2*, 633; (b) Padwa, A. *Acc. Chem. Res.* **1976**, *9*, 371; (c) Orahovats, A.; Heimgartner, H.; Schmid, H.; Heinzelmann, W. *Helv. Chim. Acta* **1975**, *58*, 2662.
64. Griffin, G. W.; Horn, K. A. *J. Am. Chem. Soc.* **1987**, *109*, 4919.
65. Kumar, C. V.; Chattopahyay, S. K.; Das, P. K. *J. Phys. Chem.* **1989**, *88*, 5639.
66. Closs, G.-L. *J. Am. Chem. Soc.* **1971**, *93*, 1546.
67. Horn, K. A.; Chaleauneuf, J. E. *Tetrahedron* **1985**, *41*, 1465.
68. (a) Grasse, P. B.; Brauer, B. E.; Zupacic, J. J.; Kaufmann, K. J.; Schuster, G. B. *J. Am. Chem. Soc.* **1983**, *105*, 6833; (b) Griller, D.; Hadel, L. M.; Nazran, A. S.; Platz, M. S.; Wong, P. C.; Savino, T. G.; Scaiano, J. C. *J. Am. Chem. Soc.* **1984**, *106*, 2227.
69. Zupancic, J. J.; Grasse, P. B.; Lapin, S. C.; Schuster, G. B. *Tetrahedron* **1985**, *41*, 1471.
70. Casal, H. L.; Tanner, M.; Werstiuk, N. H.; Scaiano, J. C. *J. Am. Chem. Soc.* **1985**, *107*, 4616.
71. (a) Criegee, R. *Angew Chem.* (Int. Ed. Engl.) **1975**, *14*, 745; (b) Criegee, R. *Rec. Chem. Prog.* **1957**, *18*, 111.
72. Wong, P. C.; Griller, D.; Scaiano, J. C. *J. Am. Chem. Soc.* **1982**, *104*, 663.
73. (a) Becker, R. S.; Bost, R. O.; Kolo, J.; Bertoniere, N. R.; Smith, R. L.; Griffin, G. W. *J. Am. Chem. Soc.* **1970**, *92*, 1302; (b) Do-Minh, T.; Trozzolo, A. M.; Griffin, G. W. *J. Am. Chem. Soc.* **1970**, *92*, 1402; (c) Trozzolo, A. M.; Leslie, T. M.; Sarpotdar, A. S.; Small, R. D.; Ferraud, G. J. *Pure Appl. Chem.* **1979**, *51*, 261.
74. Scaiano, J. C.; McGimpsey, W. G.; Casal, H. L. *J. Am. Chem. Soc.* **1985**, *107*, 7204.
75. Alberti, A.; Griller, D.; Nazran, A. S.; Pedulli, G. F. *J. Am. Chem. Soc.* **1986**, *108*, 3024.
76. McGimpsey, W. G.; Scaiano, J. C. *Tetrahedron Lett.* **1986**, *27*, 547.
77. Turro, N. J.; Cha, Y.; Gould, I. R. *J. Am. Chem. Soc.* **1987**, *109*, 2101.
78. Roth, H. D. *Acc. Chem. Res.* **1977**, *10*, 85.

79. Langford, A. O.; Petek, H.; Moor, C. B. *J. Chem. Phys.* **1983**, *78*, 6650 (and references therein).
80. Eisenthal, K. B.; Turro, N. J.; Sitzmann, E. V.; Gould, I. R.; Hefferon, G.; Langam, J.; Cha, Y. *Tetrahedron* **1985**, *41*, 1543.
81. Wasserman, E.; Yager, W. A.; Kuck, U. J. *Chem. Phys. Lett.* **1970**, *2*, 409.
82. Herzberg, G.; Johns, J. W. C. *J. Chem. Phys.* **1971**, *54*, 2276 (and references therein).
83. (a) Closs, G. L.; Closs, L. E. *Angew. Chem.* (Int. Ed. Engl.) **1962**, *1*, 334; (b) Jones, M. Jr.; Baron, W. J.; Shen, Y. H. *J. Am. Chem. Soc.* **1970**, *92*, 4745; (c) Bethell, D.; Stevens, G.; Tickle, P. *Chem. Commun.* **1970** 792; (d) Gaspar, P. P.; Whitsel, B. L.; Jones, M. Jr.; Lambert, J. B. *J. Am. Chem. Soc.* **1980**, *102*, 6108; (e) Closs, G. L. *Top. Stereochem.* **1968**, *3*, 193.
84. Moss, R. A.; Dolling, U.-H. *J. Am. Chem. Soc.* **1971**, *93*, 954.
85. (a) Jones, M. Jr.; Rettig, K. R. *J. Am. Chem. Soc.* **1965**, *87*, 4013, 3015; (b) Gaspar, P. P.; Lin, C. T.; Whitsel-Dunbar, B. L.; Mack, D. P.; Balasubramanian, P. *J. Am. Chem. Soc.* **1984**, *106*, 2128.
86. Baer, T. A.; Gutsche, C. D. *J. Am. Chem. Soc.* **1971**, *93*, 5180.
87. (a) Platz, M. S.; Savino, T. G.; Kanakarajan, K. *J. Org. Chem. Soc.* **1986**, *51*, 1305; (b) Savino, T. G.; Senthilnathan, V. P.; Platz, M. S. *Tetrahedron* **1986**, *42*, 2167; (c) Kirmse, W.; Horner, L.; Hoffmann, H. *Liebigs Ann. Chem.* **1958**, *614*, 19.
88. Hutchinson, C. A., Jr.; Kohler, B. E. *J. Chem. Phys.* **1969**, *51*, 3327.
89. (a) Brandon, R. W.; Closs, G. L.; Davoust, C. E.; Hutchison, C. A., Jr.; Kohler, B. E.; Silbey, R. *J. Chem. Phys.* **1965**, *43*, 2006; (b) Anderson, R. J. M.; Kohler, B. E. *J. Chem. Phys.* **1976**, *65*, 2451.
90. Gould, I. R.; Turro, N. J.; Butcher, J., Jr.; Doubleday, C., Jr.; Hacker, N. P.; Lehr, G. F.; Moss, R. A.; Cox, D. P.; Guo, W.; Munjal, R. C.; Perez, L. P.; Fedorynski, M. *Tetrahedron* **1985**, *41*, 1587.
91. Griller, D.; Nazran, A. S.; Scaiano, J. C. *J. Am. Chem. Soc.* **1984**, *106*, 198.
92. Bohne, C.; Kennedy, S. R.; Boch, R.; Negri, F.; Orlandi, G.; Sibrand, W.; and Scaiano, J. C. *J. Phys. Chem.* **1991**, *95*, *10*, 300.
93. Kirmse, W.; Kilian, J.; Steenkens, S. *J. Am. Chem. Soc.* **1990**, *112*, 6399.

CARBENES AND CARBORANES

Maitland Jones, Jr.

I. INTRODUCTION

Sometime in the late 1970s Paul Schleyer gave a seminar at Princeton, roughly titled "Carbon Atoms in Strange Situations." In his talk he mentioned, not much more than in passing, the icosahedral carboranes in which carbon (and boron) is hexacoordinate. He made the connection between the classical, two-dimensional aromatic compounds of organic chemistry and these icosahedra, constructed entirely of three-center, two-electron bonding, that are, nonetheless, still aromatic.

Advances in Carbene Chemistry
Volume 1, pages 161–184.

ISBN: 1-55938-545-6

o-carborane *m*-carborane *p*-carborane

Figure 1. In this and the illustrations following, the dots represent carbons; every other vertex is a boron, and there is a hydrogen at every encumbered vertex.

The icosahedral carboranes come in three flavors: *ortho*, in which the pair of carbons is adjacent; *meta*, in which they are in a 1,3 relationship; and *para*, in which they occupy opposite vertices of the icosahedron (Figure 1).

I can recall having the thought that we might have a try at using carboranes as substrates for reactions with carbenes and as platforms on which to base carbenes and other reactive intermediates. I am also certain that I treated the idea with the skepticism born of too-many years of looking at lines between atoms and thinking, "two electrons." Clearly, these molecules had an excess of bonds and were, therefore, something to be treated with caution, if not overtly feared. Visions of difficult vacuum line work and molecules bursting into green flame at unfortunate moments went through my mind, just as they doubtless did in the heads of the many potential graduate students I have (mostly) unsuccessfully attempted to seduce into boron chemistry.

Although it certainly is the case that the bonding is not of the usual (for organic chemistry) sort, instability is not what is found. Rather, these aromatic molecules are extraordinarily robust, sharing with their more conventional two-dimensional aromatic cousins the stability conferred by a completely filled set of low-energy, bonding molecular orbitals. In the icosahedral carboranes a set of 13 bonding molecular orbitals is occupied by 26 delocalized framework electrons. Three-center, two-electron bonding can be a fine way to hold even "organic" atoms together! Moreover, these molecules have other attractive properties. They are often crystalline, yet their spherical shape makes them volatile enough to pass happily through the gas chromatograph. In recent times the availability of ^{11}B NMR has simplified the problem of structure determination that must have bedeviled the pioneers in carborane chemistry.

In the late 1970s, I convinced Bob Gallucci to spend the last six months of his Ph.D. work on investigating the reaction between methylene and *o*-carborane. Eventually, an intrepid graduate student, Sarangan Chari, was convinced (bribed, really) to work exclusively in boron chemistry. To his credit he prospered, and a number of able young people followed Chari in our boron work. All too often, however, I still see the eyes of a prospective co-worker start to glaze over as it becomes clear that I expect him or her to deal with compounds containing ten boron atoms and only two carbons!

In this chapter I will outline our work on carbenes and carboranes over the last 12 years. Too many times I will have to stop short of a definitive answer, and point out what we think, not what we know. Beyond doubt, some of the ideas suggested here will turn out to be wrong. This is a progress report, not the final summation of what we have done.

II. REACTION OF CARBENES WITH CARBORANES

If the icosahedral carboranes are aromatic, might they not react with carbenes in the same fashion as benzene? Electrophilic aromatic substitution of carboranes is common,[1] with the borons most remote from the carbons (B_9) substituted first. The rate of substitution slows as the distance to the carbons decreases (B_9–H > B_8–H > B_4–H > B_3–H); the carbon–hydrogen bonds themselves are never touched. It is curious that the mechanism of this remarkable reaction seems not to have been explored, either experimentally or computationally. The critical question would seem to be the presence or absence of an intermediate analogous to the pentadienyl cation of classical aromatic substitution.

A visiting faculty member from Beijing, Guo-xiu Zheng, took up the carbene problem in 1982. She found that insertion occurred most easily in the boron–hydrogen bonds most remote from the carbons (again, B_9–H), and that the efficiency of insertion decreased as the distance to the carbons decreased (again, B_9–H > B_8–H > B_4–H > B_3–H) (see Figure 2 and Table 1).[2] A reasonable analysis would seem to relate the reaction of an electrophilic carbene and the known preferences in electrophilic aromatic substitution; however, there is one disturbing observation. The ratios of the products change with time, ultimately reaching the

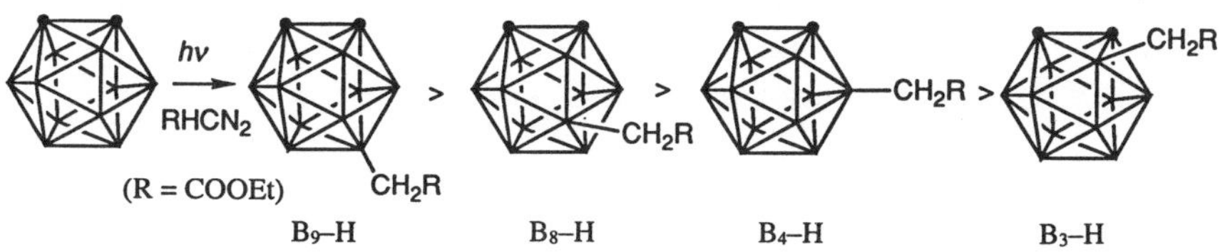

Figure 2.

Table 1. Insertion of Carboethoxycarbene into the C–H Bonds of *o*-Carborane[a]

Generation	*C–H*	*B_3–H*	*B_4–H*	*B_8–H*	*B_9–H*
	Relative Percentage of Products				
hν	0	4	19	29	47
Δ	0	3	19	27	52
Δ/CuCl	0	5	29	20	46

Note: [a]t = 1 hr.

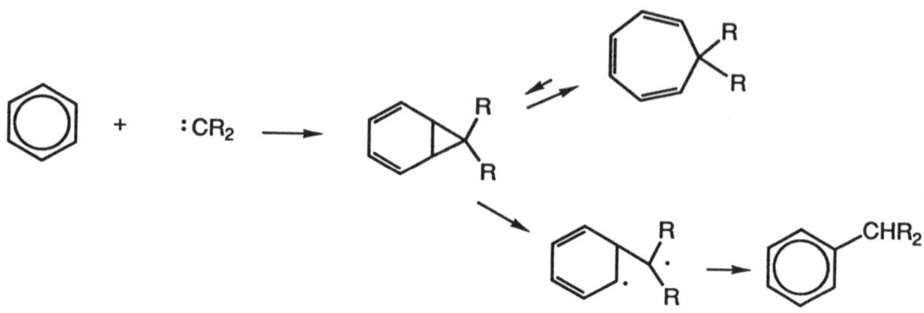

Figure 3.

statistical ratio (the C–H bonds remain unattacked). The products themselves are not isomerized under the reaction conditions. Despite our promise in 1983 to investigate this phenomenon, its cause remains a mystery. We suspect that the reaction is no straightforward insertion. In the reaction of carbenes with benzene, an intermediate norcaradiene equilibrates with a generally more stable 1,3,5-cycloheptatriene. In some cases the norcaradiene can break a "side" bond, and shift a hydrogen to give the apparent product of carbon–hydrogen insertion (Figure 3).

We think that the reaction of carboethoxycarbene with *o*-carborane may follow a similar course, with an unstable 13-vertex cage playing the part of the norcaradiene. "Walk" rearrangements could precede rearomatization and account for the randomization of positional isomers with time, but there is absolutely no firm evidence for this conjecture.

We also know that methylene reacts with *o*-carborane.[3] The positional preferences of methylene could not be worked out at the time we first did the work, although it would be a simple task now. The C–H bonds are not attacked, even by the notoriously reactive methylene.

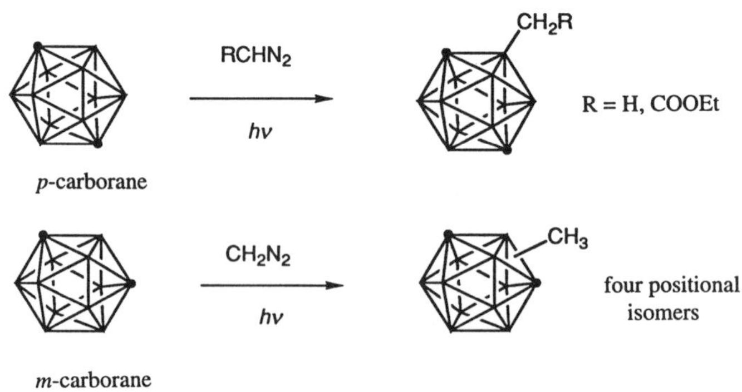

Figure 4.

There is a report in the Korean literature of the reactions of methylene, difluorocarbene, and dichlorocarbene with *o*-carborane.[4] Unfortunately, the reported properties of some of the products (^{1}H NMR spectra, melting points) are at variance with the known properties of the compounds described. The assertion that *m*-carborane and *p*-carborane did not react with methylene seemed so unlikely that we checked the data. Dr. Kaixu Yuan has shown that both methylene and carboethoxycarbene react nicely with *p*-carborane to give a single adduct, and that the appropriate number of four methyl-*m*-carboranes is produced from the photolysis of diazomethane in a hexafluorobenzene solution of *m*-carborane (Figure 4).[5] We suggest that the remainder of the aforementioned report[4] be treated with some caution.

III. PROPERTIES OF 1-*o*-CARBORANYLCARBENE

Phenylcarbene is an unexceptional, if reactive, species. It might well serve as the prototypal carbene, less reactive than the voraciously electrophilic methylene, but still reactive enough to exemplify the properties that have made carbenes so attractive to the few of us who have made their study the focus of our research efforts. Most of carbene chemistry can be encapsulated in the single diagram of Figure 5.

As with many carbenes, phenylcarbene has two reactive spin states fairly close in energy. The triplet is the ground state[6] but is essentially undetectable in chemical reactions in fluid solution. Even if one enters the equilibrium from the triplet side through photosensitized decomposition of phenyldiazomethane, it is the singlet carbene that is the reactive species. The problem is that even though the equilibrium between the two spin states must favor the ground state triplet, the rates of reaction of the singlet are many orders of magnitude greater than those of the triplet.

The triplet carbene can be detected and immobilized at low temperature. The triplet is bent, with an H–C–Ph angle of 140–155°. The zero-field-splitting parameter D (0.5098 cm^{-1}) shows that one electron is substantially delocalized over the benzene ring (Figure 6).[6]

Our original idea was simply to generate a carbon-substituted carboranylcarbene and make the obvious comparison to the classical intermediate. One problem is

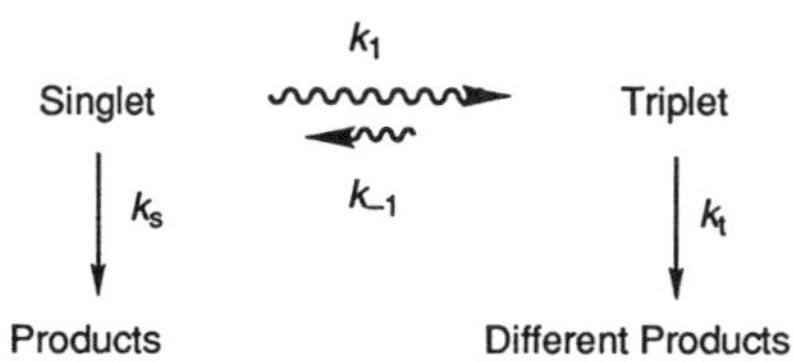

Figure 5. Carbene chemistry in a nutshell

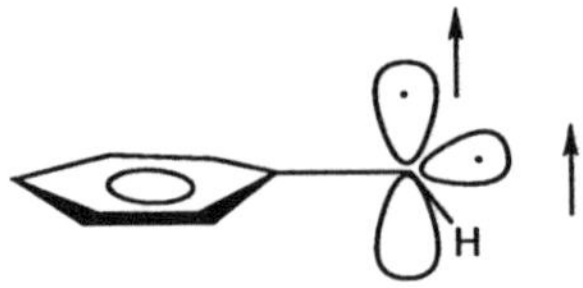

Figure 6.

how to make the precursor carboranyl diazo compound. I want to emphasize especially here that normal "organic" manipulations are possible with these carboranes. In addition, the C–H bonds of the carboranes are reasonably acidic and this offers an opportunity for further manipulation. Sarangan Chari's route to 1-diazomethyl-*o*-carborane (**1**) and 1-methyl-2-(diazomethyl)-*o*-carborane (**2**) takes advantage of the availability of 1-vinyl-*o*-carborane. Ozonolysis is followed by tosylhydrazone formation, generation of the sodium salt, and heating at about 80 °C to generate the diazo compounds (Figure 7).[7]

Irradiation of the diazo compounds **1** and **2** in frozen solutions of methyltetrahydrofuran or propylene carbonate at 5 K led to triplet EPR spectra for carbenes **3** and **4** from which the $|D/hc|$ and $|E/hc|$ values could be extracted.[8] In contrast to phenylcarbene, the $|D/hc|$ values (**3**: 0.6860, **4**: 0.6820) showed that little delocalization into the cage was possible. This is strong evidence that the frontier orbitals of the cage are poorly connected to external free valence. No evidence for hydrogen abstraction by **4** to give **5** could be seen (Figure 8).

The traditional test for the reacting spin state, not necessarily the ground state, is the stereochemistry of addition to alkenes. The presumption is that the singlet state will form both new bonds in a single step, and will therefore be unable to disturb the stereochemical arrangement of any groups on the alkene. The triplet, by

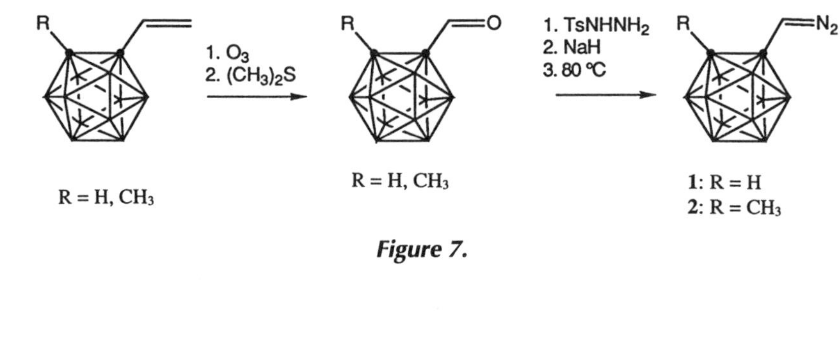

R = H, CH₃

R = H, CH₃

1: R = H
2: R = CH₃

Figure 7.

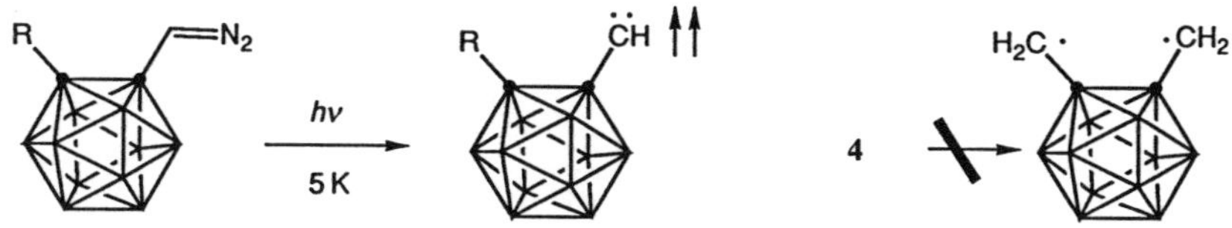

1: R = H
2: R = CH3

3: R = H
4: R = CH3

5

Figure 8.

contrast, first forms a triplet 1,3-diradical in which rotation can scramble the stereochemistry originally present in the alkene. Irradiation of **1** in *cis*-2-butene gives only a mixture of the *cis,anti* and *cis,syn* diasteriomers **6** and **7**.[7] Ten years later, Ji Li's reinvestigation of the reaction with *trans*-2-butene revealed a small amount of the *cis,anti* cyclopropane **6** not seen earlier (Figure 9).[9]

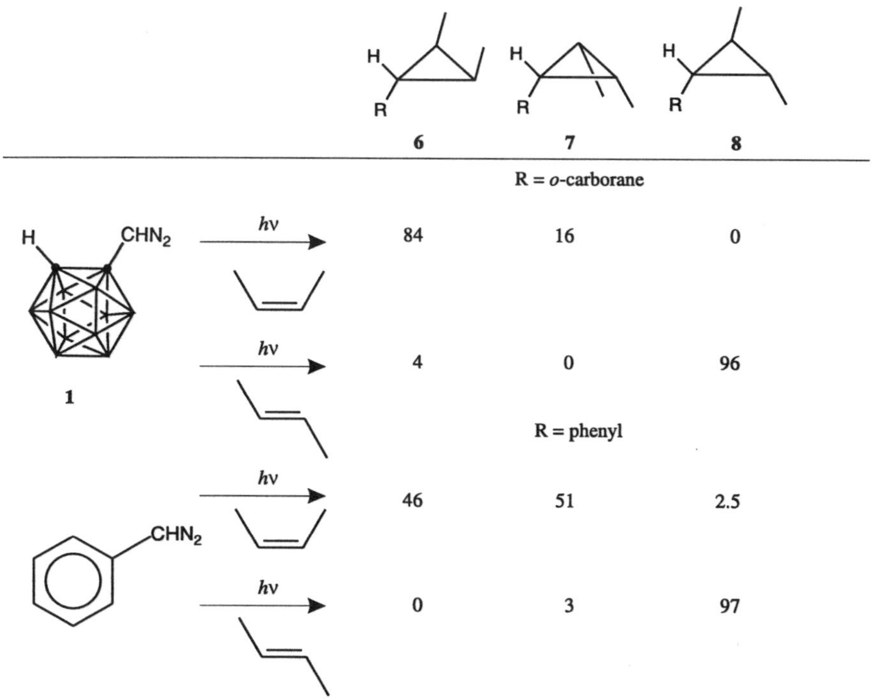

Figure 9.

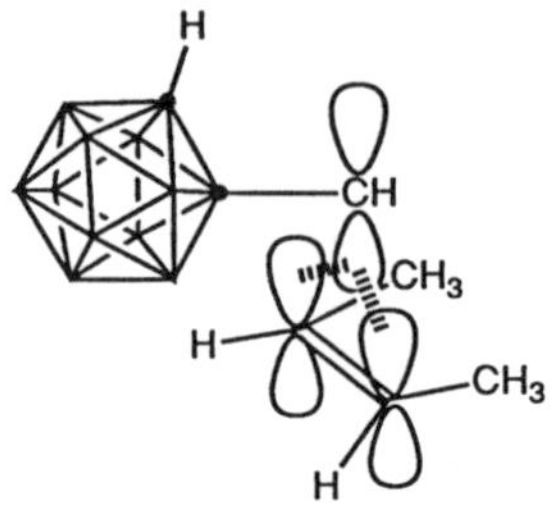

Figure 10.

The overall picture rather closely resembles phenylcarbene. Cycloaddition must be overwhelmingly through the singlet state, even though the triplet is the ground state. For the carboranylcarbene, there is a strong preference for the *cis,anti* cyclopropane over the more congested *cis,syn* compound, and this preference is not present in the less sterically demanding addition of phenylcarbene. It is always dangerous to make much of the small amounts of nonstereospecific addition that sometimes appear in otherwise stereospecific singlet carbene additions. Still, it may be that singlet addition to the *trans* alkene, in which both sides of the alkene are guarded by a methyl group, is more difficult than addition to the *cis* alkene, in which one side is open. Addition must be from above the plane of the alkene, and the carborane can point away from the methyls and toward the hydrogens in the addition to *cis*-2-butene (Figure 10). Explanations for the preference of phenylcarbene for *cis,syn* addition over *cis,anti* all depend upon orbital connection between the phenyl ring and the orbitals on the extra-ring carbon. As the EPR spectra of **3** and **4** show, there is not such a connection in the carboranylcarbenes. So, it is not surprising to see the strong preference of **3** for the *cis,anti* cyclopropane **6**.

If, as seems likely, the small amount of the *cis* cyclopropane formed in the reaction with *trans*-2-butene is the result of triplet carbene addition, it should be possible to estimate the mix of reacting spin states. In this analysis, however, it is necessary to augment the amount of *cis,anti* compound **6** with however much *trans* cyclopropane **8** derives from the triplet. In the absence of data on the stereochemical preferences of pure triplet, one can only estimate. This is especially hard in this case because the relative energies of the cyclopropane isomers are not known. Nonetheless, if the triplet is responsible for the formation of **6** from *trans*-2-butene, some **8** must be formed as well. We take, as a guess for the thermodynamic preference, a 60/40 ratio of *trans/cis*. If this is correct, the 4% of observed **6** translates into 10% triplet reaction in the addition of **3** to *trans*-2-butene. There is no sign of triplet reactivity in the addition to *cis*-2-butene.

Other reactions of **3** can be compared to those of phenylcarbene. Carbenes not only add to π systems, but insert into carbon–hydrogen bonds as well. The ability of a carbene to select among various kinds of carbon–hydrogen bonds, generally

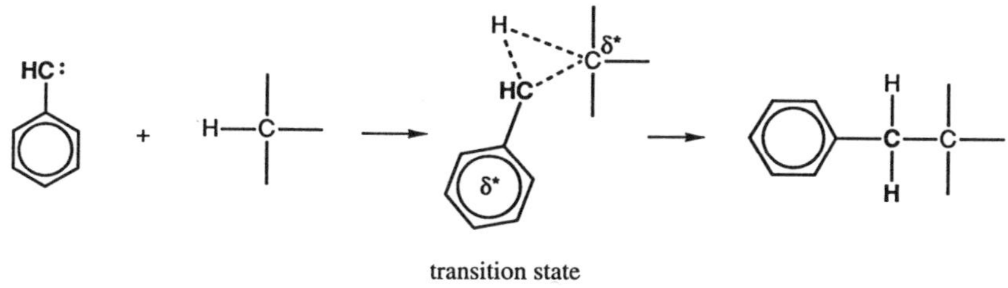

Figure 11.

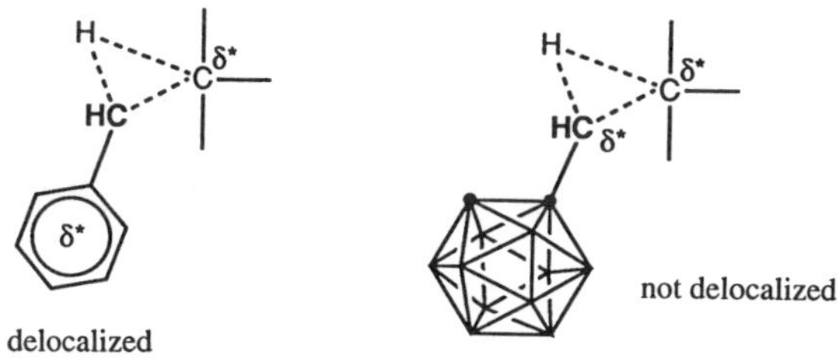

Figure 12.

primary, secondary, and tertiary, has been used as a measure of reactivity. For example, phenylcarbene, presumably singlet, generated from several sources shows the same selectivity when inserting into the different carbon–hydrogen bonds of pentane.[10] This has been taken as strong evidence that the free carbene is involved in all the reactions. For the reaction of singlet phenylcarbene with pentane, a secondary carbon–hydrogen bond is about 8.4 times as reactive as a primary carbon–hydrogen bond. In the transition state for insertion, developing free valence is stabilized both by the benzene ring and the substituents on the carbon from which the hydrogen is being plucked. Whether the transition state is polar, or, more likely, radical in nature, the more substituted the carbon, the better (Figure 11).

Carbene **3** shows similar, but slightly reduced selectivity.[7] In butane, secondary carbon–hydrogen bonds are more reactive than primary bonds by a factor of 3. The carboranylcarbene is less selective than phenylcarbene. It is risky to rationalize small energy differences, but this difference is nonetheless consistent with a pair of transition states one of which is stabilized by delocalization into the ring and another in which resonance stabilization is greatly attenuated (Figure 12).

IV. BORON-SUBSTITUTED CARBENES

A second theme in carbene chemistry has been the effect of substituents on the mix of singlet and triplet chemistry. Most attention has been paid to the rich chemistry of the narrowly separated singlet and triplet arylcarbenes and to the effects of halogen substitution. In the chemistry of arylcarbenes, reactions of both singlets and triplets have been uncovered. Indeed, it was almost 30 years ago that fluorenylidene was used to develop for the first time the reactions of well-characterized singlet and triplet states of the *same* carbene (Figure 13).[11]

Generally, halogen, oxygen, or nitrogen substitution preferentially stabilizes singlets as electrons are shared between a filled *p* orbital on the heteroatom and the empty 2*p* orbital on carbon. For example, the chemistry of the halocarbenes is completely dominated by the singlet state (Figure 14).[12]

Almost no experimental attention has been paid to the other side of the periodic table, where relatively electropositive elements dwell, and where empty orbitals tend to be more prominent than filled. However, theoretical work gives us some idea of what to expect. For an atom such as boron, the presence of an empty *p* orbital should stabilize the singlet state through overlap with the filled orbital of the divalent carbon (Figure 15).[13a,b]

In this case, however, there is a compensating effect stabilizing the triplet state. This operates through the σ system and depends on the electronegativity difference between the two atoms.[13a] The effect is to donate electrons through the σ system and, for atoms less electronegative than carbon, favors the triplet state. Thus, for a simple boron-substituted carbene such as H_2B–CH, there should be compensating effects stabilizing both the singlet and triplet states. Calculations bear this out, with

Figure 13.

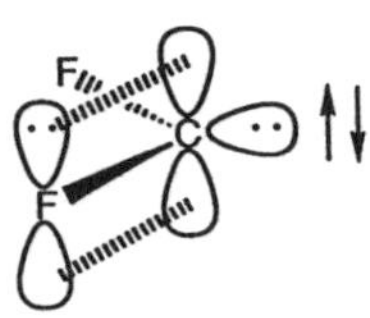

Figure 14.

Figure 15.

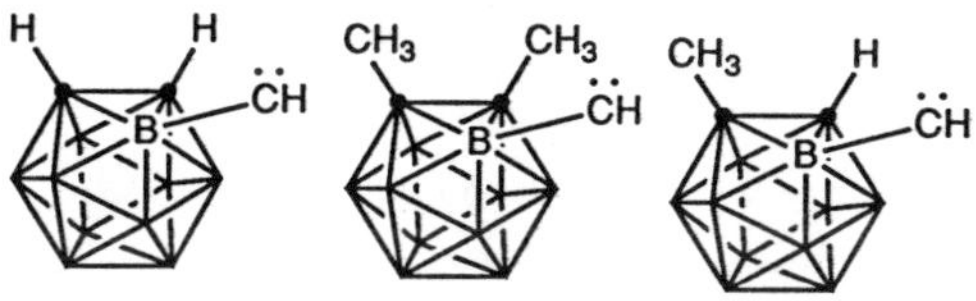

9 **10** **11**

Figure 16.

the most recent efforts predicting the two spin states to be close in energy. In the most simple example, singlet H_2B–CH is predicted to be the ground state by 4–6 kcal/mol.[14]

In 1990, Ji Li developed a route to the first boron-substituted carbenes (**9**, **10**, and **11**) through manipulations of substituted *o*-carboranes (Figure 16).[9,15]

Although Li's method does yield carbenes adjacent to a boron atom, it must be emphasized that this is no ordinary boron. In particular, the empty 2*p* orbital, so influential in the chemistry of "normal" tri-substituted boranes, is occupied in the web of three-center, two-electron bonding that makes up the icosahedral cage. Although the singlet-stabilizing π effect seems likely to be largely absent in a carboranylcarbene, the triplet-stabilizing σ effect remains. In addition, there appears to be substantial electron density at the B_3 position, and this should especially favor the triplet. A variety of ^{1}H NMR chemical shift comparisons demonstrates

Table 2. Some ^{1}H Chemical Shifts[a]

		Position of R		
	R	C_1	B_3	B_9
	C**H**3 (H underlined)	2.04	0.56	0.30
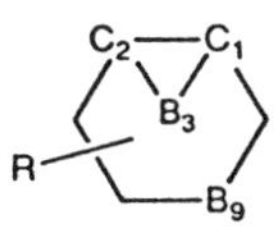	C**H**N2 (H underlined)	4.38	3.15	
	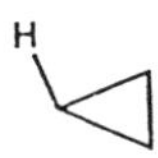	1.36	ca. −0.7	

Note: [a]δ, $CHCl_3$.

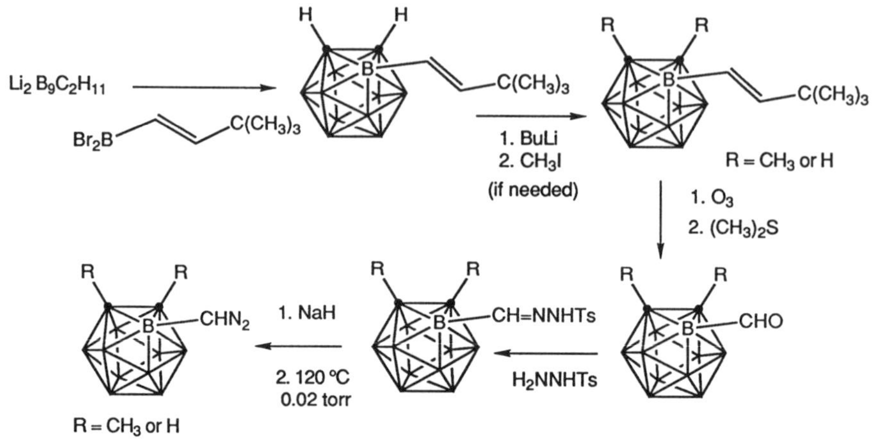

Figure 17.

the extra shielding provided by B_3 (Table 2). The B_9 position is even more shielded and should be especially effective at triplet stabilization.[16]

Thus, our preliminary expectation was that a boron-substituted carboranylcarbene might show enhanced triplet reactivity, as the triplet should be especially favored in the equilibrium of Figure 5.

The new carbenes are generated by the traditional route; namely, photolysis of a diazo compound. The diazo compounds are made from the tosylhydrazone salts, themselves available through the route shown in Figure 17. The tosylhydrazones were made by insertion of a vinyl-substituted boron into $Li_2B_9C_2H_{11}$,[17] methylation of one or two cage carbons for **10** and **11**, ozonolysis, and reaction with tosylhydrazine. Heating the sodium salts under vacuum (120 °C, 0.02 torr) led to the diazo compounds in about 80% yield.

In collaboration with Linda Bush of the Berson group at Yale, Rodney Blanch has determined that photolysis of these diazo compounds at ~11 K leads to persistent triplet EPR signals. At about 32 K the signals irreversibly decay. Analysis of the zero-field splitting parameters showed that both triplet **9** and **10** are linear ($|E/hc| \approx 0$). Carbene **9** was shown to be a ground state triplet as the Curie law was obeyed.[18] Table 3 summarizes the EPR spectra of carboranylcarbenes.

Table 3. EPR Spectra of Carboranylcarbenes[a]

Carbene	$\|D/hc\|$	$\|E/hc\|$
1-*o*-Carboranylcarbene	0.6860	0.0302
1-Methyl-2-*o*-carboranylcarbene	0.6920	0.0293
3-*o*-Carboranylcarbene	0.657	< 0.002
1,2-Dimethyl-3-*o*-carboranylcarbene	0.661	< 0.002

Note: [a]Zero-field splitting paramters (cm^{-1}).

Now it is time to see if the reactivities of the boron-substituted carbenes match our expectations of enhanced triplet reactivity.

Table 4 compares the stereochemistry of addition of **3**, **9**, and **10** to *cis*- and *trans*-2-butene. It is the results of addition to *cis*-2-butene that are most revealing. *trans*-2-Butene is likely to give overwhelming amounts of *trans* adduct no matter what the mechanism, but a stepwise addition to the *cis* isomer should lead to substantial amounts of the more stable *trans* adduct. The last column of Table 4 is calculated on the assumption that the thermodynamic preference for *trans* adduct over *cis* is about 60/40. Although only small amounts of triplet are involved in any of the cycloaddition reactions, there seems to be more reactive triplet **10** than **9**.

Table 4. Stereochemistry of Addition Reactions of Carboranylcarbenes

	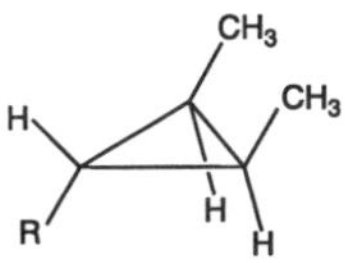	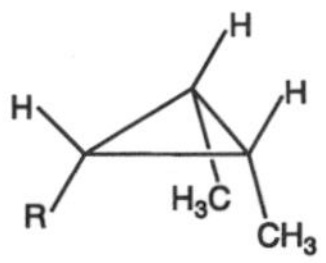	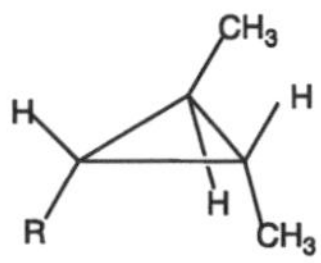	% Triplet reaction
	Relative Percentage of Products			
9 + 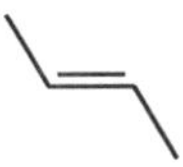	—	—	100	0
9 + 	74	24	2	3
10 + 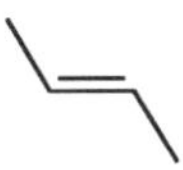	9	—	91	22
10 + 	63	25	12	20
3 + 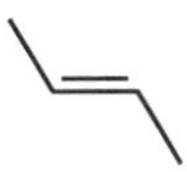	4	—	96	10
3 + 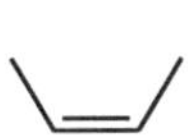	84	16	0	0

Note: R = carborane

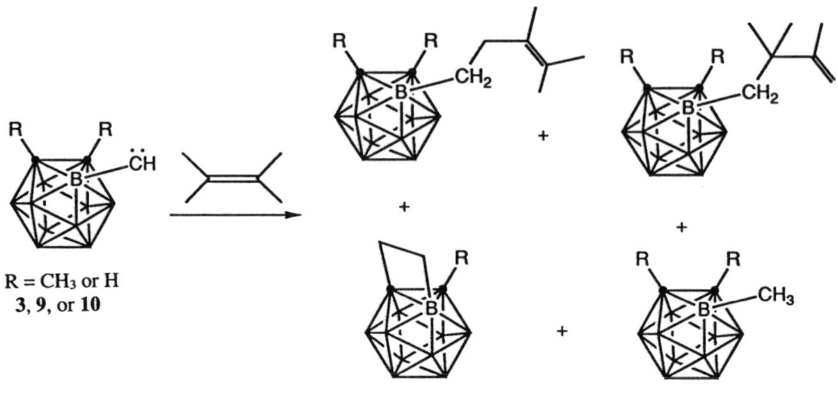

$R = CH_3$ or H
3, **9**, or **10**

12: $R= CH_3$
13: $R = H$

14: $R = CH_3, CH_3$
15: $R = H, CH_3$
16: $R = H, H$

Figure 18.

This is surely reasonable given the additional interference to two-bond-forming reactions (singlets) of the pair of methyl groups on **10**. One-bond forming reactions (triplets), taking place at the periphery of the molecule, will have an advantage for sterically congested carbenes, and this seems to be expressed here.

The reactions of carbenes **3**, **9**, and **10** with alkenes lead to products other than cyclopropanes. Both inter- and intramolecular carbon–hydrogen insertion reactions occur with alkenes, and substantial amounts of 3-methyl-*o*-carboranes are also formed (Figure 18).

One might hope to evaluate the contribution of triplet carbene in the insertion reaction by determining the amount of isomerized products in the intermolecular insertion reaction. Triplet carbene should react by hydrogen atom abstraction, and this should lead to the isolation of isomerized insertion product (Figure 19).

Unfortunately, the products of insertion into the allylic carbon–hydrogen bonds of alkenes are not stable, and isomerize to give the products of allylic rearrangement. So, even though the small amounts of observed isomerized materials correlate

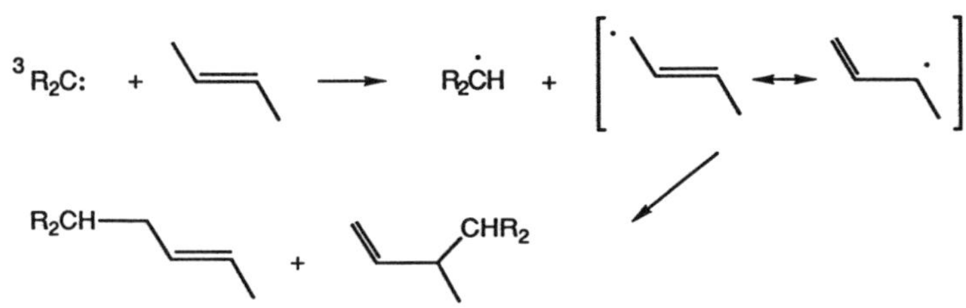

Figure 19.

rather well with the small amounts of the “wrong” stereoisomer obtained in cycloaddition reactions, these data are not reliable. Table 5 gives the relative amounts of cyclopropanes, total intermolecular insertion products, products of intramolecular insertion (the “cyclobutanes” **12** and **13**), and *B*-methyl compounds **14**, **15**, and **16** for *B*-substituted carbenes **9**, **10**, and **11** as well as for the *C*-substituted carbene **3**.

It appears as though cyclopropane formation is favored somewhat by easy access to the alkene. Thus, *cis*-2-butene, in which one side of the alkene is unguarded by

Table 5. Products from Carbenes **3**, **9**, **10**, and **11**

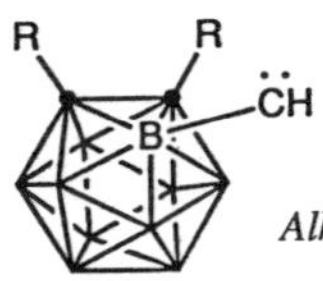	*Alkene*	*Total allylic insertion*	*Total cyclopropanes*	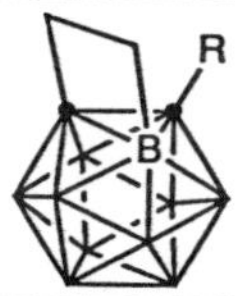	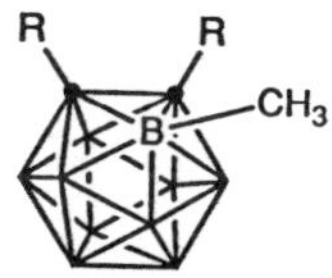
		Percentage of Products			
R = R = CH_3 **10**		36	44	4	15
	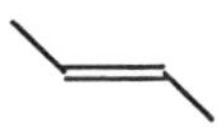	40	33	4	23
	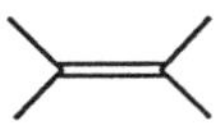	38	25	8	29
R = R = H **9**		28	50	—	19
	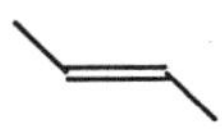	41	36	—	23
	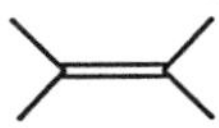	48	16	—	36
R = H, R = CH_3 **11**	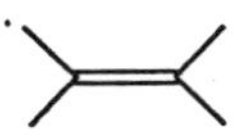	49	18	3	31
 **3**	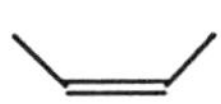	8	84	—	8
	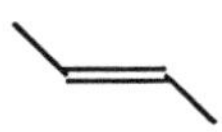	7	87	—	6
	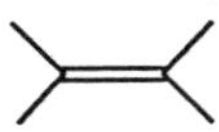	27	67	—	6

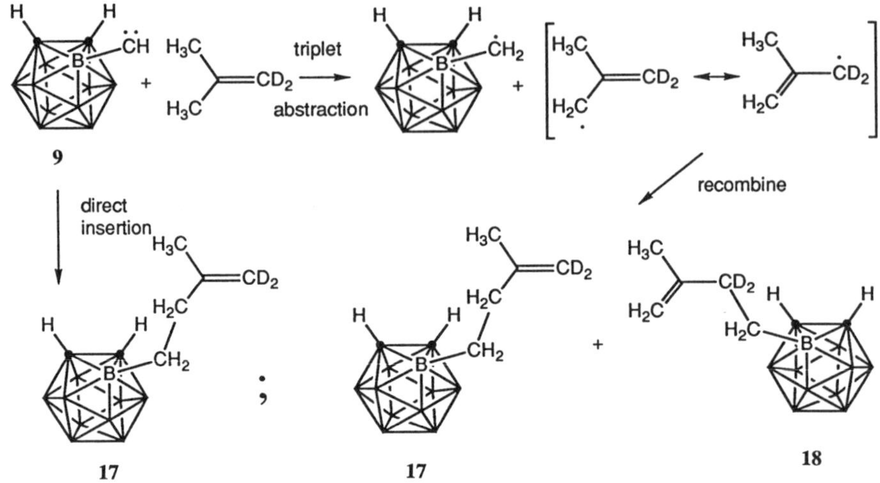

Figure 20.

methyl groups, gives the most cyclopropane, and tetramethylethylene, in which the maximum shielding of the double bond exists, gives the least.

As the products of allylic rearrangement were not stable over time, they could not be used as a diagnostic for triplet carbene activity. Therefore, a labeling experiment was undertaken in the style of Doering and Prinzbach, who in 1959 had the daring to irradiate ^{14}C-labeled 2-methylpropene in what was essentially neat diazomethane.[19] They determined that allylic insertion took place without rearrangement, and attributed this result to the singlet nature of reacting methylene. Deuterium-labeled 2-methylpropene was synthesized by Ji Li and used as a substrate for carbene **9**. As Figure 20 shows, the triplet should reveal itself through the appearance of protons in the terminal methylene position of the "insertion products" **17** and **18**.

Analysis of the crude products by ^{1}H NMR spectroscopy showed only small amounts of protium in the position of absorption of the terminal methylenes of **18** (δ = 4.74 ppm, 5–10%). Although it was not possible to make an accurate quantitative determination of the amount of triplet participation in allylic insertion, clearly the insertion product was overwhelmingly **17**, the isomer derived from the singlet, not **18**, the isomer diagnostic for the triplet. The reason seems to be that carboranyl radicals prefer abstraction of a second hydrogen to recombination. Methyl compounds **14**, **15**, and **16** are substantial products of the reactions of carboranylcarbenes with alkenes (see Figure 18 and Table 4). Similarly, **16** appears in 15% yield in the reaction of 2-methylpropene with **9**. It seems safe to assign most insertion products to the singlet.

Table 6. The Effect of Oxygen on the Product Ratio from Carbene **10**

10 +

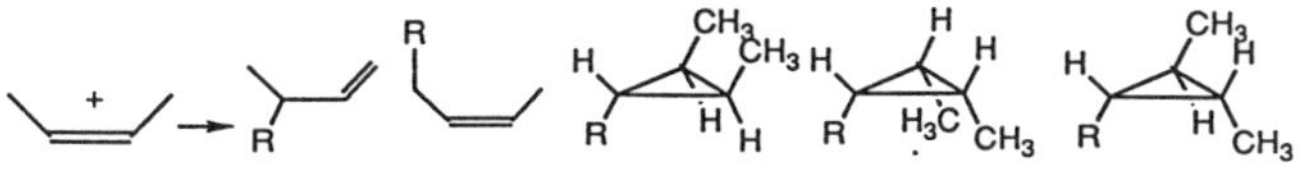

						12	**14**
No oxygen	11	25	28	11	5	4	15
In oxygen	7	32	37	21	< 1	2	trace

Note: R = carborane.

To summarize what we know at this point: It appears that the ground state triplet, boron-substituted carbenes **9**, **10**, and **11** give cyclopropanes largely through their singlet states. Table 4 reveals no great difference between these intermediates and **3**, the carbon-attached carbene. What, then, of our expectation of increased triplet reaction? Where is the triplet? The answer seems to be, "In the methyl groups of compounds **14**, **15**, and **16**." These compounds must arise by a double abstraction reaction and are most reasonably attributed to the triplet. In order to test this notion and to verify our mental separation of products into those formed from singlet carbene (cyclopropanes of retained stereochemistry and most insertion products) and triplet carbene ("wrong" cyclopropanes plus an estimated portion of the other cyclopropanes, and methyl compounds) we have repeated the reaction of **10** with *cis*-2-butene in an atmosphere of oxygen. Oxygen is a known trap for triplets,[20] and should reduce or eliminate compounds that owe their origin to a triplet reaction. Table 6 shows the products of the reaction of **10** with *cis*-2-butene in the absence and presence of oxygen.

As predicted, the products attributed to triplet carbene, the *trans* cyclopropane and methyl compound **14**, vanish. In their place are found *o*-carborane-3-carboxaldehyde, *o*-carborane-3-carboxylic acid, and 3-*o*-carboranyl formate (not shown in Table 6). The 3-aldehyde is the expected product of oxidation of the triplet carbene, and the other two compounds are known to derive from overoxidation of the aldehyde.[16] We now feel it justified to provide a new estimate of the amount of triplet reactivity present in the reactions of carbenes **3**, **9**, **10**, and **11** with *cis*- and *trans*-2-butenes (Table 7). The remaining uncertainties are the amount of triplet reactivity to assign to the allylic insertion reaction, and the amount (if any) to which triplet participates in the self-insertion reaction to give **12** and **13**. These will add only very small corrections to the numbers of Table 7, which assumes that the insertion products, like the cyclopropanes, are mainly derived from the singlet carbene. Carbenes **9**, **10**, and **11** show roughly twice the triplet reactivity as does **3**.

Carboranylcarbenes also react with alkanes. Ji Li and David Caparrelli have determined the selectivities of carbenes **9**, **10**, and **11** with a variety of alkanes.[9]

Table 7. Singlet and Triplet Reactivity for some Carbenes

Carbene	*Alkene*	*% Triplet*[a]	*% Singlet*[b]
3	*cis*-2-butene	9	91
	trans-2-butene	16	84
	2,3,-dimethyl-2-butene	> 6	
9	*cis*-2-butene	22	78
	trans-2-butene	23	77
	2,3,-dimethyl-2-butene	> 36	< 64
10	*cis*-2-butene	24	75
	trans-2-butene	30	70
	2,3-dimethyl-2-butene	> 29	< 71
11	2,3-dimethyl-2-butene	> 31	< 69

Note: [a]"Wrong" cyclopropane plus some "right" cyclopropane, plus methyl compound.
[b]"Right" cyclopropane plus allylic insertion, plus self-insertion.

Carboranylcarbenes also react with alkanes. Ji Li and David Caparrelli have determined the selectivities of carbenes **9**, **10**, and **11** with a variety of alkanes.[9] Note that these selectivities reflect a combination of direct singlet insertion combined with some triplet abstraction-recombination (Table 8).

An analysis of the reaction of **9** with cyclohexane and dodecadeuteriocyclohexane[21] was used to verify the notion that most insertion product comes from singlet carbene, and to provide an estimate for the isotope effect for the singlet insertion reaction. Carbene **9** was generated from irradiation of the diazo compound in both pure cyclohexane and dodecadeuteriocyclohexane to produce insertion products **19** and **20**. In a third experiment, **9** was allowed to react with a 1:1 mixture of cyclohexane and dodecadeuteriocyclohexane. The presumption in this experiment was that triplet **9** would be revealed through formation of "cross products" **21** and **22** formed from recombination of radicals produced by hydrogen and deuterium abstraction (Figure 21).

Adducts **19** and **20** were well-separated by gas chromatography, but the cross products **21** and **22** differ by only a single deuterium from **19** and **20** and could not be separated. Analysis of mass spectra showed that within the peak for **19** there was a small amount of a product with an extra deuterium (presumably **22**), and that within the peak for **20** was a compound with one less deuterium, (**21**). This analysis indicates that triplet abstraction could account for only about 5% of the overall insertion reaction.

When the reaction with the mixture of cyclohexanes was run in the presence of oxygen—a scavenger likely to remove all, or almost all, of the radicals formed by triplet reactions—the ratio of **19** and **20** changed from 1:1 to 1.2:1. This provides a measure of the isotope effect k_H/k_D for the singlet insertion reaction of **9** with cyclohexane.

Table 8. Carbon–Hydrogen Insertion into Alkanes

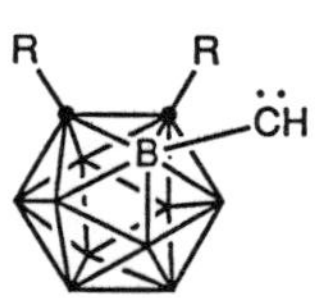	*Alkane*	*Primary (1°)*	*Secondary (2°)*	*Tertiary (3°)*	*2°/1° per H*	*3°/1° per H*	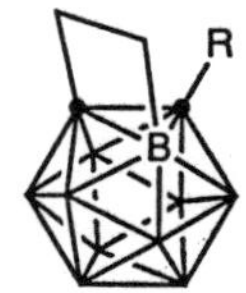 **12** *or* **13**	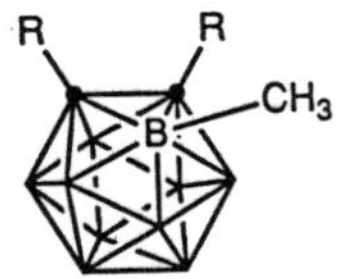 **14**, **15**, *or* **16**
$R = R = CH_3$ **10**	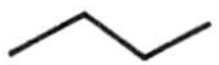	43	40		1.5		4	10
	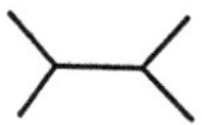	64		15		1.5	3	9
$R = R = H$ **9**		43	41		1.5			11
	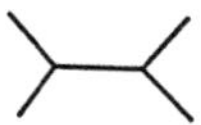	68		16		1.4		10
$R = H, R = CH_3$ **11**	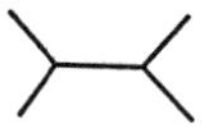	58		16		1.5	3	13
 **3**	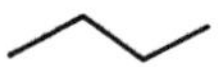				3.0			

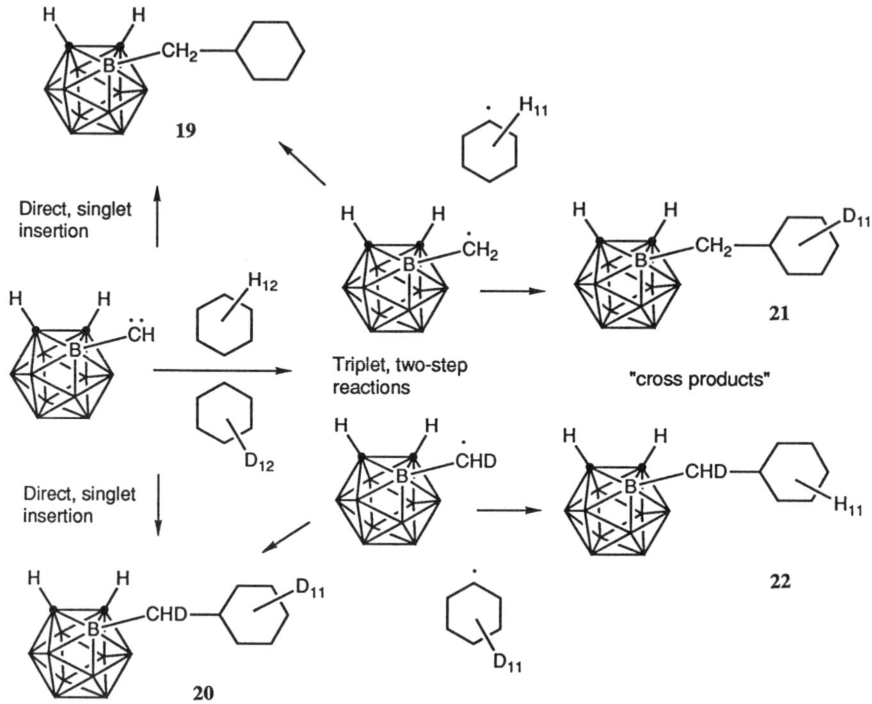

Figure 21.

V. OTHER CARBORANYLCARBENES

Almost no work has been reported on *m*-carboranylcarbenes, and exactly nothing has been published on *p*-carboranylcarbenes. The carbon-attached *m*-carboranylcarbene was shown by Chari to add in largely stereospecific fashion to *cis*- and *trans*-2-butene, to form cyclopropanes through addition to other alkenes, and to produce modest amounts (14–20%) of insertion products.[7] No detailed investigations are known, and no boron-attached *m*-carboranylcarbenes have yet been described (Figure 22).

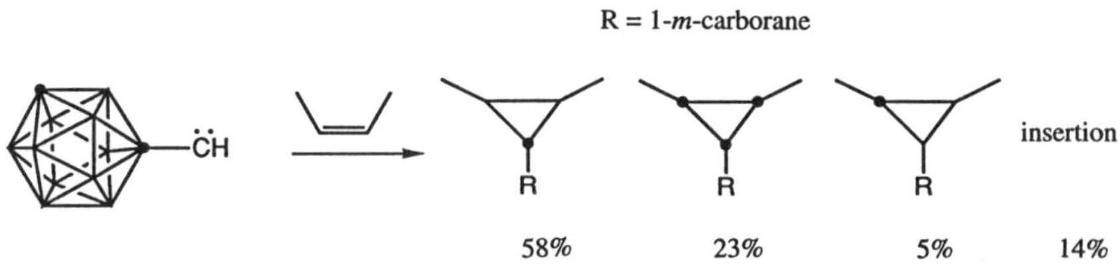

Figure 22.

VI. INTRAMOLECULAR REACTIONS OF CARBORANYLCARBENES

Shao-hai Wu took advantage of Guo-xiu Zheng's earlier observation[2] that only B–H bonds were active in the intermolecular insertion reaction of carbenes and *o*-carboranes (Table 1) to design a synthesis of the first carbon-to-boron bridged *o*-carboranes.[22] As found by Zheng, the 4 position is more reactive than the 3 position. Wu was able to produce six-, five-, and even four-membered rings (only 4% yield in the last case).[23] The mechanism of the cyclization reaction is open to question, as Wu found that Wolff rearrangement dominated the photolytic decompositions of the diazo compounds. Catalytic decomposition was by far the most effective method of forming the insertion products, and a free carbene is not likely to be involved (Figure 23).

A more exotic intramolecular reaction was found by Chiang and Chari in the gas phase decomposition of diazo compound **2**.[7] Not surprisingly, insertion took place at the proximate methyl group to give **23**. Much more unusual was the formation of 1-vinyl-*o*-carborane (**24**) in approximately equal amounts to **23** (Figure 24).

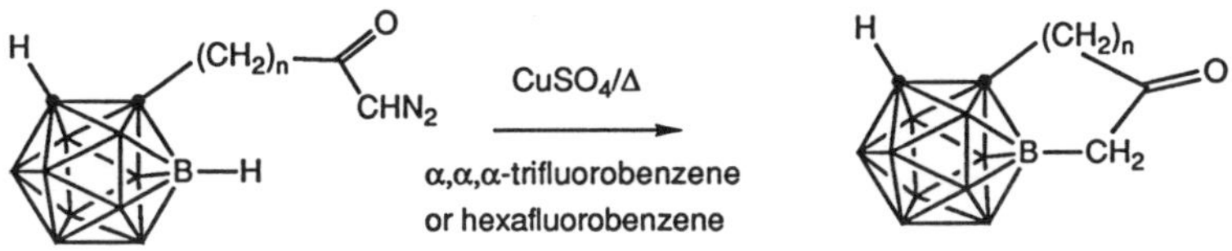

$n = 0, 1, 2$

Figure 23.

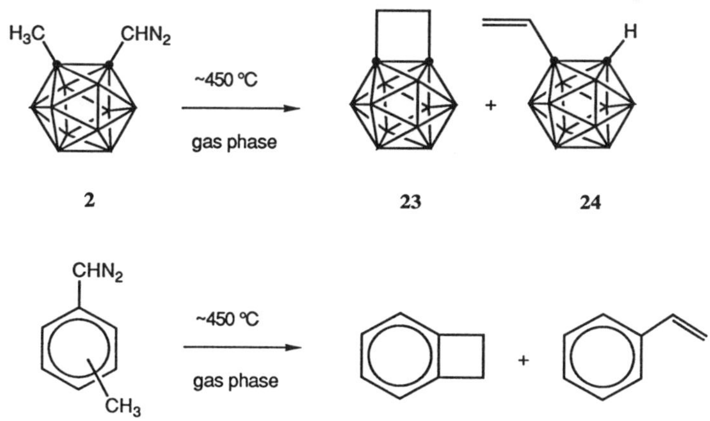

2 **23** **24**

Figure 24.

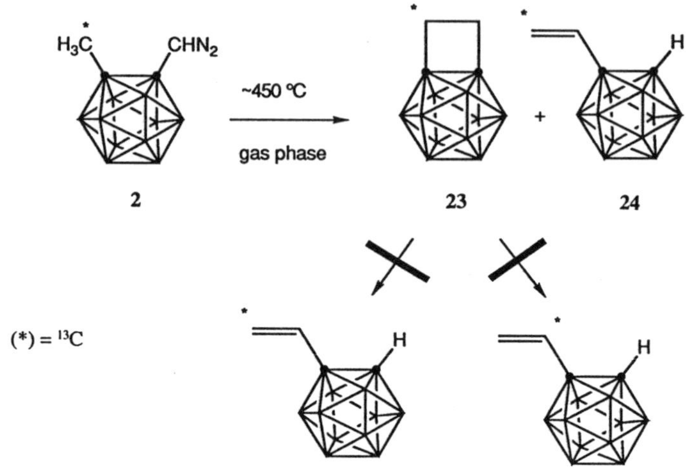

Figure 25.

A clear analogy can be made to reactions of the tolylcarbenes that rearrange in the gas phase to benzocyclobutene and styrene.[24] As in the benzene case, it was necessary to exclude rearrangement of the four-membered ring compound **23** to the vinyl compound **24**. This is a trickier business than it might seem at first, as gas-phase carbene reactions are highly exothermic and "hot molecule" rearrangements are all too likely in the initially formed products. Thus it is not a sufficient control experiment simply to resubmit **23** to the reaction conditions. Instead, it was necessary to decompose labeled diazo compound and show that the label was found only in the terminal vinyl position of **24** (Figure 25).

Unfortunately, as of this writing (April 1992), the remaining and long-overdue labeling experiments are not completed, although they are underway.[25] Accordingly, little can be said about the mechanism of this fascinating reaction at this point.

It is known that the related *meta* carbene **25** does not lead to 1-vinyl-*m*-carborane,[7] although this experiment must be treated with the caution appropriate to all reports of negative results (Figure 26).

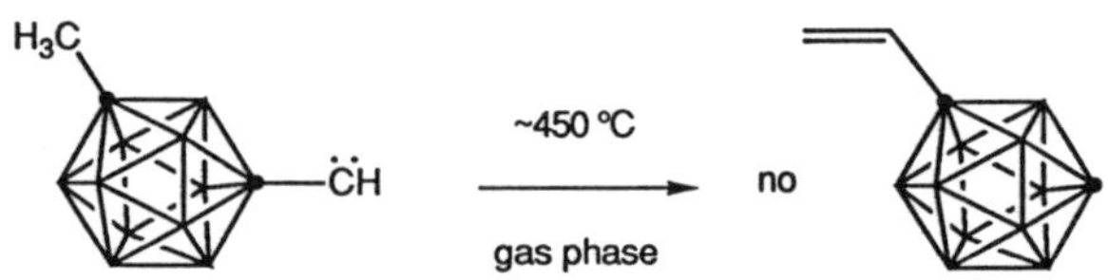

25

Figure 26.

ACKNOWLEDGMENTS

One of the blessings of working in such a non-trendy area as this one is that the students attracted to it are very likely to be doing this kind of chemistry for exactly the right reasons. The density of very able people in physical-organic, or mechanistic chemistry is very high, it seems to me. Whatever the reasons, I have been lucky to be associated with some extraordinary young people over the last several years, and I have learned much from working with them. Since the early days of Bob Gallucci and Sarangan Chari, Drs. Rodney Blanch, Sheau-hwa Chiang, Kaixu Yuan, Guo-xiu Zheng, graduate students Ji Li and Shao-hai Wu, and undergraduate David Caparrelli have all contributed mightily to our work on carboranes and carbenes. My thanks go to all these people, as well as to Paul Schleyer for provoking our initial interest.

We are also grateful for the continuous financial support from the National Science Foundation and the Petroleum Research Fund that has made this work possible.

REFERENCES

1. (a) Olah, G. A.; Surya Prakash, G. K.; Williams, R. E.; Field, L. D.; Wade, K. *Hypercarbon Chemistry*; Wiley, New York, 1987. (b) Muetterties, E. L. *Boron Hydride Chemistry*; Academic, New York, 1975. (c) Grimes, R. N. *Carboranes*; Academic, New York, 1970. (d) Onak, T. *Organoborane Chemistry*; Academic, New York, 1975. (e) Bregadze, V. I. *Chem. Rev.* **1992**, *92*, 209.
2. Zheng, G.-x.; Jones, M., Jr. *J. Am. Chem. Soc.* **1983**, *105*, 6487.
3. Gallucci, R. R. Ph.D. Thesis, Princeton University, 1976.
4. Sung, D. D.; Lee, J. D.; Choi, S. K. *Bull. Korean Chem. Soc.* **1987**, *8*, 63.
5. Yuan. K.; Jones, M., Jr. *Tetrahedron Lett.* **1992**, *33*, 7481.
6. Trozzolo, A. M.; Murray, R. W.; Wasserman, E. *J. Am. Chem. Soc.* **1962**, *84*, 4990.
7. Chari, S. L.; Chiang, S.-H.; Jones, M. Jr. *J. Am. Chem. Soc.* **1982**, *104*, 3138.
8. Hutton, R. S.; Roth, H. D.; Chari, S. *J. Phys. Chem.* **1981**, *85*, 753.
9. Li, J.; Caparrelli, D. J.; Jones, M. Jr. *J. Am. Chem. Soc.* **1993**, *115*, 408.
10. (a) Dietrich, H.; Griffin, G. W.; Pettersen, R. C. *Tetrahedron Lett.* **1968**, *153*. (b) Closs, G. L.; Moss, R. A. *J. Am. Chem. Soc.* **1973**, *86*, 4042.
11. (a) Jones, M., Jr.; Rettig, K. R. *J. Am. Chem. Soc.* **1965**, *87*, 4013. (b) Jones, M., Jr.; Rettig, K. R. *J. Am. Chem. Soc.* **1965**, *87*, 4015.
12. Regitz, M., Ed. *Carbene(oide), Carbine*, Houber-Weyl, Thieme, Stuttgart, 1989.
13. (a) Davidson, E. R. In *Diradicals*; Borden, W. T. Ed.; Wiley, New York, 1982, Chapter 2. (b) Feller, D.; Borden, W. T.; Davidson, E. R. *J. Chem. Phys.* **1979**, *71*, 4987.
14. (a) Luke, B. T.; Pople, J. A.; Krough-Jespersen, M.-B.; Apeloig, Y.; Karni, M.; Chandrasekhar, J.; Schleyer, P. v. R. *J. Am. Chem. Soc.* **1986**, *108*, 270. (b) Schleyer, P. v. R.; Luke, B. T.; Pople, J. A. *Organometallics*, **1987**, *6*, 1997.
15. Li, J.; Jones, M. Jr. *J. Am. Chem. Soc.* **1991**, *114*, 1094.
16. Li, J. Ph.D. Thesis, Princeton University, 1993.
17. Hawthorne, M. F.; Young, D. C.; Garrett, P. M.; Owen, D. A.; Schwerin, S. G.; Tebbe, F. N.; Wegner, P. A. *J. Am. Chem. Soc.* **1968**, *90*, 862.
18. Blanch, R. J.; Li, J.; Bush, L. C.; Jones, M. Jr. *J. Am. Chem. Soc.* **1992**, *114*, 9236.
19. Doering, W. von E.; Prinzbach, H. *Tetrahedron* **1959**, *6*, 24.
20. Morgan, S.; Platz, M. S.; Jones, M. Jr.; Myers, D. R. *J. Org. Chem.*, **1991**, *56*, 1351.

21. Barcus, R. L.; Hadel, L. M.; Johnston, L. J.; Platz, M. S.; Savino, T. G.; Scaiano, J. C. *J. Am. Chem. Soc.* **1986**, *108*, 3928.
22. Wu., S.-h.; Jones, M., Jr. *Inorganic Chemistry* **1986**, *25*, 4802.
23. Wu, S.-h., Ph.D. Thesis, Princeton University, 1989.
24. Gaspar, P. P.; Hsu, J.-P.; Chari, S.; Jones, M., Jr. *Tetrahedron* **1985**, *41*, 1479
25. Blanch, R. J.; Yuan, K.; Jones, M. Jr. *J. Am. Chem. Soc.* **1993**, *115*, 12597. The labeling shows that the carbons move in lockstep; the carbene carbon enters the cage, the cage carbon to which the carbene is attached slides over to the other cage position, and the other cage carbon emerges as a new carbene. Hydrogen migration then forms **24**.

THE ROLE OF CARBENES IN THE CHEMISTRY OF LOW-COORDINATED PHOSPHORUS

Holger Memmesheimer and Manfred Regitz

Advances in Carbene Chemistry
Volume 1, pages 185–213.

ISBN: 1-55938-545-6

I. INTRODUCTION*

For many years the chemist's interest in phosphorus was confined to classes of compounds such as phosphines, phosphine oxides, all types of phosphorus acid derivatives, and phosphonium salts.[1] The Wittig reaction[2] and the PO-activated olefination discovered by Horner[3] provided the foundations for the exploitation of organophosphorus chemistry in the field of organic synthesis. But then, an end of the explosive development which started in the 1950's seemed to be in sight.

The repudiation, however, of the so-called "double bond rule", which excluded the formation of $(p,p)\pi$ multiple bonds between phosphorus and elements in the first and second eight-element periods of the periodic table,[4] constituted the starting point for a completely unexpected and continuously expanding renaissance in organophosphorus chemistry.[5] The first milestones of this development were the generation and characterization of the short-lived phosphaacetylene (**1**; R = H),[6] the phosphacyanine cations,[7] and the phosphabenzenes (**3**).[8]

In the meantime, numerous phosphorous–carbon (P–C) and phorporous–phosphorous (P–P) multiple-bond systems containing low-coordinated phosphorus atoms such as **1** to **6** (λ represents the valency, and σ the coordination number of the phosphorus atom) have been synthesized and their reactivities studied. If compounds of these types are to be isolated, kinetic stabilization—that is, the introduction of sterically demanding substituents—is an indispensible prerequisite (only in a few individual cases has thermodynamic stabilization been successful). Although this has detrimental effects on the reactivity, it is hardly apparent in connection with the highly reactive carbenes or similar species. The present review is concerned mainly with the reactions between these reactive species and low-coordinated phosphorus compounds of the types **1–6**. In some cases, the preparative significance of successful carbene reactions will be discussed. Although

$\lambda^3\sigma^1$-Phosphorus
1

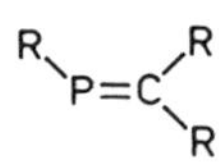

$\lambda^3\sigma^2$-Phosphorus
2

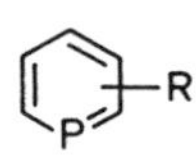

$\lambda^3\sigma^2$-Phosphorus
3

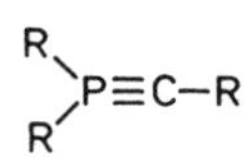

$\lambda^5\sigma^3$-Phosphorus
6a

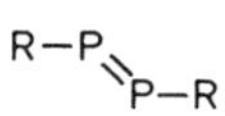

$\lambda^3\sigma^2$-Phosphorus
4

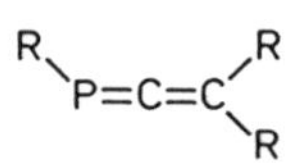

$\lambda^3\sigma^2$-Phosphorus
5

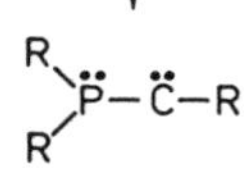

$\lambda^3\sigma^3$-Phosphorus
6b

reactions of carbenes with low-coordinated phosphorus compounds have only been investigated in the past few years, the results obtained so far are very impressive.

II. PHOSPHAALKYNES

Addition reactions of carbenes with phosphaalkynes **1**[5,9] should give rise to 1-phosphirenes, the parent compounds of the cyclic phosphaalkenes. Diazoalkanes cannot be used as the carbene source since they undergo rapid (3 + 2) cycloadditions with P–C triple-bond systems to furnish 1,2,4-diazaphospholes.[10] On the other hand, thermolysis reactions of the cyclic isomers of diazoalkanes, namely the diazirines, have proved to be a suitable source of carbenes because the starting materials themselves do not react with the phosphaalkynes.

A. Reactions with Halocarbenes

When the diazirines **7** were subjected to thermolysis conditions in the presence of excess phosphaalkyne **1** (only in the case of the crystalline 1-adamantylphosphaacetylene was benzene used as a solvent), the 1-halo-2-phosphirenes **10a–l** were unexpectedly formed in yields of 15–63%. The temperature required for the thermolysis depends on the substituent pattern of the carbene precursor: aryloxy- and alkoxydiazirines undergo extrusion of nitrogen at room temperature whereas temperatures of 90–160 °C are necessary for other substituents.

It must be assumed that the (2 + 1) cycloaddition of the carbene **8** with the phosphaalkyne **1** initially gives rise to the 1-phosphirenes **9a–l** which then undergo subsequent isomerization via (1,3)halogen shift to furnish the products **10a-l**. The

7 **8** **9** **10**

9,10	a	b	c	d	e	f	g	h	i	j	k	l
R^1	Ph	Ph	Ph	*t*-Bu	Ph	Ph	$PhCH_2$	PhO	PhO	MeO	EtO	*i*-Bu
R	*t*-Bu	*t*-Bu	*t*-Bu	*t*-Bu	1-Ad	CMe_2Et	*t*-Bu	*t*-Bu	CMe_2Et	*t*-Bu	*t*-Bu	*t*-Bu
X	Cl	Br	F	Cl	Cl	Cl	Cl	Cl	Cl	Cl	Cl	Cl

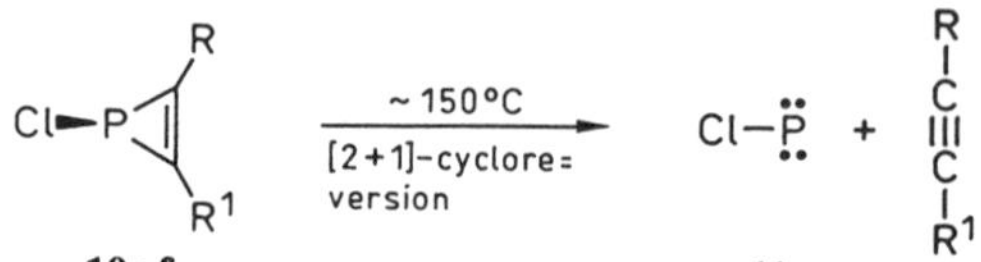

10e,f 11 12

12	a	b
R^1	Ph	Ph
R	1-Ad	CMe_2Et

thermodynamic difference in going from **9** to **10** must be enormous since even fluorine undergoes the shift in spite of the high dissociation energy of the C–F bond (**9c** → **10c**). The disappearance of the energy-rich P–C double bond in **9** in favor of a C–C double bond in **10** is also most certainly a further factor facilitating the isomerization. A crystal structure analysis of **10a** showed that the chlorine substituent is twisted out of the plane of the three-membered ring by 76°, i.e., the phosphorus atom has sp^3 hybridization so that antiaromatic interactions between the lone electron pair at phosphorus and the π-system do not occur. Just the reverse situation is valid for the 1-azirine/2-azirine system in which the latter species apparently possesses a "planar nitrogen" and has only a short life-time.[12]

Until now, 1-chloro-2-phosphirenes were only known in the form of the corresponding P–W$(CO)_5$ complexes;[13,14] 1-chloro-2,3-diphenyl-2-phosphirene can apparently be liberated from its tungsten complex (^{31}P NMR signal at $\delta = -74.0$)[15] but cannot be isolated.

The unavoidable high temperatures needed for the generation of chloro-(phenyl)carbene (**8**; X = Cl, R^1 = Ph), and for the distillation workup of the 1-chloro-2-phosphirenes **10e** and **10f**, also lead to partial fragmentation of the three-membered ring derivatives. The acetylenes **12a** and **12b** and chlorophosphinidene (**11**) are formed in this process.[11] No reliable information about the fate of the electron-deficient species **11** is available at present.

The cyclopropanations of the phosphaalkyne **13** with mono- and dihalocarbenes carried out at 0 °C also proceeded in the sense of a 1-halo-2-phosphirene fragmentation. The dihalocarbenes were generated through α-elimination from the corresponding dihalo compounds by the action of potassium *tert*-butoxide.[16] The products isolated exclusively were the 2,4,6-tris(*tert*-butyl)phenylacetylenes **16a–d**. In this case too, no information about the fate of the phosphinidine fragment **11** is available. Neither 1- nor 2-phosphirenes (**14** or **15**) could be detected.

The relatively large P–Cl bond lengths of 2.166(2) Å obtained from the crystal structure analysis of **10a** account for the easy nucleophilic replacement of the chlorine substituent, and thus also account for the sensitivity of the 1-halo-substituted 2-phosphirene system[11] toward hydrolysis.

If it were possible to remove the chloride ion from the 1-chloro-2-phosphirene, the phosphirenylium cation species **17a** would be formed. Species of this type,

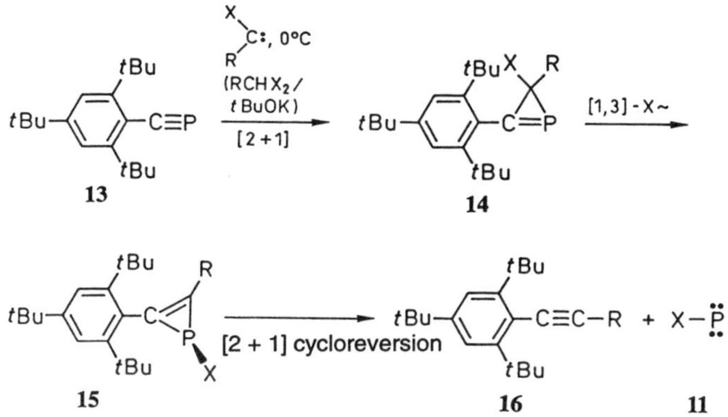

14–16	a	b	c	d
X	Cl	Br	Cl	Cl
R	Cl	Br	Ph	Me

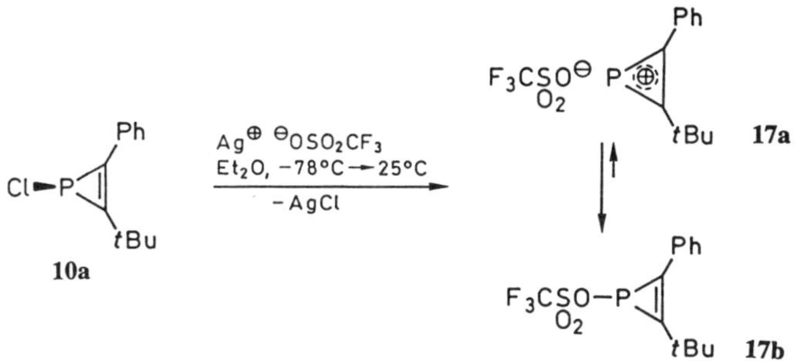

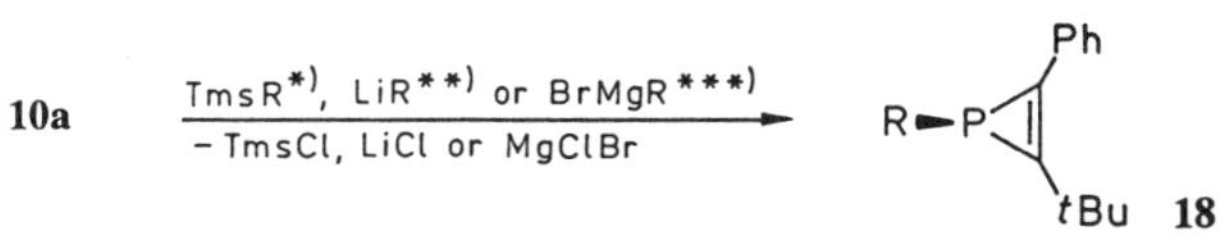

18	a*	b*	c*	d*	e**	f**	g**
R	CN	N_3	N=C(*t*-Bu)(*t*-Bu)	P=C(OTms)(*t*-Bu)	*t*-Bu	N–*i*-Pr_2	OMe

18	h**	i**	j***	k***
R	C≡C–*t*-Bu	C(=N_2)–Tms	CH=CH_2	Mes

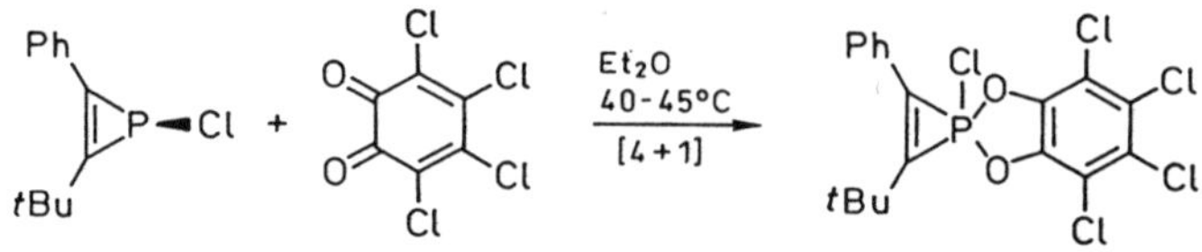

10a **19** **20** (31%)

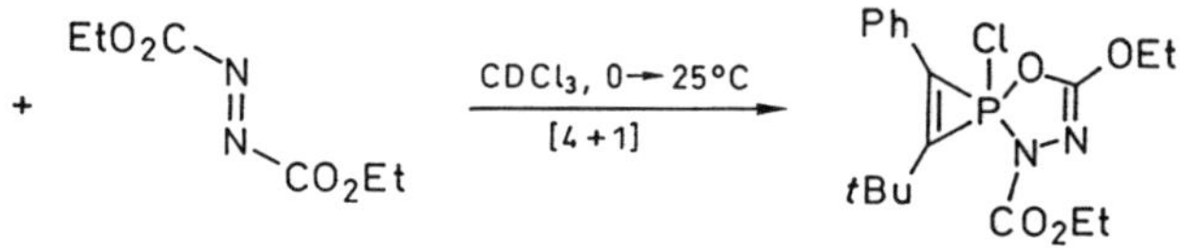

21 **22** (100%)

however, which should exhibit Hückel aromatic character, had previously not been reported in the literature. The reaction of **10a** with silver triflate seemed promising since it did indeed furnish the distillable product **17** in quantitative yield.[17] The dependence of the chemical shifts of the ^{31}P and ^{13}C NMR signals of the atoms of the three-membered ring on the polarity of the solvent was indicative, at least, of an equilibrium of the type **17a** $\rightleftharpoons$ **17b** in which the latter species presumably dominates. The fact that the ^{17}O NMR spectrum of **17** exhibited only one signal at $\delta = 159.2$ without $^1J_{O,P}$ coupling can be explained easily in terms of a rapid P/O exchange. The chlorine substituent in **10a** can be replaced by treatment with a range of *C*-, *N*-, *P*-, and *O*-nucleophiles. Silyl derivatives (→ **18a–d**), organolithium derivatives (→ **18e–i**), and Grignard compounds (→ **18j, 18k**) may be used for this purpose.[11,18]

The 2-phosphirenes **10** each contain a $\lambda^3\sigma^3$-phosphorus atom; these species also play a key role in the synthesis (by oxidative addition) of the previously unknown phosphirenes possessing pentavalent, pentacoordinated phosphorus atoms.

Thus, for example, **10a** reacted smoothly with the orthoquinone **19** in a (4 + 1) cycloaddition process to furnish the phosphorane **20**.[19] Diethyl azodicarboxylate (**21**) behaved as a hetero-1,3-diene toward **10a**; the reaction produced the spirocyclic system **22**.[20]

B. Reactions with Vinylcarbenes

It is well known that tetrachlorocyclopropene (**23**) undergoes thermal ring opening to generate tetrachlorovinylcarbene.[21] When this carbene precursor was heated with the phosphaalkyne **1** (R = *t*-Bu), the 1-chloro-2-trichlorovinyl-2-phosphirene **24** was formed. In this case also, the addition of carbene to the P–C triple bond is followed by a (1,3) chlorine shift.[22] The 1-chlorophosphirene **24** is extremely sensitive to hydrolysis and reacts with two equivalents of water to furnish the butadienylphosphorous acid **25**.[22]

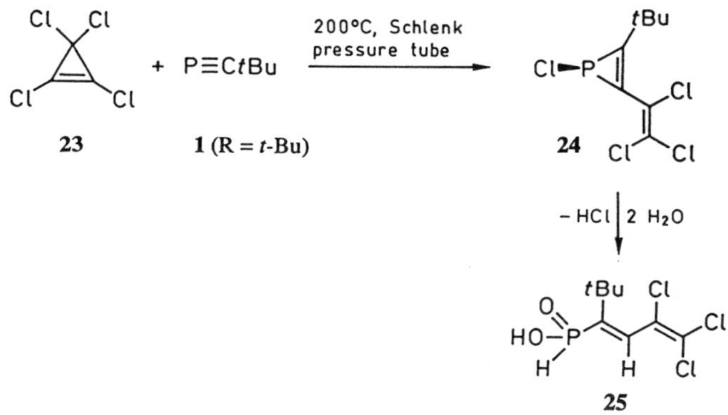

In contrast to tetrachlorocyclopropene (**23**), the 1,2-dichloro-3,3-dialkylcyclopropenes **26** exist in equilibrium with the corresponding chlorovinylcarbenes **27** even at room temperature.[23]

When this process was allowed to proceed in the presence of phosphaalkynes **1** (R = *t*-Bu, CMe_2Et, 1-Ad), the doubly functionalized 2-phosphirenes **29a–i** were obtained.[22] In these cases also, the 3-chloro-3-vinyl-1-phosphirenes **28** are most certainly the primary products although they cannot be detected, for example, by ^{31}P NMR monitoring of the reaction, since the (1,3) chlorine shifts are very rapid. The driving forces for this isomerization process are the same as those for the reaction **9** → **10**.[11]

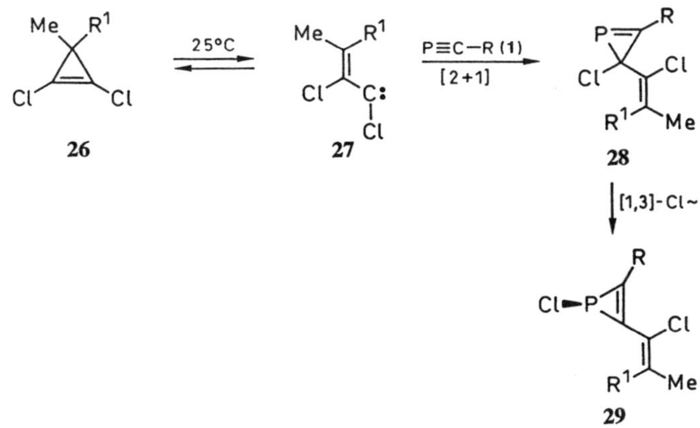

28,29	**a**	**b**	**c**	**d**	**e**	**f**	**g**	**h**	**i**
R^1	Me	CH_2Cl	CH_2OMe	Me	CH_2Cl	CH_2OMe	Me	CH_2Cl	CH_2OMe
R	*t*-Bu	*t*-Bu	*t*-Bu	CMe_2Et	CMe_2Et	CMe_2Et	1-Ad	1-Ad	1-Ad

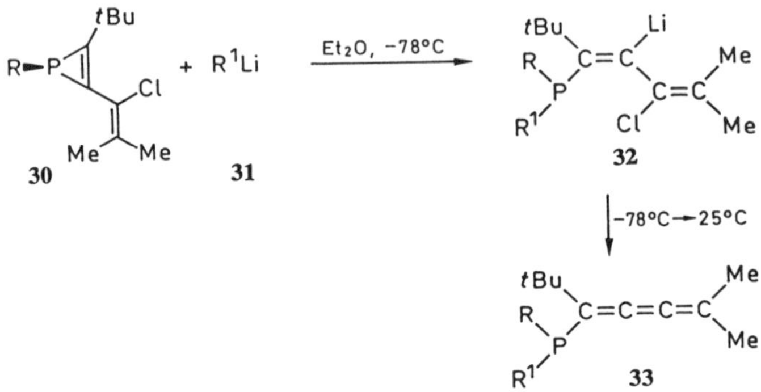

32,33	a	b	c	d	e	f	g	h	i
R[1]	Me	*n*-Bu	*t*-Bu	Me	Ph	CH═CH2	CH═CH2	C≡C–*t*-Bu	C≡C–*t*-Bu
R	Me	*n*-Bu	Me	*t*-Bu	Me	Me	*n*-Bu	Me	*n*-Bu

The chlorine substituents both at phosphorus and at the vinyl group render the 2-phosphirenes **29** interesting starting materials for ring opening reactions. For example, reactions with two molar equivalents of an organometallic derivative gave rise to the phosphino-1,2,3-butatrienes **33a–i**.[24]

This reaction can also be performed as a two-step process. Accordingly, reaction of **29a** with one equivalent of methyl- or *n*-butyllithium in diethyl ether at –78 °C gave rise initially to the *P*-alkyl-substituted 2-phosphirenes **30** (R = Me, *n*-Bu; 82 and 71% yields, respectively). Repetition of the reaction with a second equivalent of the same reagent furnished compounds **33a** and **33b**. When, on the other hand, 1-methyl-2-phosphirene was allowed to react with *n*-butyllithium, or when 1-*n*-butyl-2-phosphirene was allowed to react with methyllithium, the same product—the butadiene **33c** (= **33d**)—was obtained in each case.[24] The chlorine substituent in **29a** can also be replaced by phenyl, vinyl, and ethynyl groups. Ring opening reactions of these products then furnish the cumulenes **33e–i**.[24]

An explanation of the mechanism of this novel butatriene synthesis involves the 1,3-butadiene intermediate **32** which subsequently undergoes β-elimination of lithium chloride to furnish the products. In principle, both P–C single bonds are available for the ring opening of the phosphirene; however, for general and steric reasons, the bond system bearing the vinyl group is attacked specifically. An X-ray crystal structure analysis of the $W(CO)_5$ complex of **33a** finally clarified all doubts about the outcome of the reaction.[24]

C. Reactions with Carbene Complexes

Alkynes can be employed for the annellation of carbene ligands; this reaction opens a direct access to highly substituted hydroquinones.[25] Some parallels between the reactivity of alkynes and phosphaalkynes[5,9] prompted the question of

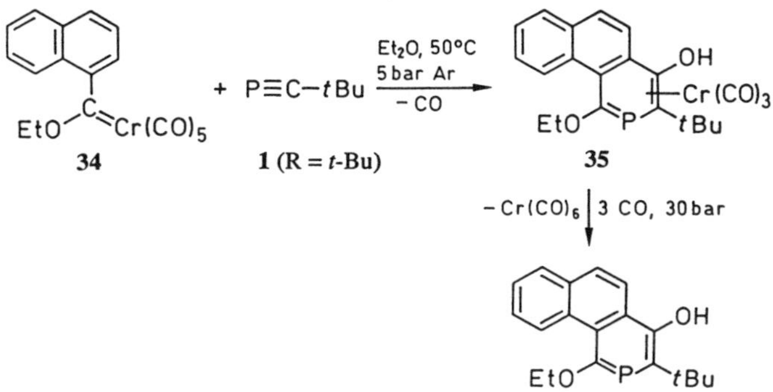

34 **1** (R = *t*-Bu) **35** **36**

whether reactions of the latter with carbene ligands could be used to prepare functionalized phosphaarenes.

When the pentacarbonyl[ethoxy(1-naphthyl)carbene]chromium complex **34** was allowed to react with **1** (R = *t*-Bu), the expected phosphaphenanthrene–hydroquinone complex **35** was formed in 82% yield, the $Cr(CO)_3$ fragment being coordinated at the phosphaaromatic ring.[26] The incorporation of the phosphaalkyne moiety proceeds regiospecifically and is apparently controlled by steric factors. The phosphaarene **36** was obtained in quantitative yield from **35** by ligand exchange under a pressure of carbon monoxide.[26,27] The concomitantly formed $Cr(CO)_6$ can be recycled in the reaction.

The reactions of the *o*-methoxyphenylcarbene complexes **37** or **38** with the phosphaalkyne **1** (R = *t*-Bu) followed a completely different route: the 1,3-oxaphospholes **39**, in which the $Cr(CO)_3$ fragment is still coordinated to the phenyl ring, were obtained.[28]

To date, this is the sole access to non-benzo-annellated 1,3-oxophospholes. Benzo-condensed heterocycles of the same type had previously been prepared by condensation of *o*-hydroxyphenylphosphines with imidic acid chlorides (see also Ref. 5, page 260).[29,30]

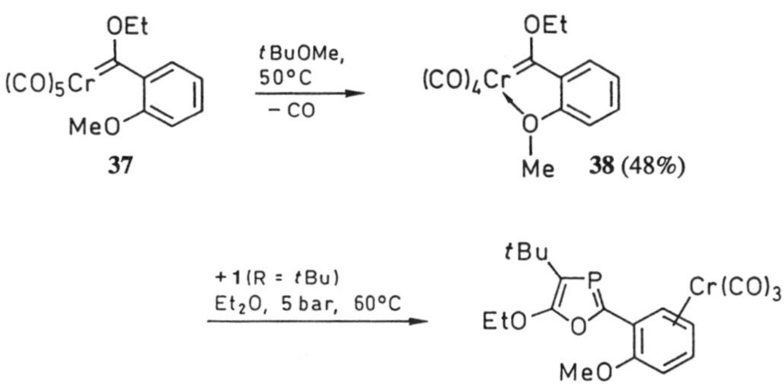

37 **38** (48%) **39** (77%)

D. Intramolecular Reactions of Phosphavinylcarbenes

The synthesis of cyclopropenes by photolysis of 3*H*-pyrazoles is well known;[31] thus, it was of interest to determine whether the photoreactions of 3*H*-1,2,4-diazaphospholes **41** would follow an analogous course. Accordingly, the phosphaalkyne **1** (R = *t*-Bu) was allowed to react with diazoalkanes possessing bulky substituents such as **40a–d**. The 3*H*-1,2,4-diazaphospholes **41a–d** were formed in quantitative yield in pentane at –40 °C.[18,32,33] The cycloaddition reactions were regiospecific and no evidence, at least within the limits of NMR monitoring, was obtained for the presence of isomers with the structure **42**. The cycloaddition is controlled by electronic factors, as was confirmed by the addition of diazomethane (**40**; R = H) to phosphaacetylene (**1**; R = H).[34]

The 3*H*-1,2,4-diazaphospholes **41a–d** are stable at –40 °C but undergo (H^+-catalyzed?) sigmatropic (1,5) R-shifts in chloroform at room temperature to furnish the 4*H*-1,2,4-diazaphospholes **43**.[33,35] In order to avoid this isomerization, which would make the formation of 1-phosphirenes impossible, the 3*H*-1,2,4-diazaphospholes must be submitted to photolysis immediately upon formation.

When the 3*H*-1,2,4-diazaphospholes **41**, prepared from the diazoalkane **40a** and the phosphaalkynes **1** (R = *t*-Bu, *i*-Pr, CH_2–*t*-Bu, 1-methyl-1-cyclohexyl), were irradiated in pentane at –40 °C, the phosphavinylcarbenes **44a–d** were generated (in these experiments, no attempts were made to detect the diazo intermediates).[18,32,33,35]

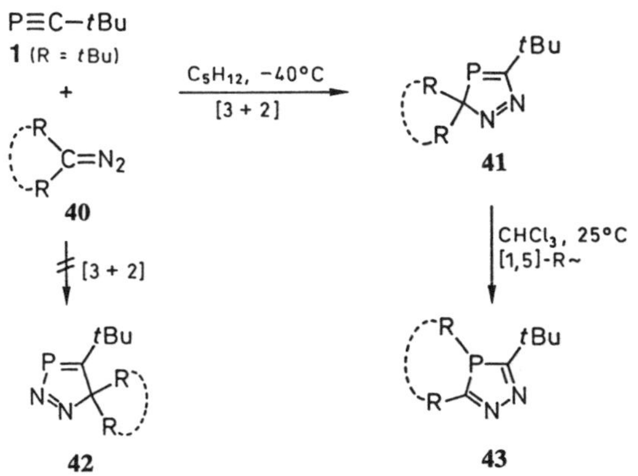

40–43	a	b	c	d
R ... R	*t*-Bu / *t*-Bu	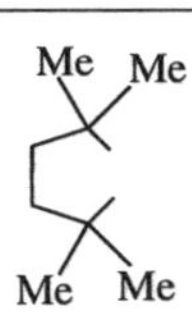	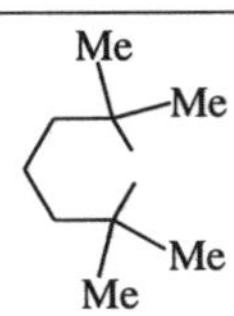	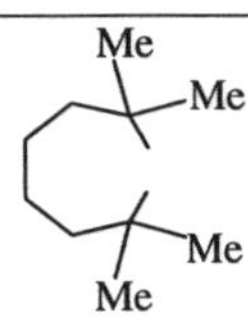

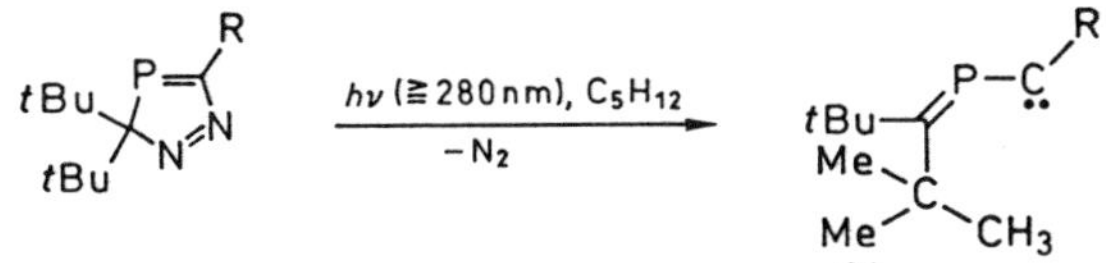

40a (R instead of *t*-Bu)

44

formula 13

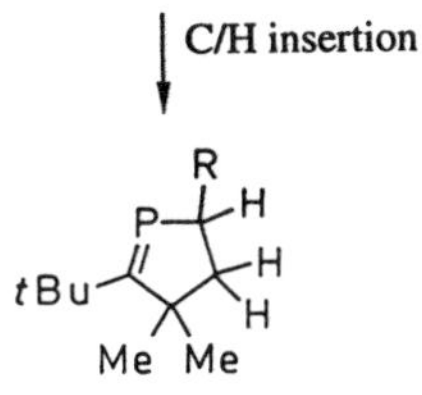

45

44,45	**a**	**b**	**c**	**d**
R	*t*-Bu	$CHMe_2$	CH_2–*t*-Bu	

Instead of the expected (1,3) ring closure of **44** to furnish 1-phosphirenes, the 1-phospha-1-cyclopentenes **45a–d** were obtained in good yields by way of intramolecular carbene insertion into one of the C–H bonds of the methyl groups.[32,35] These compounds represent the first five-membered ring systems containing a P–C double bond; they are thermally stable, but must be handled under an inert gas atmosphere on account of their sensitivity toward oxidation and hydrolysis.

The expectations with regard to the formation of the 1-phosphirene system were finally fulfilled by photolyses of the spirocyclic 3*H*-1,2,4-diazaphospholes **41b–d**.[33,35] The fact that the formation of carbenes **47a–c** follows the opening of the

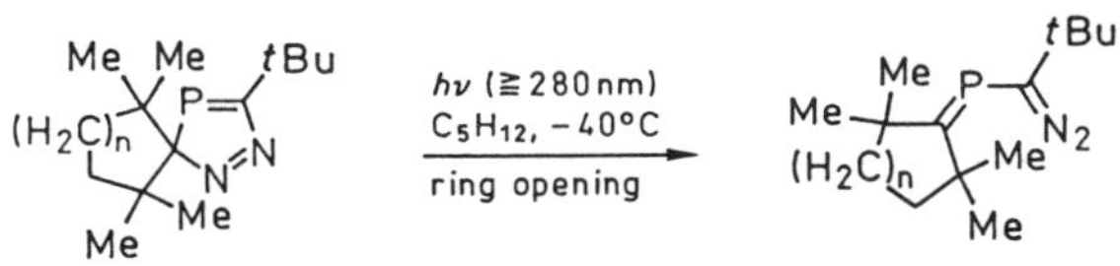

41b–d

46

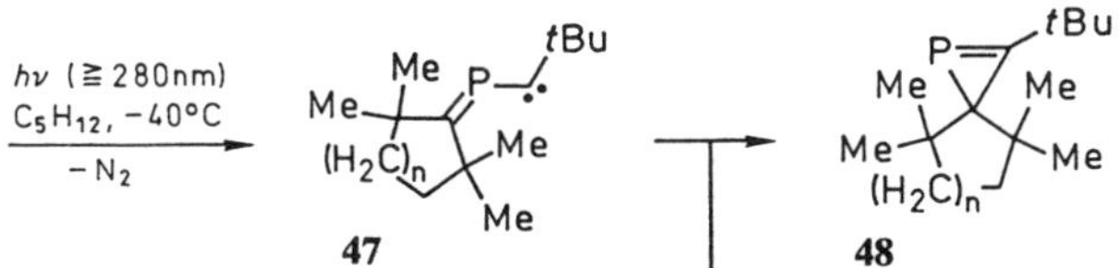

47

48

46–49	*a*	b	c
n	1	2	3
% **48**	13	11	2.5
% **49**	40	54	50

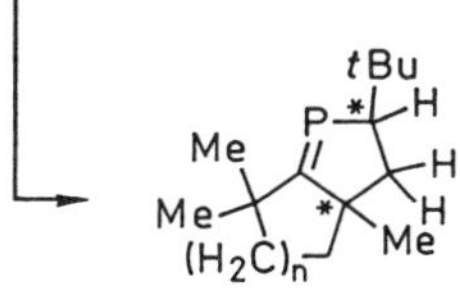

49
(mixture of diastereomers)

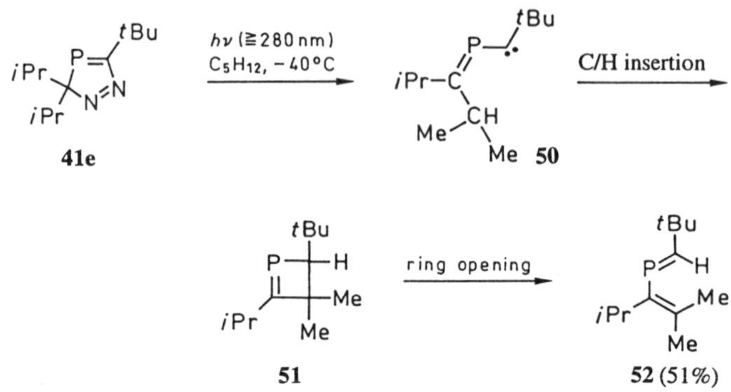

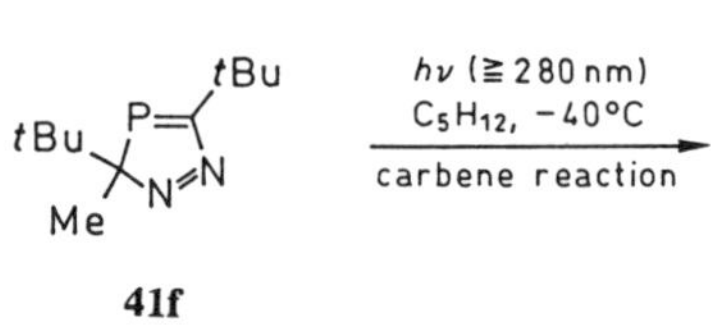

3*H*-diazaphosphole ring of **41** to give the phosphavinyldiazo compounds **46a–c** was demonstrated for the reaction of **41c** ($n = 2$) to **46b** ($n = 2$): ν ($C{=}N_2$) = 2030 cm^{-1}.[33]

Longer photolysis under otherwise identical conditions gave rise to the carbenes **47a–c**, which preferentially reacted further in one of two ways. Again, C–H insertion into a methyl group attached to the ring predominated and yielded the condensed 1-phospha-1-cyclopentenes **49a–c** (40–54%).[33,35] In all cases, the spirocyclic 1-phosphirenes **48a–c** (2.5–13%) were also isolated as thermally stable oils by a combination of bulb-to-bulb distillation and medium pressure liquid chromatography.[33,35] For unequivocal structure elucidation, **48b** was converted into an "end-on" complex at phosphorus by treatment with $W(CO)_5$·THF under exchange of the tetrahydrofuran ligand. The crystal structure of the resultant complex was determined by X-ray crystallography. The P–C double bond length was found to be 1.634 Å, that is, it is in the usual range and thus confirms the unsaturated, three-membered ring structure.[33]

Although differing with respect to the final product, the photoreactions of **41e** and **41f** are certainly comparable with regard to the actual carbene reaction.[18,35] The light-induced ring opening of **41e** with subsequent extrusion of nitrogen and carbene formation (→ **50**) is initially concluded by a methyl C–H insertion step (→ **51**); however, the fact that the 1-phospha-1-cyclobutene **51** underwent electrocyclic ring opening to the 2-phospha-1,3-butadiene **52** under very mild conditions can be attributed to the ring strain in **51**.[18,35]

The 3*H*-1,2,4-diazaphosphole **41f** reacted analogously and furnished the 2-phospha-1,3-butadiene **53** upon photolysis under identical conditions.[18,35]

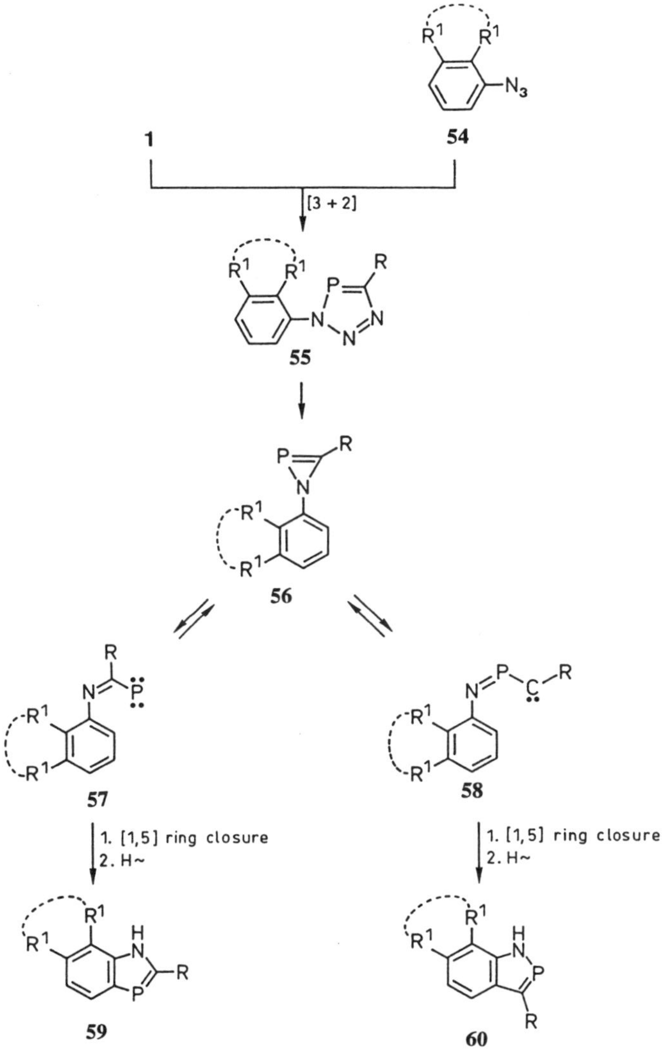

55–60	a	b	c
R	*t*-Bu	*t*-Bu	Me
R^1	H	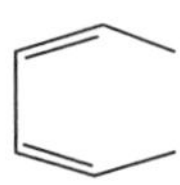	H
R^1	H		H
% **59** + **60**	95	99	60
ratio **59**:**60**	1:4	1:3	2:1

In the context of intramolecular carbene reactions for the construction of low-coordinated phosphorus compounds, flash vacuum photolyses of the 3*H*-1,2,3,4-triazaphospholes **55a–c** must also be discussed.[36,37] The heterocyclic starting materials **55** are accessible from the regiospecific cycloaddition reactions of the azides **54** with the phosphaalkynes **1** (R = *t*-Bu, Me).

When compounds **55** were subjected to flash vacuum photolysis (540–580 °C, 10^{-5}–10^{-4} mbar), they extruded nitrogen. On the basis of the product pattern obtained, it must be assumed that the azaphosphirenes **56a–c** were formed initially, and that they could, in principle, undergo two ring-opening reactions. The first involves the formation of phosphinidines of the type **57a–c** and the second the formation of carbenes of the type **58a–c**. Subsequent (1,5) ring closures with the aromatic moiety followed by *H*-shifts (which leads to aromatization) then furnished the final products **59a–c** and **60a–c**, respectively,[36,37] in high yields. In the flash vacuum photolyses of **55a** and **55b**, the 1,2-azaphospholes **60a** and **60b** were the major products whereas in the case of **55c** the 1,3-isomer **59c** was the main product.

III. PHOSPHAALKENES

Our knowledge about carbene additions to phosphaalkenes is rather sparse in comparison to our knowledge of these reactions with phosphaalkynes. The (1,5) cyclizations of phosphaalkenes with carbene centers will be discussed in this context.

A. Reactions with Halocarbenes

When the halocarbenes **8** (R^1 = Ph, OPh) were generated by pyrolysis of the corresponding diazirines **7** (R^1 = Ph, OPh) in the presence of phosphaalkenes of the type **61**, the phosphiranes **62a–d** were formed smoothly in yields of 47–89%.[38]

The phosphiranes **62b** and **62d** possess two centers of chirality while **62a** and **62c** even have three such centers; thus, the occurrence of diastereomers must be expected. This is not the case, however, as shown already by NMR spectroscopy of the crude products.[38] Nuclear magnetic resonance comparisons with other, similarly substituted phosphiranes[39] have allowed the stereochemistries at phosphorus and at the silyl-substituted carbon atom to be assigned (the "lone pair" at phosphorus and the silyl group have a *cis*-orientation), while that at the chlorine-substituted carbon atom could not yet be clarified.[38]

When the phosphiranes **62a–d** were heated at 130 °C and then subjected to bulb-to-bulb distillation, the 1-chloro-2-phosphirenes **63a–d** were isolated in good yields. Elimination of chlorotrimethylsilane at the C–C bond is almost certainly responsible for the formation of the phosphirenes; however, cleavage of chlorotrimethylsilane from the P–C bond and a subsequent chlorine shift is also feasible (see also Section II.A).

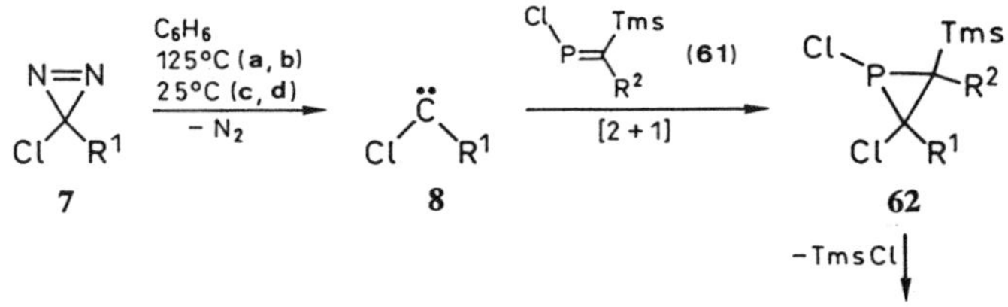

7 8 62

62,63	a	b	c	d
R^1	Ph	Ph	PhO	PhO
R^2	Ph	Tms	Ph	Tms

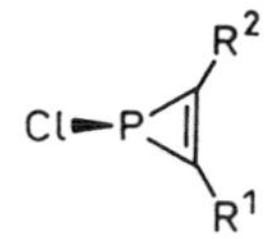

63 (47–89%)

B. Intramolecular Ring Closure Reactions

Phosphaalkenes of the structural type **66** in which the tri-*tert*-butylcyclopropenyl moiety serves to effect the kinetic stabilization, can be prepared from the cyclopropenylphosphine **64**. Lithiation to furnish **65** followed by reaction with an acyl chloride [and a subsequent (1,3) silyl shift][40] thus furnished the stable cyclopropenylphosphaalkenes **66a–d**.[41]

When compounds **66a–d** were irradiated in benzene, a highly selective isomerization to produce the 2*H*-phospholes **68a–d** took place. These compounds represent the first stable members of this class of compounds. A reasonable explanation of this isomerization process involves the carbene intermediates **67a–d** which undergo (1,5) ring closure to form the products.[41] Of course, a diradical process cannot be excluded with certainty.

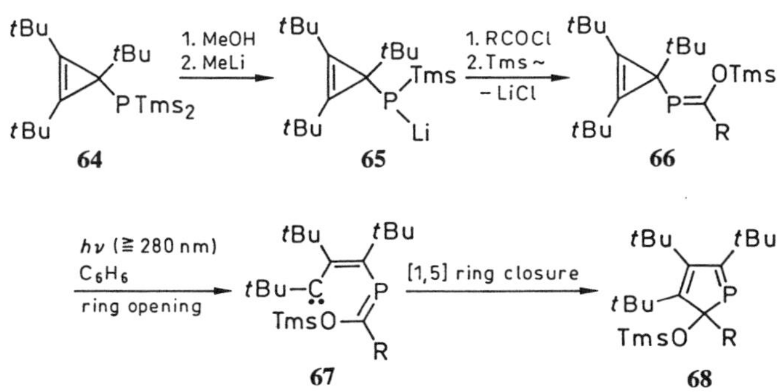

66–68	a	b	c	d
R	Ph	$C_6H_2(CH_3)_3$-2,4,6	$C_6H_2(OCH_3)_3$-2,4,6	CH_3

IV. PHOSPHABENZENES

2,4,6-Triphenylphosphabenzene (**69**; R = Ph) did not react via cyclopropanation with dichlorocarbene (**8**; R^1 = Cl), which had been generated from sodium trichloroacetate or from (bromodichloromethyl)phenylmercury. This result may be attributed to the low nucleophilicity of the phosphorus atom.[42]

In contrast, 2,4,6-triphenyl- and 2,4,6-tri-*tert*-butylphosphabenzenes (**69**, R = Ph or *t*-Bu, respectively) did react with the haloalkanes **70** (R^1 = H, Cl, Ph) to furnish the benzene derivatives **75a–f**.[42] The reaction sequence was interpreted as follows: the carbanion generated from **70** by the action of potassium *tert*-butoxide undergoes addition to the phosphabenzenes (→ **71a–f**), and **71a–f** eliminate chloride ion to furnish the phosphanorcaradienes **74a–f**. Valency isomerization of the type **74** ⇌ **73** ⇌ **72** provides the prerequisites for the extraction of phosphorus. This proceeds from **72a–f** via cheletropic fragmentation to produce the final products **75a–f** together with chlorophosphinidine. The fate of the latter species is still unknown.[42] When the pentacarbonylchromium or pentacarbonylmolybdenum complexes **76**[43] were allowed to react with diazodiphenylmethane, the phosphanorcaradiene **77** bearing the respective $M(CO)_5$ fragment was formed.[44]

It cannot yet be concluded whether the norcaradiene formation (→ **77**) should be interpreted as a carbene or carbenoid reaction (the metal complexes could catalyze the decomposition of the diazoalkane), or whether a (3 + 2) cycloaddition

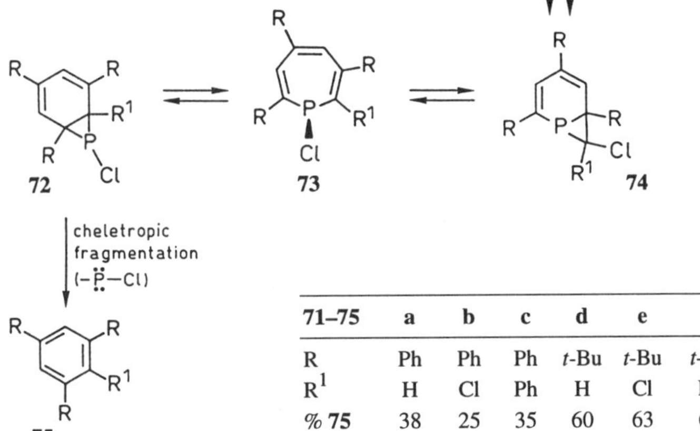

71–75	a	b	c	d	e	f
R	Ph	Ph	Ph	*t*-Bu	*t*-Bu	*t*-Bu
R^1	H	Cl	Ph	H	Cl	Ph
% 75	38	25	35	60	63	64

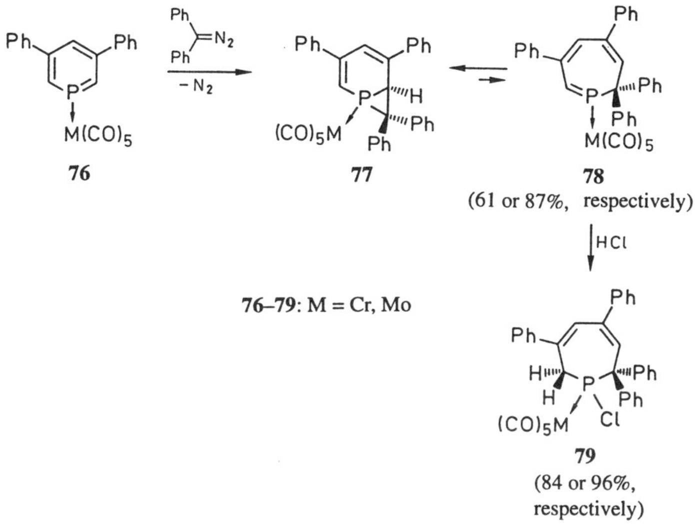

with subsequent extrusion of nitrogen is responsible for the formation of the products.[44]

If the monocyclic valency isomers **78** are compared with **77**, it may be assumed with certainty that they are thermodynamically less stable since they contain an energy-rich phosphaalkene moiety. The addition of hydrogen chloride to **77**, which gave rise to the monocyclic product **79**, may be interpreted as a first indication for the existence of a valency isomer equilibrium of the type **77** $\rightleftharpoons$ **78**.[44]

V. DIPHOSPHENES

Diazodiphenylmethane (**40a**) and 9-diazofluorene (**40b**) reacted with the *trans*-diphosphene **80** (which is sterically shielded by the presence of the 2,4,6-tri-*tert*-butylphenyl groups) under thermal or photochemical conditions to yield the diphosphiranes **82a**, **82b** (major products) and the phosphaalkenes **83a** and **83b** (minor products); the concomitant formation of the azines **84a** and **84b** could not be suppressed.[45]

The carbenes **81a** and **81b** are assumed to be responsible for the outcome of the reactions. In this context, the reaction of **80** with diazomethane, which takes place even at 0 °C in tetrahydrofuran/diethyl ether, should be mentioned since it also gave rise to a diphosphirane **82** (R = H).[46]

The halocarbenes **85**, generated by α-elimination, also underwent (2 + 1) cycloaddition with the diphosphene **80** to furnish the *C*-halo-substituted diphosphiranes **86a–d**.[45,47] In these cases as well, some formation of phosphaalkenes occurred. The tempting assumption that the latter arise through subsequent fragmentation of the diphosphirane **86** can be discounted since the carbene reaction

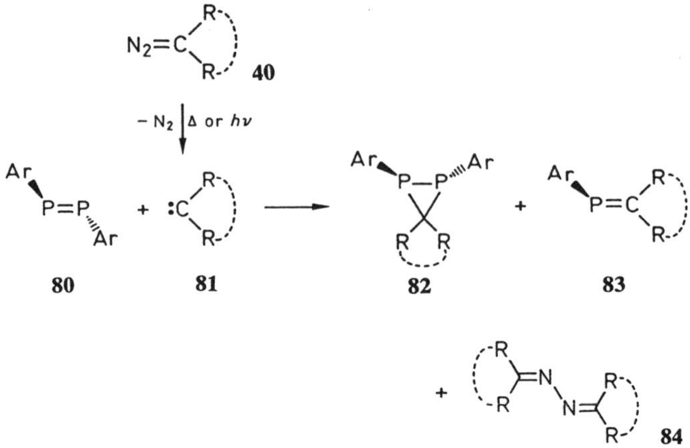

40, 81-84	a	b
R	Ph	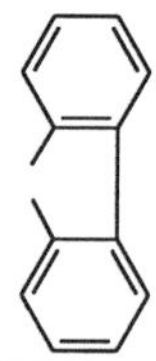
R	Ph	

is performed at 0 °C whereas the diphosphiranes are still stable at considerably higher temperatures.[45] Rather, phosphorane intermediates possessing $\lambda^5\sigma^3$-phosphorus [Ar–P(=P–Ar)(=CXR)] are considered to be responsible for the formation of the phosphaalkenes.[45]

Not all of the diphosphiranes formed by carbene addition to diphosphenes are thermally stable.

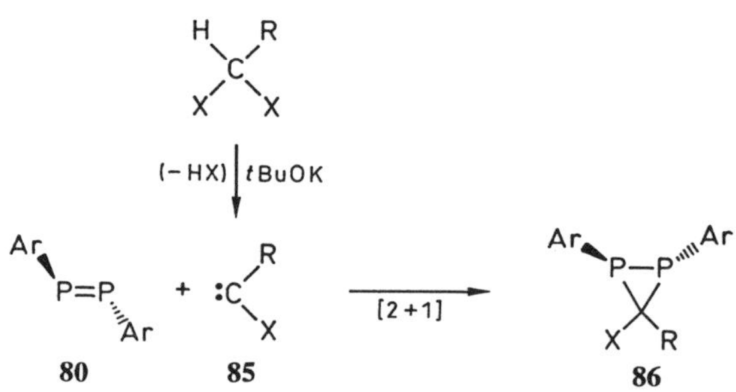

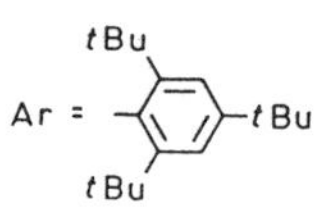

86	a	b	c	d
X	Cl	Br	Br	Cl
R	Cl	Br	Ph	CH_3

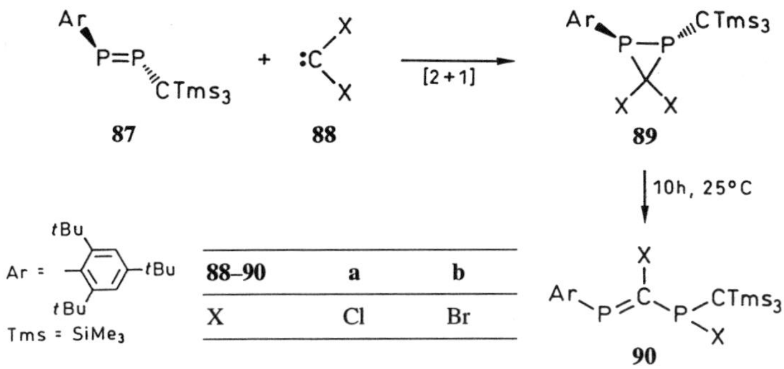

88–90	a	b
X	Cl	Br

For example, the compound **89a**, formed primarily by addition of the unsymmetrically substituted diphosphene **87** to dichlorocarbene (**88a**), underwent rearrangement to the 1,3-diphosphapropene **90a** upon attempts to purify it over silica gel. This is attributed to P–P ring opening and migration of chlorine. The stereochemistry at the P–C double bond was confirmed by an X-ray crystallographic analysis.[48] The diphosphirane **89b** underwent an identical isomerization to **90b** upon storage in pentane solution at room temperature. The $^2J_{P,P}$ coupling constants of the two 1,3-diphosphapropenes are unusually large (457 and 484 Hz, respectively).[48] In the case of **86d**, the rearrangement to the 1,3-diphosphapropene was achieved by treatment with an excess of *n*-butyllithium.[47]

1,1-Dihalocyclopropanes can be dehalogenated and transformed to allenes by treatment with organometallic reagents.[49,50] Diphosphiranes of the type **86** possess the same prerequisites and should thus function as suitable precursors for syntheses of 1,3-diphosphaallenes.[51]

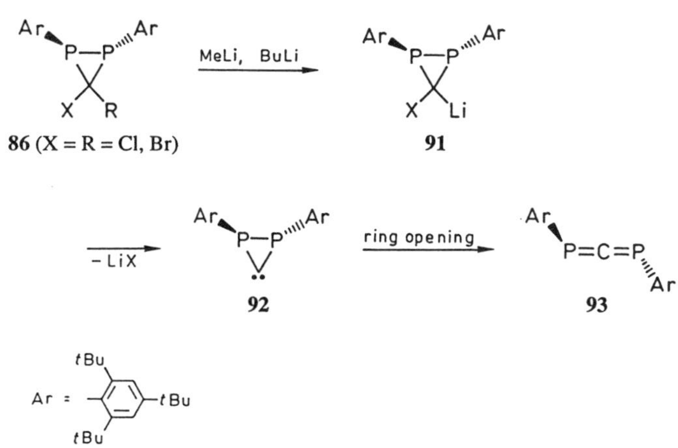

When the dihalodiphosphiranes **86** (X = R = Cl, Br) were treated with methyl- or *n*-butyllithium, the carbenoids **91**—or also the carbenes **92**—might have been formed, while underwent P–P ring opening to furnish the 1,3-diphosphaallenes **93**.[51]

VI. PHOSPHACUMULENES

Cumulenes containing at least one P═C moiety are well-known in the chemistry of low-coordinated phosphorus.[52] Investigations of the carbene reactions of these systems have only just started but promise many new and interesting results.

Thus, the 1-phosphaallene **94** underwent addition with dichlorocarbene (**88**; X = Cl) via degradation of the more energy-rich P–C double bond to furnish the methylenephosphirane **95** together with the phosphaalkene **96** as a minor product.[53] This is the key reaction for the conversion of the 1-phosphaallene **94** into the 1-phosphabutatriene **97**: exchange of chlorine for lithium in **95** generated the carbenoid **98** which, in turn, furnished the phosphacumulene **97**[53,54] either directly or after α-elimination of lithium chloride.[49,50]

The 1,3-diphosphaallene **99** reacted with dichlorocarbene (**88**; X = Cl) in the same manner as **94**: addition of the electron-deficient species yielded the phosphinylidenephosphirane **100**[53] which, however, could not be isolated. Instead, it underwent isomerization to the methylenediphospirane **101** for thermodynamic

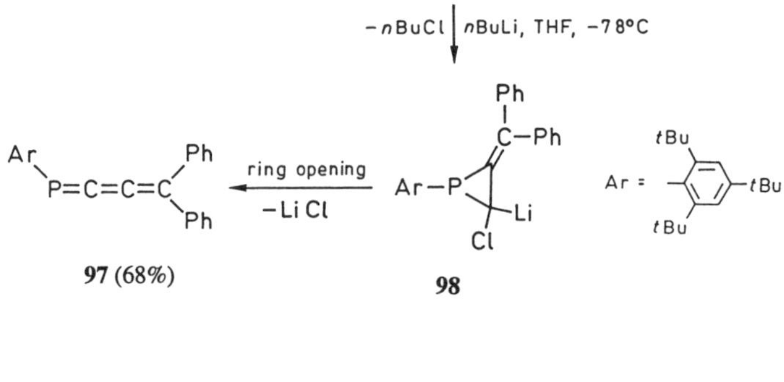

94 **88** (X = Cl) **95** (35%) **96** (10%)

97 (68%) **98**

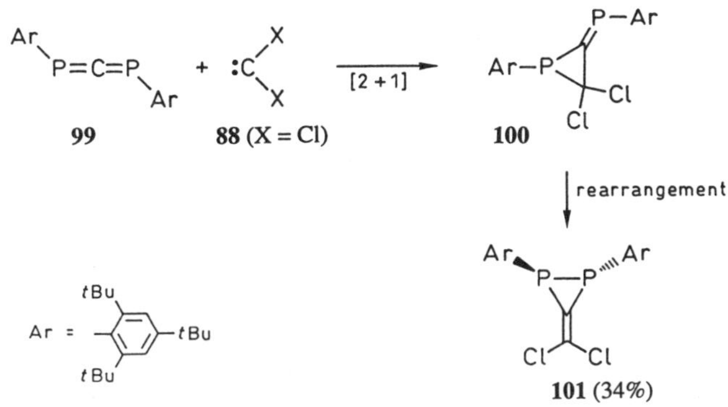

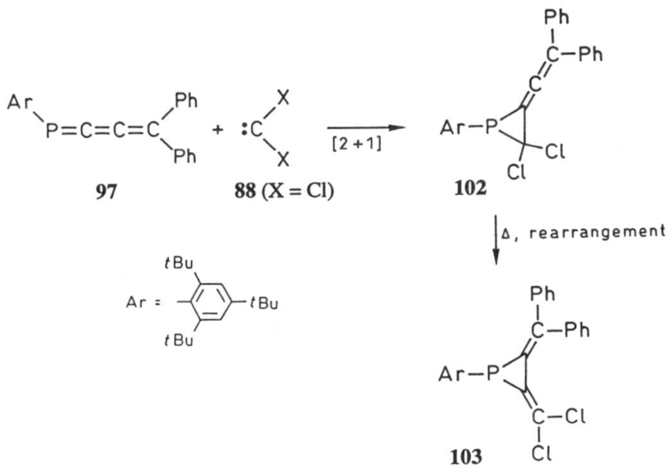

reasons. In this process, the P=C structural element is formally exchanged for a C=C system.[53]

Finally, the 1-phosphabutatriene **97** is also susceptible to cyclopropanation by dichlorocarbene (**88**; X = Cl).[55] In this case again, the phosphaalkene moiety is attacked. But pyrolysis of the vinylidenephosphirane **102** gave rise to a rearrangement to furnish the bis(methylene)phosphirane **103**.[55]

VII. THE PHOSPHINOCARBENE–λ^5-PHOSPHAALKYNE QUESTION

1,2-Dicarbenes of the type **104a** are not capable of existence since the deficit of electrons at the two carbon atoms is compensated by formation of a C–C triple bond

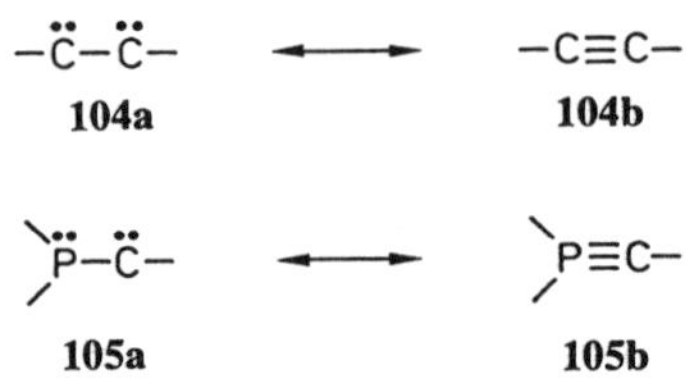

according to **104b**. As a consequence, all attempts to prepare the apparently unstable 1,2-bis(diazo) compounds resulted in the formation of acetylenes.[56,57]

A comparable situation exists in the chemistry of low-coordinated phosphorus[58]: phosphinocarbenes of the type **105a** can also be formulated as the λ^5-phosphaalkynes **105b** and this poses the question as to the generation and reactivity of these species.[59]

A possible approach to a solution of this question is provided by the phosphinodiazo compound **106**. This compound was prepared by reaction of lithiated diazo(trimethylsilyl)methane with chlorobis(diisopropylamino)phosphine.[60,61]

Thermolysis of **106** at 160 °C produced a red oil (80% yield) that was unambiguously assigned the constitution **107** on the basis of spectral and analytical data.[61] It exhibits, as will be mentioned below, characteristics of both a carbene (**107c**) and a λ^5-phosphaalkyne (**107a**) species. A preliminary balance allows the conclusion that the "lone pair" at phosphorus participates in the compensation of the electron deficit at carbon according to **107b**; the silyl group also exerts a stabilizing effect on the carbanionic center. The parallels to nucleophilic carbenes bearing amino-substituents are unmistakable.[58]

C–H-Insertion is considered to be a typical reaction of carbenes and has also been realized with **107**, although requiring unusually high temperatures. Only at 300 °C

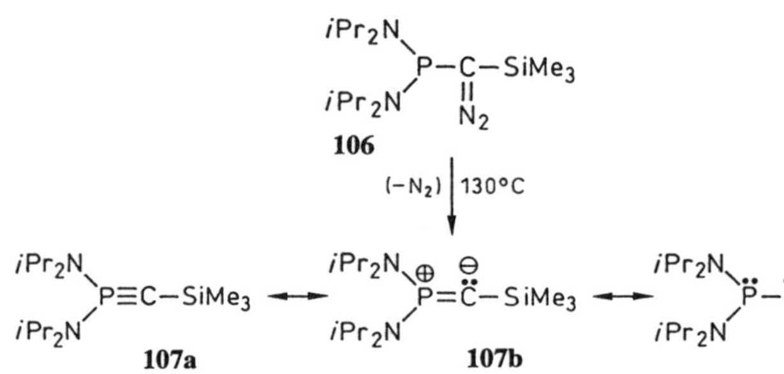

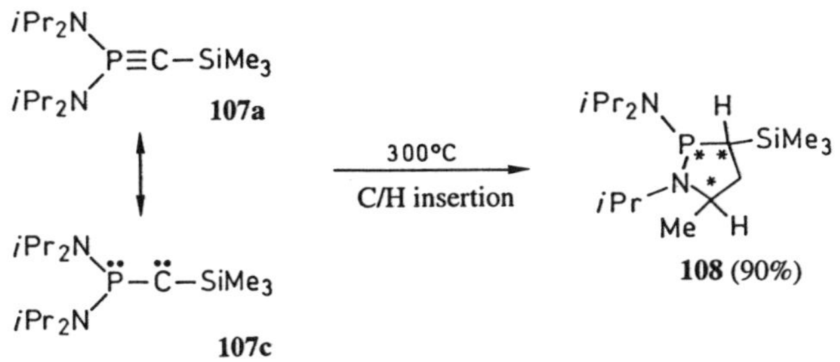

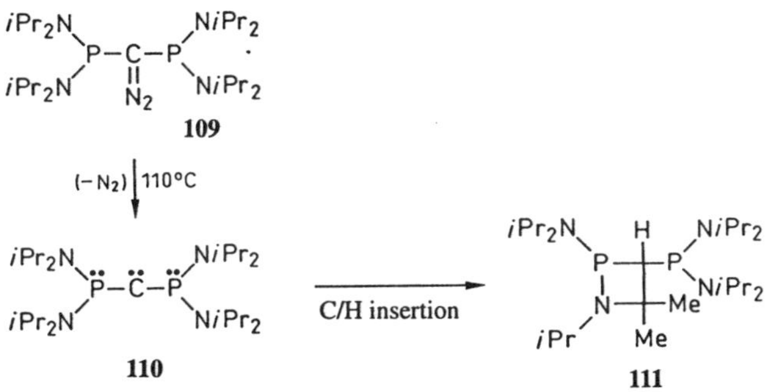

did an intramolecular insertion of **107a** ↔ **107c** into the C–H bonds of the methyl groups of the isopropyl groups to furnish **108** (as a mixture of diastereomers) take place.[61] The thermodynamic stabilization of the carbene center by the phosphino group seems to be rather large.

The carbene **110**, generated by heating the bis(phosphino)diazo compound **109** at 110 °C, behaved analogously with respect to the insertion reaction. However, the insertion did not take place at a CH_3 group of the isopropyl substituent, as is the case for **107**, but at the CH group. The result is the 1,2-azaphosphetane **111** obtained in 92% yield as a homogeneous diastereomer.[62]

In contrast to **107** with its bulky isopropylamino substituents, the λ^5-phosphaalkynes **113a** or the phosphinocarbenes **113c**, respectively, cannot be isolated.

When the species **113a** and **113b** were generated by photolysis of the respective diazo compound **112a** and **112b**,[63,64] only dimers were obtained. These are not the olefins **115** derived formally from the carbenes but rather the λ^5-phosphetes **114**.[63,64] These products are the result of a "head-to-tail" dimerization of the λ^5-phosphaalkynes. Bis(dimethylamino)-λ^5-phosphaacetylene (**113a**; H in place of $SiMe_3$), generated by elimination of two molecules of hydrogen fluoride from bis(dimethylamino)(difluoro)methylphosphorane, also underwent a similar dimerization to furnish the λ^5-diphosphete **114a** (H in place of $SiMe_3$).[65,66]

Diazo compounds with phosphino and phosphoryl substituents such as **116a–c** are thermally labile and extrude nitrogen already at 0–20 °C.[67] A short-lived

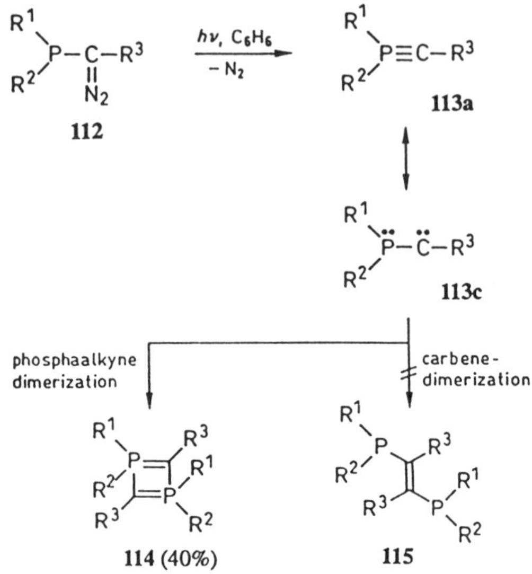

112–115	a	b
R^1	NMe_2	—N(Me)—
R^2	NMe_2	—N(Me)—
R^3	$SiMe_3$	$SiMe_3$
Yield (%)	86	40

intermediate that can be described by the limiting forms **117a** ↔ **117b** ↔ **117c** most certainly participates in this reaction. When the final products of the reaction are considered, it can be deduced that the oxygen atom originating from the R^2R^3PO group has been transferred to the R^1R^1P group; that is, a (1,3) oxygen shift must have occurred.[67] This is valid for the reactions **117a**, **117c** → **120a**, **120c** but cannot be recognized in the reaction **117b** → **120b** because of the identical substituents on both phosphorus groups. On the basis of investigations on acylphosphino(diazo) compounds in which *O*-shifts were also observed,[68,69] the ylides **118a–c** should be intermediates of the isomerizations **117a–c** → **119a–c**, which occur before the dimerizations to **120a–c** can take place.

The occurrence of (1,3) oxygen shifts can also be proposed to explain the products formed in the reaction of the diazo compound **106** with nitrosyl chloride.[70] The nitroso-diazo compound **121** is most certainly formed as the primary product, but undergoes spontaneous extrusion of nitrogen to yield the species **122b** ↔ **122c**. The *O*-shift occurs in this system to furnish the phosphoryl nitrile **123**.[70]

Finally, some reactions have been realized in which the carbene character of the decomposition products from phosphinodiazo compounds is clearly expressed. The bis(diazoalkyl)phosphines **124a–e** are thermally labile and eliminate one equivalent of nitrogen at temperatures as low as 25–60 °C.[63,64,71]

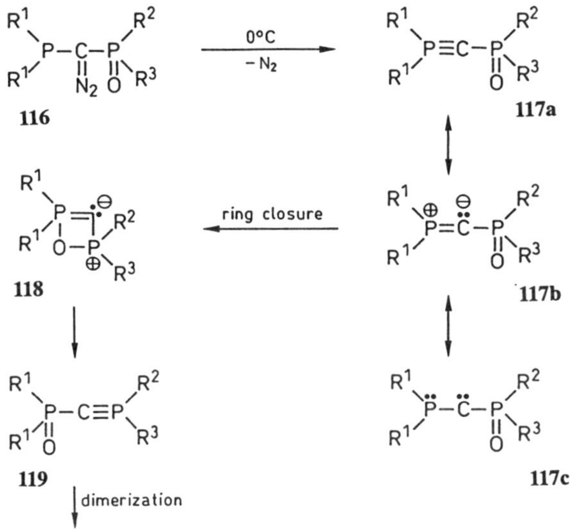

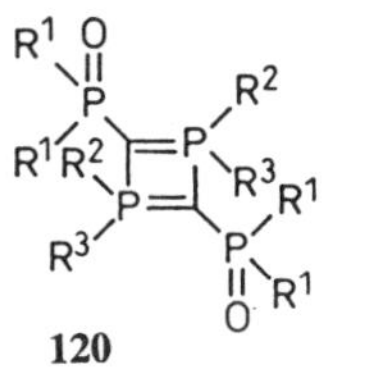

116–120	*a*	b	c
R^1	N-*i*-Pr2	Ph	N- *i*-Pr2
R^2	Ph	Ph	Ph
R^3	Ph	Ph	OMe

The 4*H*-1,2,4-diazaphospholes **126a–e** are the final products of these reactions.[63,64,71] Their formation can be explained by (1,5) ring closure of the carbene centers in **125a–e** with the remaining terminal diazo nitrogen atom; however, the intermediates **125** must not necessarily be formed. The direct transformation **124** → **126** is also feasible; that is, simultaneous cyclization and nitrogen extrusion.

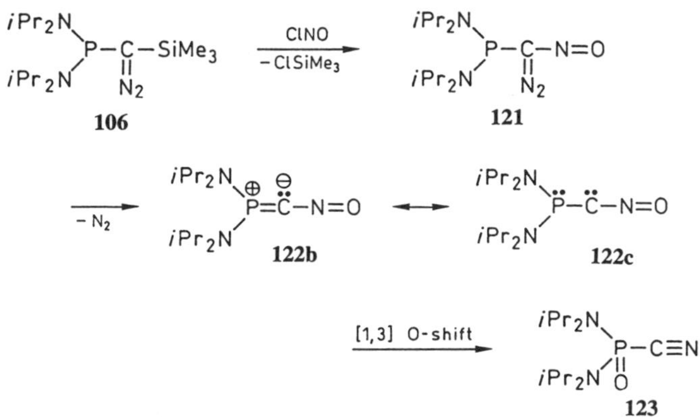

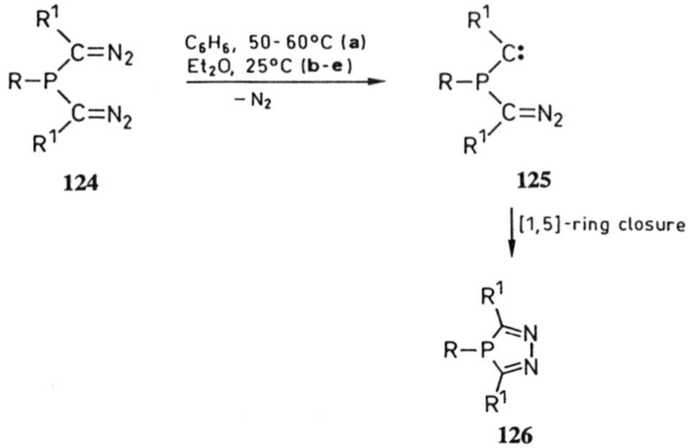

124–126	a	b	c	d	e
R	Me	*i*-Pr2N	Et2N	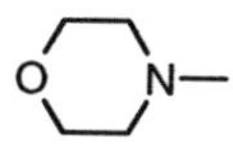	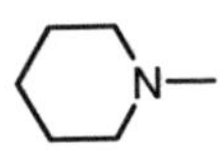
R^1	Me3Si	*t*-Bu	*t*-Bu	*t*-Bu	*t*-Bu
% **126**	70	27	35	47	37

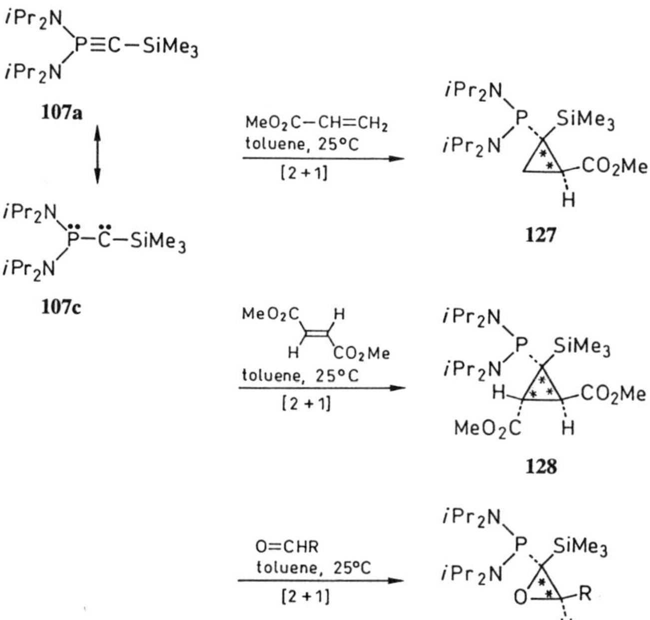

(R = Ph, Styryl; ~80%)

The reactions of the stable species **107** with electron-deficient olefins or aldehydes illustrate its unambiguous carbene reactivity. This again reflects the nucleophilic character of **107** as expressed by the limiting form **107b**.[72,73]

Thus, the reaction of **107** with methyl acrylate gave rise to the cyclopropane derivative **127** as a single diastereomer in a (2+1) cycloaddition process.[72] The reaction of **107** with dimethyl fumarate furnished **128** in practically quantitative yield. The stereochemistry of the starting olefin is retained in the cyclopropane product and, as in the above-mentioned reaction, only one diastereomer is formed.[72] The additions of **107** to benzaldehyde or cinnamaldehyde were also stereospecific and gave rise to the epoxides **129**.[72]

ACKNOWLEDGMENTS

We thank the Deutsche Forschungsgemeinschaft, the Volkswagen-Stiftung, the Fonds der Chemischen Industrie, and the Landesregierung von Rheinland-Pfalz for financial support of those investigations from our laboratories mentioned in this review.

NOTES

Organophosphorus Compounds;* Part 63. For Part 62, see: Regitz, M.; Wettling, T.; Breit, B.; Birkel, M.; Geissler, B.; Bergsträsser, U.; Barth, S.; Binger, P. *Phosphorus, Sulfur, Silicon* **1992, in press.

REFERENCES

1. Regitz, M., Ed; *Houben-Weyl, Methoden der Organischen Chemie*; Thieme: Stuttgart, 1982; Vols. E1, E2.
2. Wittig, G.; Geissler, G. *Liebigs Ann. Chem.* **1953**, *580*, 44–57.
3. (a) Horner, L.; Hoffmann, H.; Wippel, H. G. *Chem. Ber.* **1958**, *91*, 61–63; (b) Horner, L.; Hoffmann, H.; Wippel, H. G.; Klahre, G. *Chem. Ber.* **1959**, *92*, 2499–2505.
4. cf. Mulliken, R. S. *J. Am. Chem. Soc.* **1950**, *72*, 4493–4503.
5. For a survey, see Regitz, M.; Scherer, O. J.; Eds.; *Multiple Bonds and Low Coordination in Phosphorus Chemistry*, Thieme: Stuttgart, 1990.
6. Gier, T. E. *J. Am. Chem. Soc.* **1961**, *83*, 1769–1770.
7. Dimroth, K.; Hoffmann, P. *Angew. Chem.* **1964**, *76*, 433; *Angew. Chem.* (Int. Ed. Engl.) **1964**, *3*, 384.
8. Märkl, G. *Angew. Chem.* **1966**, *78*, 907--908; *Angew. Chem.* (Int. Ed. Engl.) **1966**, *5*, 846–847.
9. Reviews: (a) Regitz, M.; Binger, P. *Angew. Chem.* **1988**, *100*, 1541–1565; *Angew. Chem.* (Int. Ed. Engl.) **1988**, *27*, 1484–1508; (b) Regitz, M. *Chem. Rev.* **1990**, *90*, 191–213; (c) Regitz, M. In *Heteroatom Chemistry*; Block, E., Ed; VCH: New York, 1990; pp 295–322; (d) Regitz, M. *Bull. Soc. Chim. Belg.* **1992**, *101*, 359–379.
10. Rösch, W.; Regitz, M. *Angew. Chem.* **1984**, *96*, 898–899; *Angew. Chem.* (Int. Ed. Engl.) **1984**, *23*, 900–901; (b) Rösch, W.; Hees, U.; Regitz, M. *Chem. Ber.* **1987**, *120*, 1645–1652.

11. (a) Wagner, O.; Ehle, M.; Regitz, M. *Angew. Chem.* **1989**, *101*, 227–229; *Angew. Chem.* (Int. Ed. Engl.) **1989**, *28*, 225–226; (b) Wagner, O.; Ehle, M.; Birkel, M.; Hoffmann, J.; Regitz, M. *Chem. Ber.* **1991**, *124*, 1207–1213.
12. (a) Anderson, D. J.; Gilchrist, T. L.; Rees, C. W. *J. Chem. Soc., Chem. Commun.* **1969**, *147*; (b) Regitz, M.; Arnold, B.; Damion, D.; Schubert, H.; Fusser, G. *Bull. Soc. Chim. Belg.* **1981**, *96*, 615–631.
13. Deschamps, B.; Mathey, F. *Tetrahedron Lett.* **1985**, *26*, 4595–4598.
14. Mercier, F.; Mathey, F. *Tetrahedron Lett.* **1986**, *27*, 1323–1326.
15. Deschamps, B.; Mathey, F. *New. J. Chem.* **1988**, *12*, 755–759.
16. Yoshifujii, M. Kawai, Y.; Yasunami, M. *Tetrahedron Lett.* **1990**, *31*, 6891–6892.
17. (a) Regitz, M.; Heydt, H.; Wagner, O.; Schnurr, W.; Ehle, M.; Hoffmann, J. *Phosphorus, Sulfur, Silicon* **1990**, *49/50*, 341–344; (b) Hoffmann, J.; Regitz, M. Unpublished results, University of Kaiserslautern, 1991.
18. Wagner, O.; Regitz, M. Unpublished results, University of Kaiserslautern, 1989.
19. Ehle, M.; Wagner, O., Bergsträsser, M.; Regitz, M. *Tetrahedron Lett.* **1990**, *31*, 3429–3432.
20. Ehle, M.; Regitz, M. Unpublished results, University of Kaiserslautern, 1990.
21. (a) Weber, W.; de Meijere, A. *Angew. Chem.* **1980**, *92*, 135–136; *Angew. Chem.* (Int. Ed. Engl.) **1980**, *19*, 138–140; (b) Weber, W.; de Meijere, A. *Chem. Ber.* **1985**, *118*, 2450–2471.
22. Memmesheimer, H.; Al-Dulayymi, J. R.; Baird, M. S.; Wettling, T.; Regitz, M. *Synlett* **1991**, 433–435.
23. (a) Baird, M. S.; Buxton, S. R.; Whitley, J. S. *Tetrahedron Lett.* **1984**, *25*, 1509–1512; (b) Al-Dulayymi, J. R.; Baird, M. S.; Clegg, W. *Tetrahedron Lett.* **1988**, *29*, 6147–6148; (c) Al-Dulayymi, J. R.; Baird, M. S.; Hussain, H. H. *Tetrahedron Lett.* **1989**, *30*, 2009–2012; (d) Baird, M. S.; Hussain, H. H. *Tetrahedron* **1989**, *45*, 6221–6238; (e) Al-Dulayymi, J. R.; Baird, M. S. *Tetrahedron* **1989**, *45*, 7601–7614; (f) Al-Dulayymi, J. R.; Baird, M. S.; Clegg, W. *J. Chem. Soc., Perkin Trans. 1* **1989**, 1799–1803.
24. Memmesheimer, H.; Bergsträsser, U.; Hoffmann, J.; Baird, M. S.; Regitz, M. *Synlett* **1992**, 635–637.
25. Reviews: (a) Dötz, K. H. *Angew. Chem.* **1984**, *96*, 573–594; *Angew. Chem.* (Int. Ed. Engl.) **1984**, *23*, 587–608; (b) Dötz, K. H. In *Organometallics in Organic Synthesis: Aspects of a Modern Interdisciplinary Field*; tom Dieck, H.; de Meijere, A., Eds.; Springer: Berlin, 1988; pp 85 ff.
26. Dötz, K. H.; Tiriliomis, A.; Harms, K.; Regitz, M.; Annen, U. *Angew. Chem.* **1988**, *100*, 725–727; *Angew. Chem.* (Int. Ed. Engl.) **1988**, *27*, 713–714.
27. Dötz, K. H. *J. Organomet. Chem.* **1977**, *140*, 177–186.
28. Dötz, K. H.; Tiriliomis, A.; Harms, K. *J. Chem. Soc., Chem. Commun.* **1989**, 788–790.
29. Heinicke, J.; Tzschach, A. *Z. Chem.* **1980**, *20*, 342–343.
30. Heinicke, J.; Tzschach, A. *Phosphorus Sulfur* **1985**, *25*, 345–356.
31. cf. Franck-Neumann, M.; Buchecker, C. *Tetrahedron Lett.* **1969**, 15–19.
32. Regitz, M.; Rösch, W.; Allspach, T.; Annen, U.; Blatter, K.; Fink, J.; Hermesdorf, M.; Heydt, H.; Vogelbacher, U.; Wagner, O. *Phosphorus Sulfur* **1987**, *30*, 479–482.
33. Wagner, O.; Maas, G.; Regitz, M. *Angew. Chem.* **1987**, *99*, 1328–1330; *Angew. Chem.* (Int. Ed. Engl.) **1987**, *26*, 1257–1259.
34. Fuchs, E. P. O.; Hermesdorf, M.; Schnurr, W.; Rösch, W.; Heydt, H.; Regitz, M. *J. Organomet. Chem.* **1988**, *338*, 329–340.
35. See also the reviews cited under Refs. 5 and 9.
36. Rösch, W.; Facklam, T.; Regitz, M. *Tetrahedron* **1987**, *43*, 3247–3256.
37. Hermesdorf, M.; Birkel, M.; Heydt, H.; Regitz, M.; Binger, P. *Phosphorus, Sulfur, Silicon* **1989**, *46*, 31–41.
38. Schnurr, W.; Regitz, M. *Tetrahedron Lett.* **1989**, *30*, 3951–3954.
39. (a) Märkl, G.; Hölzl, W.; Trötsch-Schaller, J. *Tetrahedron Lett.* **1987**, *28*, 2693–2696; (b) Schnurr, W.; Regitz, M. *Z. Naturforsch.* **1988**, *43b*, 1285–1292.

40. (a) Becker, G. Z. *Anorg . Allg. Chem.* **1977**, *430*, 66–76; (b) Allspach, T.; Regitz, M.; Becker, G.; Becker, W. *Synthesis* **1986**, 31–36.
41. Zurmühlen, F.; Regitz, M. *J. Organomet. Chem.* **1987**, *332*, C1–C5.
42. Märkl, G.; Merz, A. *Tetrahedron Lett.* **1971**, 1269–1272.
43. Märkl, G.; Beckh, H. J. *Tetrahedron Lett.* **1987**, *28*, 3475–3478.
44. Märkl, G.; Beckh, H. J.; Ziegler, M. L.; Nuber, B. *Angew. Chem.* **1987**, *99*, 1214–1216; *Angew. Chem.* (Int. Ed. Engl.) **1987**, *26*, 1134–1141.
45. Elemad-Moghadam, G.; Bellan, J.; Tachon, C.; Koenig, M. *Tetrahedron*, **1987**, *43*, 1793–1797.
46. Bellan, J.; Elemad-Moghadam, G.; Payard, M.; Koenig, M. *Tetrahedron Lett.* **1986**, *27*, 1145–1146.
47. Gouygou, M.; Tachon, C.; Koenig, M. *New J. Chem.* **1989**, *13*, 315–317.
48. Gouygou, M.; Bellan, J.; Escudié, J.; Couret, C.; Dubourg, A.; Declercq, J.-P.; Koenig, M. *J. Chem. Soc., Chem. Commun.* **1989**, 593–595.
49. Skattebøl, L.; Solomon, S. In *Org. Synth., Coll. Vol. 5*; Baumgarten, H. E., Ed., Wiley: New York, 1973, pp. 306–310.
50. Brandsma, L.; Verkruijsse, H. D. *Synthesis of Acetylenes, Allenes and Cumulenes. A Laboratory Manual*; Elsevier: Amsterdam, 1981.
51. Yoshifuji, M.; Sasaki, S.; Niitsu, T.; Inamoto, N. *Tetrahedron Lett.* **1989**, *30*, 187–188.
52. cf. Appel, R., in Ref. 5, pp 181–191.
53. Yoshifuji, M.; Toyota, K.; Yoshimura, H.; Hirotsu, K.; Okamoto, A. *J. Chem. Soc., Chem. Commun.* **1991**, 124–125.
54. Independent synthesis of 1-phospha-1,2,3-butatrienes: Märkl, G.; Sejpka, H.; Dietl, S.; Nuber, B.; Ziegler, M. L. *Angew. Chem.* **1986**, *98*, 1020–1021; *Angew. Chem.* (Int. Ed. Engl.) **1986**, *25*, 1003–1004.
55. Toyota, K.; Yoshimura, H.; Mesuki, T.; Yoshifuji, M. *Tetrahedron Lett.* **1991**, *32*, 6879–6882.
56. Curtius, T. *Ber. Dtsch. Chem. Ges.* **1889**, *22*, 2161–2164.
57. Surveys: (a) Regitz, M. *Diazoalkane*; Thieme: Stuttgart, 1977; pp 99–101, 118; (b) Regitz, M.; Maas, J. *Diazo Compounds*; Academic: Orlando, Florida, 1986, pp 234–235, 260–261.
58. Regitz, M. *Angew. Chem.* **1991**, *103*, 691–693; *Angew. Chem.* (Int. Ed. Engl.) **1991**, *30*, 674–676.
59. Review: Bertrand, G., in Ref. 5, pp 443–454.
60. Baceiredo, A.; Bertrand, G.; Sicard, G. *J. Am. Chem. Soc.* **1985**, *107*, 4781–4783.
61. Igau, A.; Grützmacher, H.; Baceiredo, A.; Bertrand, G. *J. Am. Chem. Soc.* **1988**, *110*, 6463–6466.
62. Menu, M. J.; Dartiguenave, Y.; Baceiredo, A.; Bertrand, G. *Phosphorus, Sulfur, Silicon* **1990**, *47*, 327–334.
63. Regitz, M.; Keller, H. Unpublished results, University of Kaiserslautern, 1988.
64. Regitz, M.; Facklam, T. Unpublished results, University of Kaiserslautern, 1989.
65. Svara, J.; Fluck, E.; Riffel, Z. Z. *Naturforsch.* **1985**, *40b*, 1258–1263.
66. Neumüller, B.; Fluck, E. *Phosphorus Sulfur* **1986**, *29*, 23–29.
67. Keller, H.; Maas, G.; Regitz, M. *Tetrahedron Lett.* **1986**, *27*, 1903–1906.
68. Baceiredo, A.; Igau, A.; Bertrand, G.; Menu, M. J.; Dartiguenave, Y.; Bonnet, J. J. *J. Am. Chem. Soc.* **1986**, *108*, 7868–7869.
69. Bertrand, G.; Baceiredo, A.; Sicard, G.; Granier, M. *Phosphorus Sulfur* **1987**, *30*, 353–356.
70. Sicard, G.; Baceiredo, A.; Grocco, G.; Bertrand, G. *Angew. Chem.* **1988**, *100*, 289–290; *Angew. Chem.* (Int. Ed. Engl.) **1988**, *27*, 301–302.
71. Keller, H.; Regitz, M. *Tetrahedron Lett.* **1988**, *29*, 925–928.
72. Igau, A.; Baceiredo, A.; Trinquier, G.; Bertrand, G. *Angew. Chem.* **1989**, *101*, 617–618; *Angew. Chem.* (Int. Ed. Engl.) **1989**, *28*, 621–622.
73. For further reactions demonstrating the nucleophilic carbene character of **107**, see: Igau, A.; Baceiredo, A.; Grützmacher, H.; Pritzkow, H.; Bertrand, G. *J. Am. Chem. Soc.* **1989**, *111*, 6853–6854; see also Refs. 58 and 59.

INDEX